Conservation Agriculture for Africa

Building Resilient Farming Systems in a Changing Climate

This book is dedicated to the global Conservation Agriculture movement but particularly to all the pioneer farmers, researchers, educationalists and extension agents as well as all the champions in the public, private and civil sectors and in the donor community who are making Conservation Agriculture a global reality.

Conservation Agriculture for Africa

Building Resilient Farming Systems in a Changing Climate

Edited by

Amir H. Kassam

Food and Agriculture Organization of the United Nations, Rome, Italy and University of Reading, UK

Saidi Mkomwa

African Conservation Tillage (ACT) Network, Nairobi, Kenya

and

Theodor Friedrich

Food and Agriculture Organization of the United Nations, Havana, Cuba

CABI

CABI is a trading name of CAB International

CABI
Nosworthy Way
Wallingford
Oxfordshire OX10 8DE
UK

Tel: +44 (0)1491 832111
Fax: +44 (0)1491 833508
E-mail: info@cabi.org
Website: www.cabi.org

CABI
745 Atlantic Avenue
8th Floor
Boston, MA 02111
USA

T: +1 (617)682-9015
E-mail: cabi-nao@cabi.org

A catalogue record for this book is available from the British Library, London, UK.

Library of Congress Cataloging-in-Publication Data

Names: Kassam, A. H., editor.
Title: Conservation agriculture for Africa : building resilient farming systems in a changing climate / [edited by] Amir H. Kassam, Saidi Mkomwa, and Theodor Friedrich.
Description: Boston, MA : CABI, [2016] | Includes bibliographical references and index.
Identifiers: LCCN 2016022769| ISBN 9781780645681 (hbk : alk. paper) | ISBN 9781780648415 (epub)
Subjects: LCSH: Sustainable agriculture--Africa.
Classification: LCC S472.A1 C66 2016 | DDC 338.1096--dc23 LC record available at https://lccn.loc.gov/2016022769

ISBN-13: 978 1 78064 568 1

Commissioning editor: Ward Cooper
Editorial assistant: Emma McCann
Production editor: Lauren Povey

Typeset by SPi, Pondicherry, India.
Printed and bound in the UK by CPI Group (UK) Ltd, Croydon, CR0 4YY.

Contents

Contributors vii

Foreword xi

Preface xv

Acknowledgements xix

Acronyms and Abbreviations xxi

Keywords xxvii

1 **Conservation Agriculture in Africa: An Overview** 1
Saidi Mkomwa, Amir H. Kassam, Theodor Friedrich and Reynolds K. Shula

2 **Weatherproofing Agriculture with Conservation Agriculture** 10
Amir H. Kassam, Saidi Mkomwa and Theodor Friedrich

3 **Conservation Agriculture: Growing More with Less – the Future of Sustainable Intensification** 30
Patrick C. Wall

4 **Empowering Smallholder Farmers with Profitable and Sustainable Farming Using Conservation Agriculture: The Case of East Africa** 41
Saidi Mkomwa, Simon Lugandu, Peter Kuria and Weldone Mutai

5 **Implementing the Principles of Conservation Agriculture in Malawi: Crop Yields and Factors Affecting Adoption** 75
W. Trent Bunderson, Zwide D. Jere, Christian Thierfelder, Mpatso Gama, Blessings M. Mwale, Spencer W.D. Ng'oma, Richard M. Museka, John M. Paul, Brand Mbale, Obedi Mkandawire and Phillip Tembo

6 **Innovation Systems and Farmer Perceptions Regarding Conservation Agriculture in Cabo Delgado, Mozambique** 100
Baqir Lalani, Peter Dorward, Amir H. Kassam and Jose Dambiro

7 **Conservation Agriculture in North Africa: Experiences, Achievements and Challenges** 127
Oussama El Gharras, Mohamed El Mourid and Hakim Boulal

8 **How Can We Cover Millions of Hectares with Conservation Agriculture in Africa?** 139
Roland Bunch

9 **Climate-smart Push–Pull: A Conservation Agriculture Technology for Food Security and Environmental Sustainability in Africa** 151
Zeyaur R. Khan, Charles A.O. Midega, Jimmy O. Pittchar, Alice Murage and John Pickett

10 **How to Make Conservation Agriculture EverGreen** 167
Dennis P. Garrity

11 **Mechanization of Smallholder Conservation Agriculture in Africa: Contributing Resilience to Precarious Systems** 183
Brian G. Sims, Josef Kienzle, Saidi Mkomwa, Theodor Friedrich and Amir H. Kassam

12 **Conservation Agriculture in South Africa: Lessons from Case Studies** 214
Hendrik J. Smith, Erna Kruger, Jaap Knot and James N. Blignaut

13 **Extending Conservation Agriculture Benefits Through Innovation Platforms** 246
Michael Misiko

14 **Summing Up** 265
Amir H. Kassam, Saidi Mkomwa and Theodor Friedrich

Index 273

Contributors

James N. Blignaut, Department of Economics, University of Pretoria, ASSET research, and SAEON, PO Box 144, Derdepark, 0035, Pretoria, Gauteng, South Africa. E-mail: jnblignaut@gmail.com

Hakim Boulal, IPNI – North Africa Program, PO Box: 589 - 26000, Settat, Morocco. E-mail: hboulal@ipni.net

Roland Bunch, 10737 Horizon Dr., Juneau, Alaska 99801, USA. E-mail: rdbunchw@gmail.com

W. Trent Bunderson, Total LandCare, PO Box 2440, Area 14 Plot 100, Lilongwe, Malawi. E-mail: trentbunderson@yahoo.com

Jose Dambiro, Aga Khan Foundation (Mozambique), Pemba, Cabo Delgado, Mozambique. E-mail: jdambiro@gmail.com

Peter Dorward, School of Agriculture, Policy and Development, University of Reading, Whiteknights, PO Box 237, Reading, RG6 6AR, UK. E-mail: p.t.dorward@reading.ac.uk

Oussama El Gharras, CRRA de Settat, PO Box 589 - 26000, Settat, Morocco. E-mail: oelgharras@yahoo.com

Mohammed El Mourid, ICARDA – Rabat, PO Box: 6299, Rabat Institute, Morocco. E-mail: m.elmourid@cgiar.org

Theodor Friedrich, FAO Representative in Cuba, Calle 154 y 3ra, #301, Nautico, Playa, La Habana, Cuba. E-mail: theodor.friedrich@fao.org

Mphatso Gama, Machinga Agricultural Development Division, P/Bag 3, Liwonde, Malawi. E-mail: mphatso.gama88@gmail.com

Dennis P. Garrity, World Agroforestry Centre, United Nations Avenue, Gigiri, PO Box 30677 – 00100, Nairobi, Kenya. E-mail: d.garrity@cgiar.org

Zwide D. Jere, Total LandCare, PO Box 2440, Area 14 Plot 100, Lilongwe, Malawi. E-mail: zwidejere@hotmail.com

Amir H. Kassam, Plant Production and Protection Division, Food and Agriculture Organization of the United Nations, Viale delle Terme di Caracalla 00153 Rome, Italy and School of Agriculture, Policy and Development, University of Reading, Reading, RG6 6AR, UK. E-mail: amirkassam786@googlemail.com

Josef Kienzle, Plant Production and Protection Division, FAO, Viale delle Terme di Caracalla, Rome, Italy. E-mail: josef.kienzle@fao.org

Zeyaur R. Khan, International Centre of Insect Physiology and Ecology, PO Box 30772, Nairobi, Kenya. E-mail: zkhan@icipe.org

Jaap Knot, Conservation Agricultural Extension, KEL Growing Nations Trust, Private Bag X6, Zastron, 9950, South Africa. Email: consultant@growingnations.co.za

Erna Kruger, Mahlathini Organics, 72 Tatham Road, Prestbury, Pietermaritzburg, 3201, KwaZulu-Natal, South Africa. E-mail: erna@mahlathiniorganics.co.za

Peter Kuria, African Conservation Tillage Network (ACT), PO Box 10375 – 00100 Nairobi, Kenya. E-mail: peter.kuria@act-africa.org

Baqir Lalani, School of Agriculture, Policy and Development, University of Reading, Whiteknights, PO Box 237, Reading, RG6 6AR, UK. E-mail: b.lalani@pgr.reading.ac.uk; belalani@aol.com

Simon Lugandu, African Conservation Tillage Network (ACT), PO Box 14327 Dar es salaam, Tanzania. E-mail: simon.lugandu@act-africa.org; simon.lugandu@yahoo.com

Brand Mbale, Total LandCare, PO Box 2440, Area 14 Plot 100, Lilongwe, Malawi. E-mail: brandmbale@yahoo.com

Charles A.O. Midega, International Centre of Insect Physiology and Ecology, PO Box 30772, Nairobi, Kenya, E-mail: cmidega@icipe.org

Michael Misiko, Sustainable Intensification Programme, International Maize and Wheat Improvement Center, PO Box 1041-00621, Nairobi, Kenya. E-mail: m.misiko@cgiar.org

Obedi Mkandawire, Total LandCare, PO Box 2440, Area 14 Plot 100, Lilongwe, Malawi. E-mail: omkandawire@hotmail.com

Saidi Mkomwa, African Conservation Tillage Network (ACT), PO Box 10375 – 00100 Nairobi, Kenya. E-mail: saidi.mkomwa@act-africa.org

Alice Murage, Kenya Agricultural and Livestock Research Organization (KALRO), PO Box 25- 20117, Naivasha. E-mail: alicemurage@gmail.com

Richard M. Museka, Total LandCare, PO Box 2440, Area 14 Plot 100, Lilongwe, Malawi. E-mail: richardmuseka@yahoo.com

Weldone Mutai, African Conservation Tillage Network (ACT), PO Box 10375 – 00100 Nairobi, Kenya. E-mail: weldone.mutai@act-africa.org

Blessings M. Mwale, Total LandCare, PO Box 2440, Area 14 Plot 100, Lilongwe, Malawi. E-mail: blessings.mwale@gmail.com

Spencer W.D. Ng'oma, Total LandCare, PO Box 2440, Area 14 Plot 100, Lilongwe, Malawi. E-mail: ngomaspencer@gmail.com

John M. Paul, Total LandCare, PO Box 2440, Area 14 Plot 100, Lilongwe, Malawi. E-mail: krepaul@yahoo.co.uk

John Pickett, Rothamsted Research, Harpenden, Hertfordshire, AL5 2JQ, UK. E-mail: john.pickett@rothamsted.ac.uk

Jimmy O. Pittchar, International Centre of Insect Physiology and Ecology, PO Box 30772, Nairobi, Kenya, E-mail: jpittchar@icipe.org

Reynolds K. Shula, Department of Agriculture, PO Box 50291, Lusaka, Zambia. E-mail: shula.reynolds@iconnect.zm
Brian G. Sims, Engineering for Development, 3 Bourneside, Bedford, MK41 7EG, UK. E-mail: briangsims@aol.com
Hendrik J. Smith, Grain SA, PO Box 74087, Lynwood Ridge, 0040 Pretoria, Gauteng, South Africa. E-mail: hendrik.smith@grainsa.co.za
Phillip Tembo, Total LandCare, PO Box 2440, Area 14 Plot 100, Lilongwe Malawi. E-mail: phillotp@yahoo.co.uk
Christian Thierfelder, CIMMYT, International Maize and Wheat Improvement Center, PO Box MP 163, Mount Pleasant, Harare, Zimbabwe. E-mail: c.thierfelder@cgiar.org
Patrick C. Wall, La Cañada 177, Sector "O", Bahías de Huatulco, Oaxaca 70989, México. E-mail: pcwallmex@gmail.com

Foreword

World population is estimated to reach 9.7 billion by 2050. Africa will have the largest portion of this growth with a population projected to more than double by 2050, increasing from 1.1 billion today to at least 2.4 billion. Agriculture is the main livelihood for up to 80% of the population in Africa; and agriculture is essential for Africa's development. Yet for Africa to feed itself, agricultural output must grow 5% annually to meet its expected food security requirements in 2050. Every expert agrees that this will only happen through intensification, but the real debate is about what that intensification should look like. I believe it must be conservation-based and this book serves as an important guide to how it can be accomplished.

Africa is a vast continent with more diversity in agricultural production than probably any other continent in the world. It is also a continent that was short-changed on soil fertility during the periods of soil formation. In the Midwest of the United States, glaciers deposited valuable minerals and sediment that created some of the most productive soil profiles in the world. This is also true in places like the Ukraine, Argentina and other parts of the world. However, Africa is home to some of the oldest and most weathered landscapes. As a result, less than 10% of Africa has what are considered high-quality soils, including the lower third of West Africa, parts of East Africa and areas within several countries in southern Africa including Zambia, Zimbabwe, South Africa and Mozambique. The rest of Africa's 54 countries and two disputed territories did not receive the same amount of natural benefit.

The Food and Agriculture Organization (FAO) estimates that 65% of agricultural land throughout Africa has been degraded by human activity. The Montpellier Panel – a prominent group of agriculture, ecology and trade experts from Africa and Europe – estimates that these degraded soils are too damaged to sustain viable food production. Its report *No Ordinary Matter: Conserving, Restoring and Enhancing Africa's Soil* notes that Africa suffers from the triple threat of land degradation, poor yields and a growing population.

The history of Africa's soils and land use complicates the strategy for increasing productivity in agriculture. It is not as simple as introducing Western production techniques that rely heavily on synthetic fertilizer and hybrid seeds.

I propose examining Africa's potential for agriculture through a different lens: how we analyse opportunities for farmers and how soils are remediated and protected will be a key in our success or failure to support farm populations. Africa should borrow relevant lessons from the United States, Brazil, Australia and others, but also protect and embrace its unique diversity to develop a system for agriculture that is adapted to its circumstances. If it does not, the continent overall will continue its per capita decline in productivity.

This book documents how farmers in Africa are successfully adopting Conservation Agriculture as an alternative to unsustainable and unaffordable conventional farming practices and as a solution to land degradation, climate change challenges and ever increasing food insecurity. In 2013, farmers in 20 countries in Africa applied the Conservation Agriculture, or 'CA', approach on 1.2 million hectares, of which approximately 750,000 ha were added between 2009 and 2013.

Conservation Agriculture strives to achieve high and sustained production levels while conserving natural resources. More importantly for smallholder farmers, it is a system they can apply without reliance on unaffordable or inaccessible inputs. CA also reduces land and water pollution; reduces soil erosion; enhances environmental management; improves water use efficiency; and reduces emissions of greenhouse gases by reducing the need for fossil fuel, agrochemicals and reducing tillage-induced oxidation of soil organic matter. Research, extension and development professionals have been working with smallholder farmers in testing and promoting CA technologies in Africa for nearly 20 years and in some cases efforts have been ongoing since the 1980s. There is a great deal of accumulated evidence about the relevance and feasibility of CA as a climate-smart, sustainable production intensification method for use across Africa. It is now time to expand that evidence base and provide the information to farmers, investors and governments that CA is applicable to their diversified agroecological and socio-economic situations.

As the title suggests, this book is about building resilient farming systems and offers case studies and analyses of how CA is able to make agriculture climate-smart across Africa. The book also highlights the need to turn CA knowledge into practice, by linking farmers to the broader CA service providers for information, knowledge, mechanization, markets and financial services.

CA has its doubters, including academics who regularly assert that it is not a method that is appropriate for smallholder adoption. I hope this book helps to silence the nay-sayers and encourages more local, national and international development support for this climate-smart approach to sustainable agricultural intensification and food security in Africa.

As a farmer operating more than 8000 ha in the USA and Africa, I am regularly amazed at how frequently non-farmers are charged with recommending solutions that will have tremendous influence on the issues farmers face. It is the equivalent of asking a non-medical person to perform surgery. Just as trained doctors are best equipped to perform medical procedures, farmers are best equipped to understand the issues farmers face, and they are best equipped to participate in developing practical solutions that work in the real world, not in hypothetical situations. When hypothetical solutions drive policy, it is a recipe for failure. If we do not treat African soils and policies with realistic ideas based on realistic solutions, then we will fail the farmers we all want to help.

Howard G. Buffett
Chairman and CEO
The Howard G. Buffett Foundation

Preface

Tillage, irrespective of whether it is based on mouldboard ploughs, ards, harrows, hoes and spades, has led to widespread soil and ecosystem degradation globally and more particularly in the developing regions. This is especially so in Africa where conventional agriculture has become unsustainable due to severe exploitation of natural resources, with negative impacts on the environment. In addition, agriculture in Africa today faces major challenges, including increased cost of production inputs and energy, climate change and lack of an effective paradigm for sustainable production intensification. Hence, Conservation Agriculture (CA) has emerged as a major alternative approach to sustainable agricultural intensification in Africa and has already spread across many countries in Africa in the past decade, calling for increased support to farmers by the public, private and civil sectors.

To mobilize and strengthen such support, the African Conservation Tillage (ACT) Network in close liaison with partners convened the 1st Africa Congress on Conservation Agriculture (IACCA) in Lusaka, Zambia, from 18 to 21 March 2014. The Congress focused on the theme *Conservation Agriculture (CA): Building Entrepreneurship and Resilient Farming Systems*. The purpose of the Congress, which brought together 414 delegates from 42 African and other countries of the world, was to share experiences and lessons and facilitate alliances to unblock hindrances in order to facilitate the adoption and spread of CA, especially among the smallholder farming systems and related industry in Africa.

The Congress was convened against the background of 2014 having been designated by the African Union Heads of State and Government as the year for agriculture and food security, and marked the 10th anniversary of the Comprehensive African Agriculture Development Programme (CAADP) of the New Partnership of Africa's Development (NEPAD). The Congress also marked the adoption of the new CAADP results framework, which recognizes the role of climate-smart agriculture in addressing agricultural and climate-change challenges to agriculture and national development.

This is in addition to the fact that, while worldwide hunger has decreased by 132 million people in the last 20 years, it has increased by 64 million (from 175 million to 239 million) in Africa over the same period. Africa as a continent has remained a net importer of agricultural products in the past three decades. In 1980, Africa had an almost balanced agricultural trade when both agricultural exports and imports were at about US$14 billion, but by 2007 its agricultural imports exceeded agricultural exports by about US$22 billion. While global population is projected to increase to 9.7 billion people (increase by 35%) by 2050, Africa's population is projected to increase from 1.1 to 2.4 billion, an increase of 115%, by the same date, thus requiring at least a similar increase in food production on the continent, in line with the CAADP framework to which African governments have pledged their support.

CA is defined as a management system based on three interlinked principles that should be applied in a mutually reinforcing manner: (i) continuous no or minimum physical soil mechanical disturbance (no-till crop establishment); (ii) maintenance of permanent soil mulch cover with live or dead plant material; and (iii) crop diversification in space and time. Globally, CA is now spreading at the annual rate of some 10 million hectares and covers 157 million hectares worldwide, corresponding to some 11% of global annual cropland. CA is a production system paradigm that uses all appropriate best-management practices to minimize risks and ensure ecological sustainability and resilience to underpin economic, environmental and social sustainability.

The Congress reaffirmed that restoration of soil health through CA is set to become a cornerstone in transforming the way farming is done in Africa and a major contributor to achieving CAADP's goal of 6% annual growth in the agricultural sector, which employs 80% of Africa's rural population. CA was also confirmed, through the studies and presentations made, to have the ability to convert marginal semi-arid regions into grain baskets without the need for costly irrigation investments. This attribute gives CA an edge to be adopted by millions of smallholder farmers in Africa affordably and swiftly. CA has a significantly positive impact on practising farmers across Africa in their incomes, livelihoods and wellbeing and on empowerment of women farmers. It is further recognized that women and the youth play leading roles in the accelerated adoption and upscaling of CA. CA gives farmers the choice to apply CA principles to a range of rainfed and irrigated production systems, including annual cropping, horticulture, agroforestry and crop–livestock systems, among others.

The Congress participants further recognized the efforts being taken by stakeholders in support of sustainable agriculture intensification but noted that the efforts need to be stepped up to accelerate the pace of adoption of CA and subsequently upscaling. In order to achieve the CAADP goal of 6% growth of the agriculture sector, the Lusaka Declaration agreed on ten resolution points centred on: (i) policy, political commitment and leadership; (ii) private sector engagement; and (iii) training, extension, research and innovation, and knowledge support. Governments, development partners, the private sector, farmers, training institutions, research institutions, regional economic communities and non-governmental organizations are among the stakeholders called upon for implementation of the resolutions in order to enhance adoption and scaling up of CA.

This book builds on the material presented at the Congress and provides an updated assessment of trends in the adaptation and adoption of CA in different countries of Africa. It provides the latest information on the contribution of CA systems to building resilience in a changing climate while increasing productivity, profitability and ecological sustainability. The book has continental and sub-regional focus and covers information from several countries considered relevant to CA development in Africa, namely: the status of CA in Africa (Chapter 1); CA's climate-proofing characteristics (Chapter 2); productivity and efficiency of resource use, often referred to as 'producing more from less' (Chapter 3); smallholder adoption and spread of CA in different parts of Africa (Chapters 4, 5, 6 and 7); aspects related to resilience (legume cover crops, diversification for pest control, agroforestry, mechanization) in CA-based systems (Chapters 8, 9, 10, 11); and policy and communication (Chapters 12, 13). The final Chapter 14 provides a brief summing up of what the book represents in terms of the relevance and future of CA to agriculture development in Africa.

We hope that this book will serve as a source of scientific and empirical evidence: to policy-makers and institutional leaders in the public, private and civil sectors to help in decision making in support of CA; and to academics, scientists and students in formulating their strategic directions and priorities for the development of an effective CA innovation and knowledge system for agricultural and economic development in Africa.

Acknowledgements

The editors of this book would like to recognize the considerable assistance accorded to its preparation from various persons and organizations. Since the book builds on the 1st Africa CA Congress, special recognition, with much appreciation, goes to all those stakeholders who made the Congress a success. These include farmers, donors, politicians and government officials, researchers, development experts, service providers, ACT Board and staff, reviewers and those who helped to organize and run the Congress. Great merit goes to the authors of various scientific materials shared during the Congress, upon which this book has largely been built.

Most importantly, we would like to express our tremendous gratitude to the numerous contributors who dedicated time and resources to develop, update and share their scientific information and experiences that have formed the core and integral part of this book. Their contributions will eventually serve as a source of scientific and empirical evidence to various stakeholders in the agriculture sector and entice new advances on CA innovation and knowledge systems for agricultural and economic development in Africa.

We acknowledge with many thanks the external reviewers of the different chapters of this book for their immense contribution. They include: John Ashburner, John Ashley, H.R. Mloza Banda, Gottlieb Basch, Moncef Ben-Hammouda, James Breen, Roland Bunch, Ademir Calegari, Rolf Derpsch, Patrice Djamen, Jim Ellis-Jones, Alfonse Emechebe, Tom Goddard, Eric Kueneman, Rabah Lamar, Paul Mapfumo, Drake Mubiru, Joseph Mureithi, Frank Place, Jules Pretty, Reynolds K. Shula, Brian G. Sims, Hendrik J. Smith, Francis Shaxson and Haroon Sseguya. To all the sponsors of the 1st African Congress on CA and development of this book, who included Common Market for Eastern and Southern Africa (COMESA), Norwegian Agency for Development Cooperation (NORAD), Centre for Coordination of Agricultural Research and Development for Southern Africa (CCARDESA), International Centre for Agricultural Research in the Dry Areas (ICARDA), Food and Agriculture Organization of the United Nations (FAO), European Union and Alliance for a Green Revolution in Africa (AGRA), we recognize their support and acknowledge their determination to help change the way farming is done in Africa. Their sponsorship stirred up great knowledge and information sharing and triggered new advances towards sustainable farming systems in Africa as demonstrated by the new updates captured in this book. The success of the Congress was a measure of strong dedication and volunteering spirit of the members and friends of the ACT. The partnerships with the French Agricultural Research Centre for International Development (CIRAD), the International Maize and Wheat Improvement Centre (CIMMYT) and the New Partnership for African Development (NEPAD) deserves a special mention.

We express our very special and grateful thanks to Dr Howard G. Buffett for gracing the book with an inspiring and encouraging Foreword. Dr Buffett's contribution is all the more important because he is a Conservation Agriculture farmer, and his Foundation is a major donor to programmes that support the spread of Conservation Agriculture in the developing regions, including Africa.

Finally, we end by registering our appreciation to any other persons or organizations not mentioned above that have contributed in one way or another to the build-up and compilation of the material presented in this book. We deeply recognize their contribution.

Acronyms and Abbreviations

AAAID	Arab Authority for Agricultural Investment and Development
ABACO	Agro-ecology Based Aggradation Conservation Agriculture
ACIAR	Australian Centre for International Agricultural Research
ACT	African Conservation Tillage Network
ADD	Agricultural Development Division
AFD	French agency for development
AGRA	Alliance for a Green Revolution in Africa
AIP	Agricultural Innovation Platform
AIS	Agricultural Innovation Systems
AKF	Aga Khan Foundation
AKF-CRSP	Aga Khan Foundation Coastal Rural Support Programme
AKIS	Agricultural Knowledge and Information System
AMA	Associação Meio Ambiente
AMCEN	African Ministerial Conference on the Environment
ANA	Agencia Nacional de Águas
APAD	Association pour la Promotion d'une Agriculture Durable (Tunisia)
ARC	Agricultural Research Council
ARI	Agricultural Research Institute
ARS	Agricultural Research Service
ATCC	Agricultural Technology Clearing Committee
AU	African Union
BCR	benefit–cost ratio
BRCC	Building Resilience to Climate Change
BUSOFIP	Bungoma South Farming Innovation Platform of SIMLESA (western Kenya)
C	carbon
CA	Conservation Agriculture
CA4NRM	Conservation Agriculture for Natural Resources Management

CAADP	Comprehensive Africa Agriculture Development Programme
CANA	Conservation Agriculture for North Africa
CARE	Cooperative for Assistance and Relief Everywhere
CARP	Conservation Agriculture Regional Project
CATs	Conservation Agriculture Technologies
CAWT	Conservation Agriculture with Trees
CBE	community-based extension
CBO	community-based organization
CCARDESA	Centre for Coordination of Agricultural Research and Development for Southern Africa
CFGB	Canadian Foodgrains Bank
CFU	Conservation Farming Unit (Zambia)
CGAP	Consultative Group to Assist the Poor
CH_4	methane
CIMMYT	International Maize and Wheat Improvement Centre
CIRAD	Agricultural Research Centre for International Development
CLCA	Crop–Livestock Conservation Agriculture
CO_2	carbon dioxide
COMESA	Common Market for Eastern and Southern Africa
CRT	conventional ridge tillage
CSA	climate-smart agriculture
CT	conventional tillage
CV	conventional agriculture
DAES	Department of Agricultural Extension Services
DAFF	Department of Agriculture, Forestry and Fisheries
DAP	diammonium phosphate
DARS	Department of Agricultural Research Services
DARTS	Department of Agricultural Research and Technical Services (Malawi)
DELEP	Desert Legumes Program
DFID	Department for International Development (UK)
DRDLR	Department of Rural Development and Land Reform
EA	East Africa
EAC	East African Community
ECA	EverGreen Conservation Agriculture
EFS	Eastern Free State (South Africa)
EPAGRI	Empresa de Pesquisa Agropecuária e Extensão Rural De Santa Catarina
ESAK	Higher School of Agriculture of Kef
EU	European Union
EVA	EverGreen Agriculture
FACASI	Farm Mechanization and Conservation Agriculture for Sustainable Intensification
FAO	Food and Agriculture Organization of the United Nations
FFS	farmer field school
FIP	Farmer Innovation Programme

FMNR	farmer-managed natural regeneration
FP	farmers' practice
FSR	farming systems research (model)
FTT	fertilizer-tree technology
GDP	gross domestic product
GEF	Global Environment Fund
GHG	greenhouse gas
GIZ	German Development Cooperation
GMCC	green manure cover crops
GOM	Government of Malawi
Grain SA	Grain South Africa
GSDs	growing seasons days
HEI	high external input systems
HI	Heifer International
HP	horsepower
HPI	Heifer Project International
HSGOC	Heads of State and Government Orientation Committee
HYV	high-yielding variety
IAASTD	International Assessment of Agricultural Knowledge, Science and Technology for Development
IACCA	First African Congress on Conservation Agriculture
ICARDA	International Centre for Agricultural Research in the Dry Areas
icipe, ICIPE	International Centre of Insect Physiology and Ecology
IC–LS	integrated crop–livestock system
ICRAF	World Agroforestry Centre
ICRISAT	International Crops Research Institute for the Semi-Arid Tropics
ICT	information and communication technologies
IFAD	International Fund for Agricultural Development
IFPRI	International Food Policy Research Institute
IIRR	International Institute of Rural Re-construction
ILRI	International Livestock Research Institute
INGC	National Field Crops Institute
INRA	Institut National de la Recherche Agronomique (National Agronomic Research Institute)
IP	innovation platform
IPCC	Intergovernmental Panel on Climate Change
IPM	integrated pest management
IS	innovation systems
ISCW	Institute for Soil, Climate and Water
ISFM	integrated soil fertilizer management
ITGC	Technical Institute for Field Crops
ITOC	*icipe* Thomas Odhiambo Campus
KARI	Kenyan Agricultural Research Institute
KENDAT	Kenya Network for Dissemination of Animal Traction Technologies
KSh	Kenyan shilling

KZN	KwaZulu-Natal
LAMP	Land Management and Environment Programme
LEISA	low external input sustainable agriculture
LRNP	Legume Research Network Project
LSD	least significant difference
MDG	Millennium Development Goals
MEA	Millennium Ecosystems Assessment
MIP	Maize Innovation Platform (of Nyagatare, Rwanda)
MoAIWD	Ministry of Agriculture, Irrigation and Water Development
MOU	Memorandum of Understanding
MRR	marginal rate of return
N	nitrogen
N_2O	nitrous oxide
NAIP	National Agriculture and Food Security Investment Plans
NAMA	Nationally Appropriate Mitigation Actions
NAPA	National Adaptation Programmes of Action
NARS	National Agricultural Research System
NASFAM	Malawian Smallholders Farmers' Association
NCATF	National CA Task Force
NDL	no-till with *Dolichos lablab*
NDP	no-till with dolichos and pigeon peas
NEPAD	New Partnership for Africa's Development
NGO	non-governmental organization
NLP	National LandCare Programme
NNC	no-till with no cover crops
NORAD	Norwegian Agency for Development Cooperation
NPP	no-till with pigeon peas
NPV	net present value
NRM	Natural Resources Management
NT	no tillage, or no-till
NYAMIG	Nyagatare Maize Investment Group Ltd (of MIP, Rwanda)
OECD	Organization for Economic Cooperation and Development
OVK	Orange Free State Cooperative
P_2O_5	phosphate
PCA	principal component analysis
PDARI	Integrated Agricultural and Rural Development Project
PES	Payment for Ecosystems Services
PGRCU	Plant Genetic Resources Conservation Unit
PICC-PMV	Project of Integration of Climatic Changes into the Green Moroccan Strategy
PID	participatory innovation development
PLAAS	Poverty, Land and Agrarian Studies
PM&E	participatory monitoring and evaluation
PMV	Green Morocco Strategy
PTO	power take off
R&D	research and development

RAB	Rwanda Agricultural Board
RCBD	randomized complete block design
REC	Regional Economic Community
RELMA	Regional Land Management Unit
RTI	Royal Tropical Institute
SADC	Southern African Development Community
SAFIM	South African Farm Implement Manufacturers
SARD	Sustainable Agricultural and Rural Development
SARI	Selian Agricultural Research Institute
SCAPA	Soil Conservation and Agroforestry Programme for Arusha
SDAE	Serviços Distritais das Actividades Económicas (District Services of Economic Affairs)
SDG	sustainable development goals
SED	standard error of the difference
SFIP	Smallholder Farmer Innovation Programme
SIDA	Swedish International Development Agency
SIMLESA	Sustainable Intensification of Maize and Legume Systems for Food Security in Eastern and Southern Africa
SNA	social network analysis
SOC	soil organic carbon
SOM	soil organic matter
SPSS	Statistical Package for Social Sciences
SRI	System of Rice Intensification
SSA	Sub-Saharan Africa
TCP	Technical Cooperation Project
TLC	Total LandCare
TOT	transfer of technology (model)
UNDP	United Nations Development Programme
UNEP	United Nations Environment Programme
UNICEF	United Nations Children's Emergency Fund
USAID	United States Agency for International Development
USDA	United States Department of Agriculture
VDO	Village Development Organization
VSA	visual soil assessment
WFS	Western Free State (South Africa)
WT	wheeled tractor (as in 2WT and 4WT)
WUE	water-use efficiency
WWF	World Wildlife Fund

Keywords

Numbers indicate Chapter(s) in which keywords are used.

Adaptation 2
Adoption 1
Adoption challenges 5
Africa 1, 14
Agricultural efficiency 3
Agricultural intensification 9
Agroforestry 10
CA Policy 12
Case Studies 13
Case study 6
Climate change 1, 2, 4, 5, 9
Climate resilience 4,
Climate-smart 1, 2, 9, 14
Climate-smart agriculture 10, 11
Conservation Agriculture 1, 2, 3, 4, 5, 6, 7, 8, 9, 10, 11, 12, 13, 14
Constraints to adoption 11
Cost savings 3
Cover crops 4
Desmodium 9
Direct-seeding drill 7
Discovery learning 12
Diversification 5
East Africa 4
Economically attainable yield 3
Ecosystem services 2
Education and research support 14
Environmental service 1
EverGreen Agriculture 10

EverGreen Conservation Agriculture 10
Faidherbia albida 10
Farmer empowerment 14
Farmer Field School 6, 12
Farmer Innovation 12
Farmer Perceptions 6
Farmer-centred Innovation Systems 12
Grain yield 7
Green manure/cover crops 8
Ground cover 5
In situ mulching 8
Innovation Platform 6, 12
Innovation Systems 6
Institutional bottlenecks 3
Landscape restoration 10
Local resources 8
Machinery need 11
Mechanization inputs 11
Mixed crop–livestock systems 9
Natural regeneration 5
Nitrogen fixation 8
North Africa 7
No-till 14
No-tillage 4, 7
On-farm experimentation 12
Paradigm change 14
Participatory methods 6
Productivity 5
Push-pull technology 9
Residues 7
Resilience 1, 2, 5, 14
Risk 3
Rural Institutions 13
Scaling-up 10
Smallholder farmers 4, 9
Smallholders 13, 14
Soil health 9
Soil nutrients 8
Soil organic matter 8
Sustainability 5
Sustainable agriculture 9
Sustainable intensification 11, 13
Sustainable production intensification 14
Uptake and spread of CA 14
Water services 2
Weather-proofing 2
Weed control 8

1 Conservation Agriculture in Africa: An Overview

Saidi Mkomwa,[1]* Amir H. Kassam,[2] Theodor Friedrich[3] and Reynolds K. Shula[4]

[1]African Conservation Tillage Network, Nairobi, Kenya; [2]University of Reading, UK; [3]Food and Agriculture Organization of the United Nations, Havana, Cuba; [4]African Conservation Tillage Network, Nairobi, Kenya

1.1 What is Conservation Agriculture (CA)?

Conservation Agriculture (CA) is not a single technology but a systems approach to farming based on a set of three linked complementary practices formulated locally and based on the following interlinked principles as defined by the Food and Agriculture Organization of the United Nations (FAO) (www.fao.org/ag/ca):

- **Avoiding or minimizing mechanical soil disturbance**. Sow seed or plant crops directly into untilled soil in order to: maintain soil organic matter; promote soil biological processes; protect soil structure and porosity and overall soil health; and enhance productivity, system efficiency, resilience and ecosystem services.
- **Enhancing and maintaining a permanent mulch cover with organic matter on the soil surface.** Use crop residues (including stubble) and cover crops to: protect the soil surface; conserve water and nutrients; supply organic matter and carbon to the soil system; and promote soil biological activity to enhance and maintain soil health (including structure and aggregate stability), contribute to integrated weed, pest and nutrient management, and enhance productivity, system efficiency, resilience and ecosystem services.
- **Diversification of species.** Use diversified cropping systems with crops in associations, sequences or rotations that will contribute to: enhanced crop nutrition; crop protection; soil organic matter build-up; and productivity, system efficiency, resilience and ecosystem services. Crops can include annuals, trees, shrubs, nitrogen-fixing legumes and pasture, as appropriate.

*saidi.mkomwa@act-africa.org

These three locally formulated practices should be implemented in combination with other good technologies and practices by the farmers in order to obtain full productivity, socio-economic and environmental benefits from CA. These practices cover a large range of expertise, from equipment and machinery to soil management, residue management and cover crops, to pest and diseases management, to nutrient and water management, including crop and cropping system management (FAO, 2011, 2016). In addition, each country and sub-region in Africa has its own unique resource endowment, socio-economic conditions, range of production and farming systems, and agricultural and economic development opportunities. Further, each country and sub-region has its particular level of adoption and spread of CA, depending on its national level of commitment towards CA.

This state of affairs calls for flexibility and adaptability according to the specific biophysical and socio-economic situation in each country and sub-region. Given this understanding, therefore, the CA principles need to be translated into a number of locally devised and adapted practices that can work simultaneously through contextualized crop–soil–water–nutrient–pest–ecosystem management at a variety of scales to address the major challenges in Africa and globally. These include: (i) pervasive food insecurity and poverty; (ii) high environmental impact of tillage-based agriculture; (iii) relatively high-cost tillage–seed–fertilizer–pesticide–credit approaches; (iv) absence of agroecologically-based production systems that are environmentally stable; and (v) natural and man-made disasters and crises.

CA has been shown to be relevant and appropriate for small- and large-scale farmers at all levels of farm power and mechanization, from manually operated hand tools to equipment drawn by animals to operations performed by heavy machinery. However, its spread in Africa has been slow compared with other continents, and the reasons for this lower-than-desired spread of CA can be attributed to include: (i) continued promotion and development support of tillage-based agricultural systems by national and international, public and private institutions and sector industries; (ii) weak policies and regulatory frameworks and institutional arrangements to support the promotion and mainstreaming of CA; (iii) inadequate awareness, knowledge and expertise of CA systems and the process of their adoption and spread among policymakers, academic, research, extension and technical staff; (iv) inappropriate CA technology packaging and dissemination; (v) inadequate CA-based enterprise diversification and integration in farming systems; (vi) inability of smallholders to diversify crop rotations, sequences and combinations; (vii) inadequate skills and competencies among farmers and other CA practitioners; (viii) farmers' inability to maintain year-round soil cover through the use of specially introduced cover crops, intercrops and crop residue; (ix) poor availability and access to the required CA equipment, machinery and inputs; and (x) absence of a strong continental body and strategic policy framework to guide the promotion and mainstreaming of CA across Africa.

1.2 Global and Regional Level Environment for CA

CA is gaining acceptance in all continents and many countries of the world as an alternative to other forms of agriculture as a means to enable small and larger farmers to intensify production sustainably. Due to increased environmental concerns, there is renewed interest in the formulation of alternative strategies for enhancing development activities, including agriculture, that take cognizance of the needs of smallholder farmers and of long-term investment in environmental sustainability that is also affordable and accessible to smallholders. In recent times, CA has demonstrated its potential in addressing climate-change challenges and it has also been shown to contribute to climate-change adaptation and mitigation, biodiversity conservation and delivery of ecosystem and societal services such as clean water, carbon sequestration and control of soil erosion and degradation. Consequently, CA is referred to as being climate-smart and is included as being part of climate-smart agriculture (CSA) options (FAO, 2013; Gonsalves *et al.*, 2015).

At the international level, CA relates directly to the United Nations Framework Convention on Climate Change, the International Convention on Biodiversity, the United Nations Convention to Combat Desertification and the various agreements on international waters. Under the Kyoto Protocol, various initiatives in national and international marketing of carbon offsets based on CA systems are currently operating that involve financial mechanisms that ensure environmental benefits provided by CA are recognized by society at large and that financial benefits accrue to CA practitioners. These and other future initiatives provide opportunities for environmental service payments on 'niche products' and have the potential to encourage farmers to shift to CA and other supporting sustainable agricultural practices.

As the next section shows, the challenges and constraints mentioned earlier are being surmounted across Africa for CA to be adopted by farmers in all regions. Most of these farmers in recent years are smallholders and such farmers offer opportunities for commercialization of agriculture and for service providers to provide services to support field operations with equipment and machinery that the smallholder farmer would be reluctant to invest in (Sims and Kienzle, 2015).

1.3 Where is CA Practised in Africa?

According to Kassam *et al.* (2015), based on the FAO AquaStat data (www.fao.org/ag/ca/6c.html), CA in 2013 was practised on nearly 157 million hectares of global cropland, and since 2008/09 it had increased at a rate of some 10 million hectares per annum. Africa has about 1.2 million hectares, an increase of some 750,000 ha since 2008/09. Most of this spread has been through smallholder farmers. Although latest figures for Africa are not available for all the countries where CA adoption is occurring, South Africa has the highest area (368,000 ha), followed by Zimbabwe (332,000 ha), Zambia (200,000 ha), Mozambique (152,000 ha) and Malawi (65,000 ha) (Fig. 1.1). Other notable

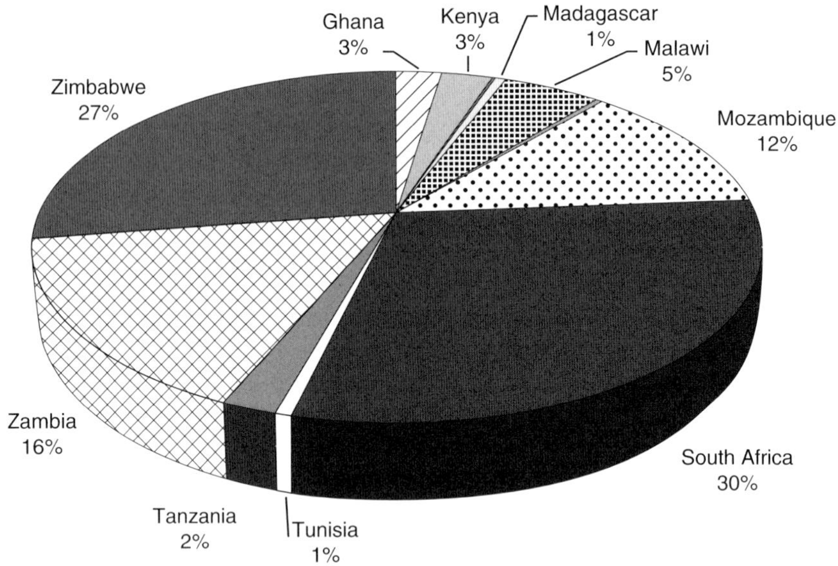

Fig. 1.1. Conservation Agriculture (CA) area in African countries as a percentage of the total 1.2 million hectares (Kassam *et al.*, 2015).

countries include Kenya, Ghana, Tanzania, Tunisia, Madagascar, Morocco, Lesotho, Namibia, Sudan and Burkina Faso.

Some large-scale farmers have been able to adopt profitable mechanized CA in several countries, such as South Africa, Zambia, Zimbabwe, Kenya, Tanzania, Morocco and Tunisia. However, in much of Africa, agriculture is dominated by smallholder farmers. They have different sets of drivers and challenges compared with large-scale farmers and they need greater support to adopt and practise CA (Derpsch *et al.*, 2015). They use manual labour, animal traction or tractor power. Several participatory approaches to CA adoption and scaling up have been tested successfully. These include Farmer Field Schools, Lead Farmer Networks and No-Till CA Associations. Where mechanization is introduced, a service provider model or a group ownership approach, or a combination, can be suitable depending on local constraints and the nature of the overall development support, including training and access to technical expertise, affordable supply chains and markets.

1.4 Key Stakeholders in the Evolution of CA in Africa

The introduction and promotion of CA for smallholder agricultural and livelihood development in Africa has been championed by the Food and Agriculture Organization (FAO) of the United Nations since the mid-1990s in partnership with non-governmental organizations (NGOs), national governments and research and development partners. Some of the earlier countries where the FAO Technical Cooperation Projects (TCPs) were implemented in partnership with national governments are Burkina Faso, Kenya, Tanzania, Uganda, Ghana, Sudan, Zambia, Swaziland, South Africa, Mozambique, Lesotho, Eritrea and Egypt. The Soil Fertility Initiative launched by FAO and the World Bank in 1996 supported CA in several African countries.

Promotion and spread of CA in Africa has relied on funding mainly from donor support through specific time-bound projects. One of the exceptionally longer-term programmes, since 1996, has been the Royal Norwegian Government-initiated CA support programme in Zambia with the Conservation Farming Unit (CFU). The programme, which is ongoing and backed by government, has attained remarkable achievements. Other interventions include the Norwegian support to the South African Development Community (SADC) countries through FAO for a period of 4 years from 2008. The Norwegian Agency for Development Cooperation (NORAD), the UK's Department for International Development (DFID) and other donors backed the Common Market for Eastern and Southern Africa (COMESA) climate-change project efforts, extending from the SADC to the East African Community (EAC). Other projects include: support to the smallholder CA project in West Africa (Burkina Faso, Guinea and Niger) funded by the International Fund for Agricultural Development (IFAD) and implemented by the African Conservation Tillage network (ACT), the World Agroforestry Centre (ICRAF) and the Agricultural Research Centre for International Development (CIRAD); IFAD-funded support to CA in four SADC countries implemented by the International Maize and Wheat Improvement Centre (CIMMYT); and ACT's capacity-building to research workers and extension officers from SADC countries with support from SADC, European Union (EU), German Development Cooperation (GIZ), FAO, Regional Land Management Unit (RELMA) and LandCare South Africa.

ACT and partners organized the FAO-sponsored Third World Congress on Conservation Agriculture in Nairobi, Kenya, in 2005 (proceedings at http://www.fao.org/ag/ca/doc/IIIWCCA.pdf). The event created the much needed awareness about CA for policy makers, research and development practitioners, farmers and the private sector. Other prominent international NGOs and development organizations that are promoting CA in Africa include: CONCERN Worldwide, Canadian Food Grains Bank, CARE International, Total LandCare, Howard Buffett Foundation and Aga Khan Foundation. The Alliance for a Green Revolution in Africa (AGRA) sponsored by the Gates Foundation and the Rockefeller Foundation began supporting CA in partnership with ACT from 2012 through their Soil Health Projects in Kenya and Tanzania. In addition, there are several national-level NGOs that are promoting CA, namely: Kwa-Zulu Natal No-till Association in South Africa; Conservation Farming Unit (CFU) in Zambia; Foundation for Development in Zimbabwe; and Association pour la Promotion d'une Agriculture Durable (APAD) in Tunisia.

The focus of most CA initiatives has been on: food security and livelihood development; participatory adaptive research with smallholder farmers for technology development for sustainable production; and advocacy for public and private sector support. Such initiatives are bound to have significant implications for adoption and spread of CA in the region, and need to be supported and encouraged.

The private sector has also contributed significantly to the current status of CA in Africa. The main stakeholders include large-scale farmers (e.g. in South Africa, Kenya, Tanzania, Zambia and Zimbabwe), CA equipment manufacturers and distributors, and agricultural input suppliers. Their successful implementation of CA, especially in marginal and diverse conditions, has provided useful learning platforms for other farmers, policy makers and development organizations. Some large-scale farmers have even introduced outreach programmes to support smallholder neighbouring farmers.

Engagement of Regional Economic Communities (RECs) across Africa in the promotion and uptake of CA is considered to be essential but currently it is weak. RECs are expected to ensure the existence of a conducive development environment for all stakeholders to play their respective roles. Good policy environment, commitment of national governments, and public and private sector institutional support are key to successful implementation of CA and climate-smart agriculture (CSA) programmes in Africa. It is therefore necessary to have a regional platform where regional bodies can share evidence-based CA information to enable the formulation and implementation of policies and institutional strategies that can attract investments to support the introduction, adoption and spread of CA as a core component of CSA initiatives.

The importance and role of CA in sustainable agricultural growth and economic development has been clearly documented and can, therefore, be considered as the most appropriate entry point in transforming agricultural production in Africa. However, as noted already, the adoption and spread of CA in Africa has been slow, due to the challenges and constraints outlined above and in this book. Addressing these challenges and constraints requires interventions at higher levels to ensure development of appropriate policies and regulatory frameworks to support adaptation, adoption, scaling up and mainstreaming of CA and to attract the private sector to invest and develop CA supportive businesses around agricultural commodity value chains and input supply chains.

1.5 New Partnership for Africa's Development (NEPAD)

NEPAD is a programme of the African Union (AU). It was established in 2001 to create institutions and mechanisms that will work to achieve Africa's development objectives (Landsberg, 2008). Realizing that Africa as a region can only take its proper place in the international community if it gains economic strength, the AU Assembly of Heads of State and Government (HSG) set an ambitious target of 7% annual growth rate in gross domestic product (GDP) over the next 20 years to eradicate poverty, achieve food security and build the foundations of sustainable economic development on the continent. The AU and its NEPAD programme intend to mobilize domestic and external resources and to establish new forms of partnership with domestic and international organizations, thus creating development opportunities for regional and sub-regional institutions.

At continental level, AU-NEPAD hopes to eradicate poverty in Africa and to place its countries on a path of sustainable growth for the next 15 years. It is a commitment of African countries, both individually and collectively, to the former Millennium Development Goals (MDGs) and the new Sustainable Development Goals (SDGs). NEPAD believes that improvement of the performance of the agricultural sector is a prerequisite in meeting its set targets for economic development. It further recognizes that improving the productivity of agriculture rests on dealing effectively with a number of critical constraints such as climate variability, poor rural infrastructure, unsupportive policies and weak institutional and regulatory frameworks. Regional and sub-regional institutions are expected to play a lead role in spearheading science and technology in the development of sustainable food and agriculture systems.

1.6 Comprehensive African Agriculture Development Programme (CAADP)

CAADP was established by AU-NEPAD in July 2003 as the highest-level policy framework for the coordinated transformation and development of agriculture in Africa, as well as for wealth creation, food security and nutrition, economic growth and prosperity for all. The overall goal of CAADP is to 'help African countries reach a higher path of economic growth through agriculture-led development which eliminates hunger, reduces poverty and food insecurity, and enables expansion of exports'. CAADP is a growth-oriented agriculture development agenda aimed at increasing agricultural growth rates to 6% per year to create the wealth needed for rural communities and households in Africa to prosper. To achieve this goal, CAADP focuses its interventions in four key pillars to achieve measurable outcomes:

- Pillar 1: extending the area under sustainable land management and reliable water control systems;
- Pillar 2: improving rural infrastructure and trade-related capacities for market access;
- Pillar 3: increasing food supply, reducing hunger and improving responses to food emergency crises; and
- Pillar 4: improving agriculture research, technology dissemination and adoption.

CA, as a climate-smart concept and approach to sustainable agricultural intensification and land management, has a major contribution to make to all four CAADP pillars. At the continental level CA has been incorporated into the regional agricultural policies of NEPAD as part of the responses to food insecurity and rising food prices, as well as embracing climate-change adaptation and mitigation measures, in sustainable land management practices.

The AU Assembly meeting in June 2014, in Malabo, Equatorial Guinea, adopted the Malabo Declaration on CAADP and commitment to accelerate 'Agricultural Growth and Transformation for Shared Prosperity and Improved Livelihoods' (AU Assembly, 2014a). The Declaration included: (i) commitment to ending hunger by 2025; (ii) commitment to halving poverty by 2025 through inclusive agricultural growth and transformation; and (iii) commitment to enhancing resilience in livelihoods and production systems to climate variability and other shocks. A decision from the same meeting on the report of the chairperson of the NEPAD Heads of State and Government Orientation Committee (HSGOC) endorsed the NEPAD Programme on Agriculture Climate Change, with its components on gender empowerment, support to smallholder farmers and the establishment of an African Climate-Smart Agriculture Coordination Platform (AU Assembly, 2014b). It is through the Platform that the Planning and Coordinating Agency (NPCA) will collaborate with partners, including RECs and NGOs, targeting 25 million households by 2025, branded as Vision 25 × 25.

Around a common set of standards, the AU-NEPAD Agriculture Climate Change Programme will rally and coordinate efforts in four interrelated impact and action areas. These are: (i) training, nurturing and skills development (formal and informal) in both technological knowledge and skills as well as management; (ii) improved availability and accessibility to relevant and appropriate CSA practices and CSA support technologies for enhanced community-level adaptive capacity in the face of climate variability, shocks as well as long-term changes; (iii) strengthening and aligning evidence-based policies and institutions to foster enabling environment for CSA development and scaling up, to be pursued through, among others (a) data and analytical support to policy processes, (b) strengthening and informed public voice, (c) local farmer/practitioners associations and purpose-clubs, and (d) champions in various constituencies; and (iv) markets and trade facilitation – from determining the commodities through to marketing and possibly local processing. This dimension is intended to facilitate examination of the viability of the farm production systems in which CSA practices are applied.

The Africa Climate-Smart Agriculture Alliance is Africa's leading initiative to catalyse and foster championing of result-oriented, on-the-ground implementation support towards the attainment of the Africa Union Vision. Operationalized within National Agriculture and Food Security Investment Plans (CAADP NAIPs), National Adaptation Programmes of Action (NAPA) and Nationally Appropriate Mitigation Actions (NAMAs), the Alliance aims to rally public, political, technical and financial action to catalyse and support local systems for expanded and accelerated scaling up of CSA. The 1st Africa CSA Alliance Forum took place in May 2015 in Addis Ababa, Ethiopia.

All CA stakeholders in the public, private and civil sectors, and particularly ACT, who are promoting the adoption and spread of CA in Africa for sustainable production intensification have a major contribution to make to the achievement of the agricultural growth and transformation commitment of the Malabo Declaration on CAADP as well as to the NEPAD Agriculture Climate Change Programme.

References

AU Assembly (2014a) Malabo Declaration on Accelerated Agricultural Growth and Transformation for Shared Prosperity and Improved Livelihoods. Assembly/AU/Decl.1(XXIII) In: *Decisions, Declarations and Resolution of the Assembly of the Union, Twenty-Third Ordinary Session, 26–27 June 2014, Malabo, Equatorial Guinea*. African Union, Addis Ababa, Ethiopia.

AU Assembly (2014b) Decision on the Report of Heads of State and Government Orientation Committee (HSGOC) on NEPAD. Doc. Assembly/AU/9(XXIII) In: *Decisions, Declarations and Resolution of the Assembly of the Union, Twenty-Third Ordinary Session, 26–27 June 2014, Malabo, Equatorial Guinea*. African Union, Addis Ababa, Ethiopia.

Derpsch, R., Lange, D., Birbaumer, G. and Moriya, K. (2015) Why do medium- and large-scale farmers succeed practicing CA and small-scale farmers often do not? Experiences from Paraguay. *International Journal of Agricultural Sustainability* 14(3), 269–281. doi: 10.1080/14735903.2015.1095974.

FAO (2011) *Save and Grow: A Policymaker's Guide to the Sustainable Intensification of Smallholder Crop Production*. FAO, Rome, 102 pp.

FAO (2013) *Climate-Smart Agriculture Sourcebook*. FAO, Rome, 545 pp.

FAO (2016) *Save and Grow in Practice: Maize, Rice and Wheat*. FAO, Rome, 112 pp.

Gonsalves, J., Campilan, D., Smith, G., Bui, V.L. and Jimenez, F.M. (eds) (2015) *Towards Climate Resilience in Agriculture for Southeast Asia: An overview for Decision-Makers*. International Center for Tropical Agriculture (CIAT), CGIAR Research Program on Climate Change, Agriculture and Food Security (CCAFS), Hanoi, Vietnam, 450 pp.

Kassam, A.H., Friedrich, T., Derpsch, R. and Kienzle, J. (2015) Overview of the worldwide spread of conservation agriculture. *Field Actions Science Reports 8*. Available at: http://factsreports.revues.org/3966 (accessed 15 September 2016).

Landsberg, C. (2008) The birth and evolution of NEPAD. In: Akokpari, J., Ndinga-Muvumba, A. and Murithi, T. (eds) *The African Union and Its Institutions*. Jacana, Johannesburg, pp. 207–226.

Sims, B. and Kienzle, J. (2015) Mechanization of conservation agriculture for smallholders: issues and options for sustainable intensification. *Environments* 2, 139–166, doi: 10.3390/environments2020139.

2 Weatherproofing Agriculture with Conservation Agriculture

Amir H. Kassam,[1]* Saidi Mkomwa[2] and Theodor Friedrich[3]

[1]*University of Reading, UK;* [2]*African Conservation Tillage Network, Nairobi, Kenya;* [3]*Food and Agriculture Organization of the United Nations, Havana, Cuba*

2.1 Introduction

Weatherproofing agriculture generally refers to protecting agriculture from the variability in weather, thereby avoiding the negative effects of fluctuations or variations, including extreme events, and of longer-term trends or changes over time and space in weather or climate. It also refers to making agriculture 'climate-smart' so that agriculture has the resilience to stand up to climate variability and climate change, and to be able to cope with extreme events and recover from shocks, while at the same time reducing the contribution of agriculture towards climate change or even helping to mitigate it.

Weather and climate (i.e. long-term weather) include parameters such as precipitation (rainfall, snow, hail and their amount, intensity, timing and duration, distribution in time and space, seasonality, droughts, floods), temperature (minimum, maximum, diurnal range, seasonality, heat waves, cold spells and frost), air humidity (level, duration, timing) and wind (speed, timing, duration, typhoons, cyclones, hurricanes). These parameters, and others such as solar radiation and characteristics of soil and terrain or landscape as well as crop and cropping system, define the reference agroecological potential of land at a particular location for specific crops, cropping systems and farming systems. When considered over large areas or landscapes and ecosystems, they define the reference agroecological as well as socio-economic potentials and suitability for agriculture systems development. Weather and climate parameters and their fluctuations and longer-term trends, as well as temporal and spatial patterns, have significant

*amirkassam786@googlemail.com

influence on the kind of agriculture (in terms of crop production or farming systems or food and agriculture systems) that can be practised or developed and sustained in an area or territory, the systems' agroecological potentials and their actual performance.

In its broader sense, weatherproofing agriculture would need to cover all aspects of the agroecosystems, including: crops and livestock at the molecular, cellular and tissue level, to whole-plant and whole-animal level, and beyond into cropping systems (including livestock) and farm system levels; and all components of the natural infrastructure comprising the soil and terrain, water, nutrients, carbon, vegetation and biodiversity, in space and time, and all the soil- and landscape-mediated ecosystem functions that define the ecological and agroecological land potentials and actual performance of crops, farms and farming systems, and the food and agriculture systems or sectors of which they form a constituent part.

Given the difficulty in predicting weather and extreme climatic events, it is important that as much resilience as possible is built into the design and management of production systems. However, it is equally important to understand the weaknesses and vulnerabilities to climate variability and change in conventional tillage-based agriculture.

This chapter elaborates on why no-till Conservation Agriculture (CA) – comprising no or minimal soil disturbance, permanent soil cover, and cropping system diversity, plus other good agricultural practices – is better able to weatherproof agriculture compared with tillage agriculture. This is elaborated by examining soil management in relation to the water-related ecosystem services for CA and tillage-based agriculture. The chapter focuses conceptually on the narrower topic of the adaptation of farming practices to address climatic variability and change. This focus relates to the adaptation of production systems and farming systems as well as of ecosystems and landscapes to address climatic variability and change, including devising solutions at the level of farmer groups and community-based actions within watersheds and beyond. The adaptation of the food and agriculture systems is relevant to this discussion because some of the weatherproofing actions that may be taken at the sector level can have a direct impact on the weatherproofing of production systems and practices at the farm and field level.

Other water-related aspects such as soil formation and its water-related functions, agricultural land use and water services at the watershed level and climate-change adaptability involving adaptation and mitigation are briefly touched upon. The chapter highlights the paradigm shift that is currently under way in which conventional tillage-based production systems are being replaced by climate-smart CA systems to achieve sustainable production intensification.

2.2 The Scope of Weatherproofing Agriculture

The inclusion of soil and terrain (landscape) and the natural resource base in the discussion and objectives of weatherproofing agriculture makes the task of explaining the scope of adaptations of farming practices incredibly complex. This is because the effects of a weather event, such as the occurrence of rainfall on crop growth and on crop and cropping system performance, depend not only on the rainfall characteristics, including the frequency of occurrence of dry periods and their timings in relation to crop phenology and development of individual crops in the cropping system, but also on the dynamics of soil water balance across the cropping system. The latter in turn depends on several other factors, including: soil quality and health; whether the soil is protected with a layer of mulch against the erosive force of rainfall and evaporation loss from soil surface; whether soil management favours the maintenance of high infiltration rates, or whether the soil is poor or good in terms of its carbon content, structure or depth; and whether the soil, including the subsoil, is compacted and its surface sealed and unable to let water infiltrate and be retained in the soil for root and crop growth and for soil life and functions.

Crop response to weather fluctuations and longer-term trends would also depend on whether the soil management practices applied by the producer favour symbiotic endophytes, which benefit plant growth and health, and the myriad other soil microorganisms such as mycorrhizae and mesofauna such as earthworms, which can flourish in the soil and contribute to the productivity potential and resilience of the soil system as well as that of plants, crops, farming systems and of larger agroecosystems.

There is clear evidence that there are symbiotic relationships between microorganisms and crops that have coevolved; and in the case of endophytic symbiotants there is evidence that they improve not only growth and development of crops but also their resilience to biotic and abiotic stresses. Mycorrhiza is a well known case but there is evidence related to rust resistance in wheat, as well as improved photosynthesis in rice (Chi *et al.*, 2005, 2010; Rodriguez *et al.*, 2009).

Indeed, it is realized that weatherproofing of agriculture requires, first and foremost, a deeper understanding of how crop and land potentials, as well as agricultural development potentials upon which they rely, can be underpinned ecologically. Such ecological underpinning is necessary to minimize or even avoid the impacts of weather or climate variability and change the performance of the production system and its component subsystems and practices, and the response of production inputs applied to raise output.

Similarly, the weatherproofing objective should be seen at all levels of farming systems and of the food and agriculture systems, covering the landscape- and ecosystem-level aspects of sustainability and resilience as well as the food and agriculture sector level aspects, and across all the related sectors such as water, energy, environment, industry, trade, manufacturing, research, education, transport, marketing, catering and communication.

At the level of food and agriculture systems, weatherproofing requires the weatherproofing of all parts of the greater agricultural system, including all the service providers on the input supply side as well as on the output value chain side, and all aspects of cropping and farming systems, from the molecular, cellular and tissue level in plants and animals to crops and cropping systems to farming systems and beyond to area, landscape and territorial levels (Bunting and Kassam, 1986). This means that weatherproofing agriculture would require that biotic and abiotic stresses arising from climatic variability and extreme events are dealt with at the genetic level, at the agronomic (phenotypic) level and at the cropping system level as well as at the whole agriculture system level.

Where conventional tillage agriculture is being practised, the natural infrastructure of land, water, land cover, biodiversity and ecosystem functions and services generally would be in a degraded state, at any scale and level of development. In some instances the natural resource base would be severely degraded to the point where land has been abandoned or has been desertified, often leading to local and international conflict. Montgomery (2007) provides a rich account of how humankind throughout the ages has been degrading the very land resources base upon which agricultural productivity and food security depend, through unsustainable tillage-based agricultural practices and systems at the farm level.

Climate represents long-term weather but, in a changing climate, weather events and patterns are in a state of flux in time and space, as has been seen from the increase in the occurrence of extreme events and also in the intensity of extreme events and in their spatial distribution. 'Climatic variability' refers to the fluctuations in climatic parameters around the mean, and 'climate change' refers to the longer-term trends in the various climatic parameters, so that the mean as well as the spread around the mean are in a state of flux. Global climate-change models attempt to predict the likely large-scale spatial and temporal changes in the weather and climate characteristics but it is not possible to do so with any precision. However, local weather forecasting models that combine historical analysis with real-time satellite weather data and images are able to offer reasonably accurate daily weather forecasts that would have the potential to contribute to weatherproofing agriculture strategies, especially if such forecasts can be made to improve farming decisions in, for example, CA systems related to timing of farm operations such as sowing and harvesting, irrigation water management, application of plant nutrients and management of insect pests, pathogens and weeds. Thus weatherproofing agriculture often means the incorporation of various aspects of resilience at all levels of the food and agriculture system to enable the society and economy as a whole to cope with increased variability and extreme events, as well as longer-term changes in climate. This really is a very tall order and requires attention to be paid to all aspects of food and agriculture systems.

2.3 Weatherproofing Agriculture: Adaptation of Farming Practices

Adaptation of farming practices to weather and climate variability has always been of serious concern to farmers of all types simply because of its effect on crop and farm biological, environmental and economic performance – output and factor productivity, including area productivity or yield, but also how crops and cropping systems can cope with weeds, insect pests and pathogens, and how land- and crop-related operations of crop establishment, weed management, nutrient management, water management, harvesting and drying as well as storage and supply chains are organized and performed. This is relevant also at the landscape and ecosystem level where there is a mosaic or a mixture of land uses, all coexisting within a common watershed whose macro-level ecosystem services performance depends on the performance of the contiguous micro-level farming systems.

Adaptation of farming practices refers to changes that can be made in farming practices to counteract, or reduce or mitigate, the likely negative effects of variability and change, and with this to reduce the risk for the farming families. If an existing farming practice can be changed to make farming more adapted to suit a new agroecological situation caused by increased variability or climate change, then it has a continuing value to agriculture. On the other hand, if further adaptation of prevailing farming practices or systems to address the new climatic variability and change situations is not possible, then they must be modified or replaced with something different that can succeed.

Agroecosystems comprise agricultural lands with farming systems and households that also have their respective socio-economic and environmental relationships with the wider surroundings in the landscape, including all the ecosystem resources and processes within the natural infrastructure. Thus the flow of ecosystem services from farmland depends on the extent to which farming practices enable or disrupt such flows and the underlying ecosystem functions.

Provisioning ecosystem services such as fresh water supplies, biological nitrogen fixation and biological products, and regulatory ecosystem services such as regulation of groundwater and streamflow and control of runoff and erosion, can only function adequately provided that the supporting ecosystem services such as water, nutrient and carbon cycling and soil formation are operating sustainably (MEA, 2005). If they are not, then the provisioning and regulating services also become suboptimal. Ideally, farming system sustainability and resilience would depend on the extent to which the system is managed to allow its ecological functions to operate optimally, including its ability to self-repair and self-regulate based on the protective ability of system buffers against climatic variability, extreme events and longer-term climate changes.

Since the First World War, the scientific formulation that has been driving the evolution of production and farming systems as well as production intensification has been based on an agricultural paradigm that assumes that, to achieve more output, more production inputs must be applied and that, for this to occur, specially bred input-responsive crop cultivars and animal breeds must be used.

This approach views crop and agricultural production systems to be closed systems and presumes that all or at least the best production inputs must be manufactured and purchased. This input–output system concept also assumes that production systems must involve intensive mechanical tillage of the soil to prepare the seedbed and to manage weeds. This approach has also involved increased standardization of agronomy and inputs, reduction in diversification and reliance on genetically altered cultivars and animal breeds as a basis for production intensification in rainfed and irrigated production systems. It was this approach that formed the ideological or doctrinal basis for the Green Revolution agricultural paradigm after the Second World War, with its heavy dependence on modern cultivars, mineral fertilizers, pesticides, intensive tillage and more and more reliable use of irrigation water: the Green Revolution was a hungry and thirsty technology.

Adaptation of farming practices and systems to climate change must first ask the questions: what is the production system paradigm that one is dealing with, and what are the root causes of ecological weaknesses and vulnerabilities of the production systems and of the overall agricultural system against climate variability and climate change? To be weatherproofed and climate-smart involves, to start with, the ability of plant tissues and whole plants and field crops, cropping systems and farm systems and ecosystem services to be able to cope with the prevailing 'normal' pattern of climate variability in space and time. If this coping ability is in place, then it becomes possible to deal with increased variability due to climate change, including extreme events of greater intensities and any longer-term trends in the climatic parameters and agroclimate and agroecological conditions as a whole. However, this 'normal' coping ability of production systems depends on the kind of agriculture that is under consideration; for example, whether it is conventional tillage-based Green Revolution agriculture or whether it is the alternative no-till agriculture such as CA described in Section 2.4.

One of the major weaknesses in the paradigm of the conventional tillage agriculture approach to production intensification is that it is not climate-smart, because it shows meagre concern for soil ecological health and for agroecosystem health. This weakness is reflected in the continued acceptance and promotion of intensive mechanical soil disturbance, no mulch cover (including stubble mulching) to protect the soil surface and to serve as substrate for sustaining soil life, and inadequate cropping system diversification (including legumes) as an integrated strategy for minimizing soil degradation and intensifying production. This weakness is exacerbated by the strident focus on yield as the main or only output of concern in an input–output production sense, not paying much attention to the ecological basis of high factor productivity (factor efficiencies) and agroecosystem resilience to biotic and abiotic stresses, including climate variability and extreme events. Worst, even less attention is paid to soil health and ecosystem-related functions and services on-farm and from the landscape and agroecosystem as a whole.

Since the Second World War the drive for intensification has been through even more intensive tillage and mechanization and standardization of agronomy and farm operations – more mineral fertilizers and more pesticides, less agrobiodiversity – while responding to market demands that have over the years led to reduced crop diversification and increased vulnerability to pests, pathogens and weeds, and to climate variability and climate change. Major effects of this kind of intensification on soils have been: decrease in soil porosity and structure, loss of soil organic matter (SOM), decreased infiltration, decreased water retention capacity and falling water tables, increased soil erosion and soil compaction, which in turn have led to agroecological, environmental, economic and social vulnerability.

By definition, climatic variability is to do with fluctuations of climatic events such as daily rainfall over time. Agricultural landscapes and ecosystems need to have their climatic resources characterized so that farming systems are devised and practised that match the prevailing climate in terms of water balance, length of growing season, start and end of the growing season, surplus rainfall above evapotranspiration, etc., and the variability of these parameters for planning of cropping systems (FAO, 1978–81; Higgins *et al.*, 1982), as well as the possibilities of integrating cover crops to take advantage of prevailing variability in the length of growing season from year to year, particularly during those years when the seasons are longer than average.

Climatic events in individual future years are not easy to predict, thus the characterization helps to define the nature of the prevailing climatic resources, including the variability in each of the important weather parameters and the associated risk of each in terms of drought, floods, heat waves and frosts. For this reason the cropping system at micro levels has to be able to cope with such diversity of possible climatic scenarios. However, also at macro levels, it will be impossible to predict exactly which type of climate and weather pattern will prevail for the next season and hence which would be the best crops to grow. A traditional coping strategy against such risks has been the diversification of cropping pattern and of production areas. This, however, was replaced during the Green Revolution with a trend to specialize, resulting in farming units producing a single product and in expansion of monocropping cultures.

2.4 Why Conservation Agriculture is Able to Weatherproof Agriculture

CA practices help to weatherproof production systems and build resilience in a number of ways, namely by: (i) rehabilitating and conserving natural resources, including reduction in soil erosion and compaction, and enhancing soil health; (ii) sequestering carbon; (iii) improving water cycling and efficiency of water and nutrient use; (iv) timeliness in operation; and (v) improved cropping ratio. All these aspects, along with genetically adapted and improved cultivars, operate together and synergistically in CA systems to weatherproof them and build and maintain system-level resilience from ecological and socio-economic viewpoints. These are expressed in a number of ways, such as: improved and more stable yields and profit; reduced production inputs of seeds, water, nutrients, biocides and energy; and greater farm output and return to investment (Kassam *et al.*, 2009, 2013; Rockstrom *et al.*, 2009; Thierfelder *et al.*, 2013, 2015a,b). See Chapter 11 for further discussion on this aspect. At the same time, since diversification of the production system is an integral component of CA, it also helps to reduce the overall production risks related to monocropping in a climate change scenario.

CA practices essentially provide ecological underpinnings to any land-based production system to enable the ecosystem services to function so that land-use systems behave with minimum disturbance to the ecosystem (Jat *et al.*, 2014; Farooq and Siddique, 2015). In addition, all three CA core practices (see Chapter 1, Section 1.1) facilitate a dynamic set of biological processes that can enable other important practices such as integrated management of pests (insects, pathogens and weeds), or integrated nutrient management, or integrated water management to function effectively. Soil biological processes operate when soil life and health are encouraged to flourish by the three core CA practices. One key contribution to soil life and health by CA practices is the return and availability of organic matter with minimum soil disturbance, allowing the soil to develop and function biologically, which is necessary for building soil structure and porosity as well as aggregate stability. This is possible when soil microorganisms, especially bacteria, fungi and mesofauna such as earthworms, termites and other arthropods, are allowed to establish themselves as part of the soil food web and perform their functions (see Chapter 11).

This is why the first CA principle of no or minimum mechanical soil disturbance (Kassam *et al.*, 2009; FAO, 2015) is not a production input in the conventional sense but, when put into practice through no-till seeding, enables a set of ecosystem processes to function *in situ* (in the field) and on-farm (across the whole farm) (Table 2.1) that can contribute to system resilience and thus to productivity and its stability.

The same can be said of the second principle of CA, i.e. maintaining soil mulch cover (including stubble mulching) using crop residues, stubble and cover crops; and the third principle, i.e. maintaining species diversification through rotations, associations and sequences, including intercropping, whose greater resilience characteristics have been reported by many (Andrews and Kassam, 1976).

These three CA principles, when put into practice together through locally formulated adapted practices by farmers, are synergistic and this is why it is preferable that they are implemented with the objective of introducing all three sets of practices over time, along with other good agricultural practices of integrated crop, nutrient, water, pest and energy management. The three CA principles have been shown to be applicable to all land-based production systems, whether rainfed or irrigated, including systems with annual or perennial crops, horticulture systems, orchards and plantations, organic production systems, rice-based systems including System of Rice Intensification (SRI), agroforestry systems, crop–livestock systems, crop–livestock–pasture–tree systems

Table 2.1. Effects of production system components fully applied together on sustainability and ecosystem services. From Friedrich *et al.* (2009).

System component ▶ To achieve ▼	No tillage (no or minimal soil disturbance)	Mulch cover (crop residues, cover-crops, green manures)	Crop diversification (for several beneficial purposes)	Legumes (as crops for fixing nitrogen and supplying plant nutrients)
Simulate optimum 'forest-floor' conditions	√	√		
Reduce evaporative loss of moisture from soil surface		√		
Reduce evaporative loss from soil upper soil layers	√	√		
Minimize oxidation of soil organic matter, CO_2 loss	√			
Minimize compactive impacts by intense rainfall, passage of feet, machinery	√	√		
Minimize temperature fluctuations at soil surface		√		
Provide regular supply of organic matter as substrate for soil organisms' activity		√		
Increase/maintain nitrogen levels in root-zone	√	√	√	√
Increase CEC[a] of root-zone	√	√	√	√
Maximize rain infiltration, minimize runoff	√	√		
Minimize soil loss in runoff, wind	√	√		
Permit/maintain natural layering of soil horizons by actions of soil biota	√	√		
Minimize weeds	√	√	√	√
Increase rate of biomass production	√	√	√	√
Speed soil porosity's recuperation by soil biota	√	√	√	√

Continued

Table 2.1. Continued.

System component ▶ To achieve ▼	No tillage (no or minimal soil disturbance)	Mulch cover (crop residues, cover-crops, green manures)	Crop diversification (for several beneficial purposes)	Legumes (as crops for fixing nitrogen and supplying plant nutrients)
Reduce labour input	√	√		
Reduce fuel-energy input	√		√	√
Recycle nutrients	√	√	√	√
Reduce pest-pressure of pathogens and insects	√	√	√	√
Re-build damaged soil conditions and dynamics	√	√	√	√
Pollination services	√	√	√	√

[a]CEC, cation exchange capacity.

and agrosilvopastoral systems. All crops, including root and tuber crops, can be seeded or planted into untilled soils. Thus, the weatherproofing qualities of CA can be integrated into any land-based production system, offering a range of weatherproofing advantages to a variety of production and farming systems (Table 2.1).

The transformation of agroecosystem equilibrium from a conventional tillage production system into a CA system takes time, depending on the starting set of constraints and level of degradation that must be addressed. This is because a new set of ecosystem processes and functions in the agroecosystem have to be (re-)established at the micro and macro levels to enable ecosystem services to re-establish themselves. This is why some benefits, such as improvements in soil moisture regime, come through sooner and improve over time while others, such as increase in SOM and recuperation of soil structure and soil biopore networks, may take longer, because the relevant soil biological processes must be re-established and take some time to do so. Similarly, to re-establish a more protective insect pest–predator dynamic cannot be done in one season, particularly when starting from a production system situation in which there has been regular intensive mechanical and chemical disturbance of the agroecosystem over many years in which habitats of natural enemies of pests and food webs have been damaged, or destroyed altogether. In general it is helpful to think of the CA transformation process in terms of 15–20 years. This is because the starting level of degradation is often quite severe and it can take up to 10 years of transition time for the soil to rebuild its biological health and perform the soil-mediated ecosystem functions that will support enhanced productivity (Sá, 2004).

Table 2.1 provides a glimpse of a range of outcomes or services that are desirable to have in a weatherproofed production system. The three

CA principles help farmers to integrate practices that can allow such outcomes and services to function within the CA cropping system and CA farm system, and indeed from CA-based landscapes at large-area scales of watersheds and regions to harness ecosystem service benefits to society as well as benefits related to productivity, resilience and system self-recuperation at multiple levels in the landscape – the field, the farm, the local community, the watershed/basin, the greater society, etc.

This is why the three interlinked CA principles are considered, first and foremost, as serving as an ecological foundation of sustainability to the production systems and permit systems to operate at a much higher agroecosystem performance level than at the level of production input such as mineral nitrogen fertilizer or a pesticide. Clearly, this issue cannot be fully discussed here, but it is important to elaborate the roles and functions of farming practices that are related to ecosystem health and ecosystem functions and services that enable production systems to remain fundamentally climate-smart.

Ecosystem services have been categorized as being either mostly supporting services, regulatory services or provisioning services. These services are all related in the natural infrastructure and cannot operate normally unless their interconnections are maintained. Thus, 'blue water' as a provisioning service depends on how well the water-related regulatory services are functioning; and both these services depend on how the supporting services of the water cycle are functioning. So, a serious interest in the water elements of weatherproofing agriculture or production systems must ensure that these sets of relationships are not debilitated by the production practices being used or promoted.

Of course, the total resilience of a production system in terms of its water-related climate-smartness would depend on a range of characteristics from molecular and cellular to plant level to crop and cropping system level. It will also depend on how well the agroecological requirements of crops and cropping systems and of farming systems match the prevailing weather and climate conditions. Additionally, it will depend on how well weatherproofing corresponds to the management of the cropping and production system in space and time, and how weatherproofed are the production inputs and service providers. Further, it will depend on the whole knowledge system, including the education, research, extension and participatory learning and innovation systems. Thus, climate-smart agriculture is not a single-cause–single-effect topic, but rather a complex dynamic system situation that must be understood, managed and serviced for sustainable production intensification and sustainable land management at all levels within the agroecosystem. This would also involve larger-scale soil and water conservation measures to rehabilitate and stabilize the landscape and improve watershed hydrology. In addition it would involve the adoption of CA to enhance soil health and achieve sustainable production intensification and, where applicable, to rehabilitate degraded marginal or abandoned agricultural lands.

2.5 Water Cycle and Water Services for Weatherproofed Production and Society

Production practices, if they are to be weatherproofed, must enable water cycling to continue unhindered as much as possible, otherwise both the provisioning water services such as available soil water to crops and aquifer recharge, as well as the regulatory services such as groundwater and streamflows, are impeded. The most vulnerable time in the water cycle on agricultural land is at the point when rainfall water drops must infiltrate into the soil quickly to maximize rainfall capture. The second point in time is in terms of the ability of the soil to retain the maximum amount of available moisture; and third is the ability of the plant and soil conditions to allow the establishment of the largest possible root system and for a plant's root system to ensure maximum access to the available soil moisture. However, it is not only that the infiltration of water into the soil as such is important, but also that the pathways in which the water is moving in the soil have important repercussions on the contamination of water resources with mineral soil nutrients or other chemicals that could be in the soils.

During soil water-deficit periods, other soil and plant elements are important but the point being highlighted here deals with the management of infiltration and soil water-holding capacity. Both infiltration and soil moisture-holding capacity depend on soil porosity and structure; and the total volume of water available for plants depends on the soil depth. In the case of infiltration, mulch cover on the soil surface helps to dissipate the energy load of rainfall, so it is less erosive and its momentum is broken, and then the infiltration of rainwater into the soil depends on the soil's surface characteristics, its moisture-holding capacity and the internal drainage characteristics of the soil system.

Under tillage-based mechanical soil and weed management, there is continuous de-structuring of the soil system and destruction of soil biological life and functions, which leads to loss in soil carbon and structure, increased runoff and soil erosion and soil compaction. All of these processes seriously reduce the surface infiltration rate, soil moisture-holding capacity and soil drainage. There is also the destruction of habitats of the soil microbiomes and mesofauna responsible for soil health and soil biological processes. The latter include humus formation and carbon sequestration, building and repairing soil structure and networks of soil biopores. These processes are important for symbiotic (including endophytic) relationships among microbiomes, and with the plant root system and the plant as a whole, that impart resilience to withstand abiotic and biotic stresses, but they also create the internal drainage structure in soils and subsoils with a network of continuous macro- or biopores (Chi *et al.*, 2005, 2010; Redman *et al.*, 2011).

The adaptation of farming practice to enhance infiltration and soil moisture-holding capacity, thus imparting weatherproofing to the production system, means that farmers and service providers must take responsibility to minimize or avoid mechanical soil disturbance, cover the soil surface with mulch, and enable soil biology to drive the soil's self-recuperating processes of building structure, networks of biopores, processes of humification and carbon sequestration, and provide substrates for soil microorganism and mesofauna. Soil tillage not only disrupts the continuous biopores in the tilled horizon but also, by destroying the habitat of the soil meso- and macrofauna, it also eliminates the capacity of soil life to penetrate through the macropores into deeper horizons of the subsoil, far beyond the reach of any subsoiler. With such impairment, in the long term, the overall infiltration capacity of tilled soils is seriously reduced.

2.6 Soil Development and its Water Functions

While soil parent material weathers from the bottom at the rock surface and in the subsoil layers to produce mineral soil particles, biologically active soil formation occurs from the top down and leads to the development of the topsoil with its 'O' and 'A' horizons. These horizons degrade rapidly under the influence of mechanical disturbance such as tillage, which not only breaks up the soil aggregate structure, destroys biopores and stokes the fire (burning of SOM) that decomposes SOM but also leads, as explained above, to a decrease in the infiltration of rainfall water into the soil (including the subsoil), an increase in soil surface sealing and an increase in runoff and water erosion. In addition it leads to a decrease in soil moisture-holding capacity and soil water retention, water percolation and drainage and poor groundwater recharge. In turn, this leads to poor response to applied and available nutrients (mainly due to leaching and erosion of soil nutrients), weak plant growth and crop development and lower productivity. Over time, tillage leads to the formation of plough pans or hoe pans in the subsoil which can cause other problems such as waterlogging, surface flooding or droughts.

Thus with time, the soil degradation described above leads to the collapse of the soil layers, causing compaction and reduced subsurface drainage. At least two sealed layers are created due to tillage: the surface seal from the pulverized surface soil and the subsurface seal from the hoe or plough pan. Rates of infiltration can reduce down to 10–20 mm/h or less in agricultural soils that are being managed with mechanical tillage practices, compared with 100–300 mm/h or more for soils that are under no-till soil management practices (Landers, 2007). Similarly, under no-till soil management the moisture content of the soil has been shown to be higher compared with that under tillage management, often during the entire season (Derpsch *et al.*, 1991; Shaxson *et al.*, 2008).

Due to soil erosion and land degradation that occurs over a long period of time, many soils do not have adequate depth for good crop production; and in many instances soils have coarse material from the subsoil appearing closer at the surface. Where the land is on slopes, this kind of degradation is more severe and requires land stabilization and rehabilitation management with soil and water conservation measures. In general, a healthy porous loamy soil with above 2% SOM can have up to 60% air space, of which some

50% would be available to hold moisture at field capacity. More specifically, a soil with a moderate depth of 50 cm (500 mm) that has 50% air space and 50% moisture-holding capacity would hold some 125 mm of water. Similarly, a soil that is 1 m in depth would hold some 250 mm of water. Available water for the crop would be some 75%, offering a great deal of buffer against drought and heat (when soil water is available to plants).

Under tillage agriculture, this is not so. Due to poor soil structure and compaction, soil moisture-holding capacity is greatly reduced and soils and crops suffer water stress relatively quickly. Thus, the aim of good soil management practice should be to ensure that the soil, regardless of its depth, is able to hold the maximum amount of moisture at any given time to serve as a buffer against dry weather. A compacted soil of 50 cm depth that has less than 0.5% carbon and is managed under tillage soil practice would probably have a pan below 20–30 cm and the soil above it would have poor structure. Not only will it not have the ability to hold its full potential amount of moisture, but also it will not allow water to go down to deeper soil layers to be retained. It is no wonder that such soils, and the crops on them, suffer drought (and heat) stress much faster than no-till soils with mulch cover.

Unless soil tillage practices are stopped, agricultural soils will continue to degrade and erosion losses will continue unabated until the topsoil and much of the soil below are gone, and farmers are forced either to abandon their land or to introduce temporary measures of ridging and tie-ridging to cope with the hard soil pan below and with the remaining de-structured topsoil above. Under such conditions, crop response to heavy (and heavier) application of mineral fertilizers will be low and will become ever lower with time. Diminishing returns set in, because the soil system is not functioning with its active and diverse biological residents. In such soils, if mineral fertilizers are being applied, it is likely that nutrient uptake efficiency will be very low. Even if organic matter is also added as a source of nutrients with or without mineral fertilizer, the poor soil condition will not be able offer the desired weather-smart water and nutrient productivity performance.

Further, even without the presence of a marked plough pan, the subsoil in tillage-based farming over time forms a solid unstructured filter body of mineral material with no larger pore volumes. In annual crops very few roots will find their way into this material, which has a significantly higher density than the tilled A horizon, at least in the beginning of the cropping cycles. Those crop plants will depend mostly on the water reserves of the A horizon and will be very susceptible to drought spells. Water infiltration capacity of the subsoil in the absence of biopores will depend only on the texture of the base mineral material. It will be high to moderate in gravel or sandy subsoils, limited in loamy materials and close to zero in clay soils.

The following is an elaboration of an example that is important in terms of weatherproofing agriculture to make it climate-smart from the viewpoint of water services in a watershed with a reservoir and a dam for generating hydroelectric power.

2.7 Climate-smart Water Services at the Watershed Level

Climate-smart agriculture must also operate at the larger area or watershed level so that the watershed services can also operate in a climate-smart manner in agricultural watersheds. The best example of this is the way in which agriculture and land use in the Parana Basin III in Brazil has been transformed into a climate-smart land management system to reduce erosion, sediment load, pollution and land degradation. This has been done through a programme called 'Cultivating Good Water' (Mello and van Raij, 2006; ANA, 2011; Itaipu, 2011) in which tillage-based agriculture has been replaced by CA. The programme has led to cleaner and less polluted water draining into the Itaipu Dam reservoir, increasing the dam life manifold, increasing agricultural productivity, improving community livelihoods and fostering a multi-stakeholder partnership that is weatherproofing the climate-smart agriculture at several nested levels.

The Itaipu example also highlights the fact that green water and blue water or water in general are part of the same greater ecosystem. What happens to water when it hits the ground as rainwater depends very much on how we take care of the soil and landscape system, which in turn can have an impact on the soil water balance, groundwater recharge, streamflows and the quantity and quality of blue-water resources available for other uses such as domestic, irrigation, hydroelectric, recreational and environmental purposes.

2.8 Policy and Institutional Support at the Sector Level

In line with what has been stated above, weatherproofing or climate-smartness must permeate all aspects of the production system, including those that are responsible for servicing the whole farming sector. Development of a commensurate research, knowledge and innovation system is a key requirement if it is to contribute to the transformation of conventional tillage agriculture towards greater climate-smartness based on, for example, CA. This is because experiential and scientific evidence will always be required in order to accelerate and strategically direct the necessary changes. It is becoming increasingly clear that, due to climate change, optimal sowing and harvest dates in many locations are changing and so are the agroclimatic characteristics of rainfed and irrigated growing environments. This calls for greater attention to be paid to characterizing any ongoing shifts in growing period zones and in their temperature and moisture regimes, and indeed in watershed hydrology.

While adoption and spread of CA across Africa is essential for building greater resilience, on its own it is not likely to be enough to lead to maximum benefits to rural communities and to the nation as a whole. The development of climate-smart agriculture must be backed up by effective policy and institutional support, so that all stakeholders in the private, public and civil sectors become aligned to the common goal of mainstreaming the weatherproofing of agriculture over the coming decades. This task cannot be left to individual sectors – science or industry or private sector or farmers or the education system or government. All must become engaged as is beginning to happen in certain parts of the world, e.g. Brazil, Paraguay, Canada, Australia, Zambia, Zimbabwe, Kazakhstan and China.

Adaptation to climate change at the overall development goal level must include both adaptation and mitigation at sector and subsector levels. For example, in the case of agriculture, weatherproofing the production system by adopting CA makes it climate-smart in terms of adaptation to climate variability and climate change, but also in terms of mitigating climate change. Climate-change mitigation at the production level is largely to do with farming practices that allow the decrease in greenhouse gas (GHG) emissions of CO_2, N_2O and CH_4 from soil and animals as well as from heavy fuel use. At the farming system level, it is also to do with GHG emissions from the manufacture of production inputs of mineral fertilizer, machinery and equipment, the size and life of machinery and equipment, etc. Thus, climate-smart agriculture must also be supported by climate-smart life cycles that service agricultural production and the input-supply and output-value chains with the smallest environmental footprint.

With intensive tillage production systems supported by conventional Green Revolution practices of high agrochemicals and heavy machinery, it is difficult to become a net mitigator of climate change simply because the GHG emissions are high and the net GHG balance contributes to greater global warming potential; whereas in CA production systems there is a net mitigation outcome because not only is there the possibility of carbon sequestration in the soil but also there is a decrease in fuel use and eventually in horsepower required because of no tillage and lower agrochemical use. Thus, the carbon footprint of CA systems can be expected to be much smaller than that of conventional tillage agriculture systems and it can contribute to the development of a low-carbon economy.

Similarly, a carbon offset trading scheme involving agriculture production is not possible with tillage-based agriculture but, following a strict protocol, it is possible under CA systems as has been done by the Alberta government in Canada for the farming sector in the prairies of Alberta Province (Haugen-Kozyra and Goddard, 2009; Marketwired, 2014). Similarly, CA systems are more water efficient and have higher biological and water productivity, so less water needs to be pumped for a given required output.

2.9 Concluding Remarks

Business as usual is no longer tenable, because much higher levels of resilience need to be built into all agricultural land-use systems to achieve the required sustainable intensification and contribute to sustainable development. This chapter does not provide a complete picture of what is required to weatherproof agriculture and what is involved to develop climate-smart agriculture at all levels. Weatherproofing agriculture for sustainable agriculture development is very much a work in progress, because changes that are required in the agriculture paradigm are revolutionary in many ways. The scope of what must change is large and some of the changes that must occur are at the foundation level of the way that we think about agriculture and the stakeholders who contribute to it. Agricultural science and practices linked to tillage agriculture and Green Revolution agriculture are becoming less and less relevant for the future because of their inability to respond adequately to the local, national and global challenges of climate variability and climate change and also of food insecurity, poverty alleviation, degradation of land, biodiversity and environment, resource scarcity, rising costs of inputs, and demand for ecosystem services and quality food.

The new ecologically based agricultural paradigm, as reflected, for example, by the science and practice of CA, is now replacing the conventional Green Revolution agricultural sciences and practices, as well as the research, education and extension systems that have been supporting them for the past several decades.

This transformation is a paradigm change of a magnitude corresponding to 'flat earth versus round earth' beliefs. For agricultural development to meet future human and environmental needs, the agronomic and genetic sciences that underpinned the Green Revolution in the 20th century need to accommodate the new scientific knowledge that is coming in now from cutting-edge work in areas such as soil microbiology, soil ecology, soil health and functions, epigenetics, molecular and cellular sciences, plant physiology, science of ecosystem functions and services, sustainable landscape management, and in farmer and societal participation in agriculture development and land resources stewardship. The practices that the Green Revolution agricultural world view supported and promoted need to be reconsidered and transcended, as they are taking agriculture and society down a destructive path of nominal development that offers bleak prospects under the changing and more challenging conditions of the 21st century. There is no more need for developing countries to follow the Green Revolution agriculture paradigm of the 20th century, as it is being replaced in every continent by a much more effective, efficient and resilient production paradigm of CA upon which to build future food and agriculture systems as well as sustainable natural resource management and economic growth.

CA is now spreading at the annual rate of some 10 million hectares and covers more than 157 million hectares globally, equivalent to some 11% of cropland (FAO, 2015; Kassam *et al.*, 2015). There are also permanent pasture-based CA systems and tree-based perennial CA systems (including plantation systems in the tropics and outside) which are expanding in several parts of the world. As Chapter 1 shows, Africa already has more than 1 million hectares of CA across many countries, involving hundreds of thousands of farmers, and much of the area has entered into the CA transformation process in recent years as more development resources and attention are being applied to support its adoption and spread.

While this chapter focuses on adaptation of farming practices to address climatic variability and change, weatherproofing agriculture as a whole in the context of equitable development and sustainable economic growth requires the engagement of all immediate and distant stakeholders. This includes farmers and their rural communities and civil societies, and all the business sector and public sector service providers in the greater food and agriculture system of the economy. Knowledge and innovation systems must be open to incorporating fresh thinking and approaches offered by the agroecological paradigm of sustainable agriculture that is now spreading globally with the engagement of farmers and their supporters in the development community who know that 'business as usual' is no longer the river down which we can continue, given the amount of devastation left in its wake.

With its three interlinked principles – no or minimal soil disturbance, permanent soil cover, and cropping system diversity – CA not only improves the resilience of a particular cropland but also enhances the entire production system and land management towards a more inclusive agricultural land-use system for rural development. This leads to the potential opening up of agricultural production for the integration of other production areas such as livestock and forestry and facilitating an economic growth model away from economies of scale and specialization towards diversity, local development and increased resilience (Landers, 2007).

References

ANA (2011) *Programme for Water Producers (Programa Produtor de Água)*. Available at: http://www.ana.gov.br/produagua/ (accessed 16 September 2016).

Andrews, D.J. and Kassam, A.H. (1976) The importance of multiple cropping in increasing world food supplies. In: *Multiple Cropping*, ASA Special Publication No. 27. American Society of Agronomy, Madison, Wisconsin, pp. 1–10.

Bunting, A.H. and Kassam, A.H. (1986) Principles of water use and dry matter production and distribution affecting choice of crops and farming systems. In: *Proceedings of the Consultant's Workshop on Drought Research Priorities, November 1986*. ICRISAT, Hyderabad, India (published 1988), pp. 43–62.

Chi, F., Shen, S.H., Chang, H.P., Jing, Y.X., Yanni, Y.G. and Dazzo, F.B. (2005) Ascending migration of endophytic rhizobia, from roots to leaves, inside rice plants and assessment of benefit to rice growth physiology. *Applied and Environmental Microbiology* 71, 7271–7278.

Chi, F., Yang, P.F., Han, F., Jing, Y.X. and Shen, S.H. (2010) Proteomic analysis of rice seedlings infected by *Sinorhizobium melioloti* 1021. *Proteomics* 10, 1831–1874.

Derpsch, R., Roth, C.H., Sidiras, N. and Köpke, U. (1991) *Controle da erosão no Paraná, Brasil: Sistemas de cobertura do solo, plantio direto e preparo conservacionista do solo*. GTZ, Eschborn, Germany.

FAO (1978–81) *Agroecological Zones Project Report. Methodology and Results for Africa (Vol. 1), West Asia (Vol. 2), South and Central America (Vol. 3), Southeast Asia (Vol. 4)*. World Soil Resources Report 48. FAO, Rome.

FAO (2015) CA adoption worldwide. FAO Conservation Agriculture Website/FAO AQUASTAT. Available at: http://www.fao.org/ag/ca/6c.html (accessed 16 September 2016).

Farooq, M. and Siddique, K.H.M. (2015) *Conservation Agriculture*. Springer International, Basel, Switzerland.

Friedrich, T., Kassam, A. and Shaxson, F. (2009) Case study: Conservation Agriculture. Annex 2 in: *Final Report Agricultural Technologies for Developing Countries* (STOA Project 'Agricultural

Technologies for Developing Countries'). European Technology Assessment Group (ETAG). ITAS, Karlsruhe, Germany.

Haugen-Kozyra, K. and Goddard, T. (2009) Conservation agriculture protocols for green house gas offsets in a working carbon market. Paper presented at the *IV World Congress on Conservation Agriculture*, 3–7 February 2009, New Delhi.

Higgins, G.M., Kassam, A.H., Naiken, L., Fischer, G.W. and Shah, M.M. (1982) *Potential Population Supporting Capacities of Lands in the Developing World: Land Resources for Population of the Future.* FAO/UNFPA/IIASA. International Institute for Applied Systems Analysis (IIASA), Laxenburg, Austria.

ITAIPU (2011) Cultivando agua boa (Growing good water). Available at : https://www.itaipu.gov.br/meioambiente/cultivando-agua-boa (accessed 14 November 2016).

Jat, R., Sahrawat, K. and Kassam, A. (eds) (2014) *Conservation Agriculture: Global Prospects and Challenges.* CAB International, Wallingford, UK.

Kassam, A., Friedrich, T., Shaxson, F. and Pretty, J. (2009) The spread of Conservation Agriculture: justification, sustainability and uptake. *International Journal of Agricultural Sustainability* 7(4), 293–320.

Kassam, A., Basch, G., Friedrich, T., Shaxson, F., Goddard, T., Amado, T., *et al.* (2013) Sustainable soil management is more than what and how crops are grown. In: Lal, R. and Stewart, R.A. (eds) *Principles of Soil Management in Agro-ecosystems*. Advances in Soil Science, CRC Press, Boca Raton, Florida.

Kassam, A., Friedrich, T., Derpsch, R. and Kienzle, J. (2015) Overview of the worldwide spread of Conservation Agriculture. *Field Actions Science Reports 8*. Available at: http://factsreports.revues.org/3966 (accessed 16 September 2016).

Landers, J. (2007) *Topical Crop-Livestock Systems in Conservation Agriculture: The Brazilian Experience*. Integrated Crop Management, Vol. 5. FAO, Rome, 92 pp.

Marketwired (2014) Alberta's Climate Change Central to close. Available at: http://www.marketwired.com/press-release/albertas-climate-change-central-to-close-1916234.htm (accessed 14 November 2016).

MEA (2005) *Ecosystems and Human Well-Being: Synthesis.* Millennium Ecosystem Assessment. Island Press, Washington, DC.

Mello, I. and van Raij, B. (2006) No-till for sustainable agriculture in Brazil. *Proceedings of the World Association Soil and Water Conservation* P1, 49–57.

Montgomery, D. (2007) *Dirt: The Erosion of Civilizations.* University of California Press, Berkeley, California.

Redman, R.S., Kim, O.Y., Woodward, C.J.D.A., Greer, C., Espino, L., Doty, S.L. and Rodriguez, R.J. (2011) Increased fitness of rice plants to abiotic stress via habitat adapted symbiosis: a strategy for mitigating impacts of climate change. *PLoS One* 6(7), e14823, doi: 10.1371/journal.phone.0014823.

Rockstrom, J., Kaumbutho, P., Mwalley, J., Nzabi, A.W., Temesgen, M. Mawwenya, L., *et al.* (2009) Conservation farming strategies in East and Southern Africa: yields and rainwater productivity from on-farm action research. *Soil & Tillage Research* 103, 23–32.

Rodriguez, R.J., Freeman, D.C., McArthur, E.D., Kim, O.Y. and Redman, R.S. (2009) Symbiotic regulation of plant growth, development and reproduction. *Communicative & Integrative Biology* 2(3), 1–3.

Sá, J.C.M. (2004) Adubacao Fosfatada no Sistema de Plantio Direto. In: Yamada, T., Abdalla, S.R.S. and Sao Pedro, S.P. (eds) *Fosforo na Agricultura Brasileira*. Associacao Brasileira para a Pesquisa da Potassa e Fosfato, Piracicaba, SP, Brazil, pp. 201–222, 726 pp.

Shaxson, F., Kassam, A., Friedrich, T., Boddy, B. and Adekunle, A. (2008) Underpinning the benefits of Conservation Agriculture: sustaining the fundamental of soil health and function. Main document for the *Workshop on Investing in Sustainable Crop Intensification: The Case of Soil Health, 24–17 July 2008*. FAO, Rome.

Thierfelder, C., Chisui, J.L., Gama, M., Cheeseman, S., Jere, Z.D., Bunderson, W.T., *et al.* (2013) Maize-based Conservation Agriculture systems in Malawi. Long-term trends in productivity. *Field Crop Research* 142, 147–157.

Thierfelder, C., Rusinamhodzi, L., Setimela, P., Walker, F. and Eash, N.S. (2015a) Conservation Agriculture and drought-tolerant germplasm: reaping the benefits of climate-smart technologies in central Mozambique. *Renewable Agriculture and Food Systems*, published online 30 September 2015, 15 pp. doi: 10.1017/S1742170515000332.

Thierfelder, C., Bunderson, W.T., Jere, Z.D., Mutenje, M. and Ngwira, A. (2015b) Development of Conservation Agriculture (CA) systems in Malawi: lessons learned from 2005–2014. *Experimental Agriculture*, published online 20 November 2015, 26 pp. doi: 10.1017/S0014479715000265.

3 Conservation Agriculture: Growing More with Less – the Future of Sustainable Intensification

Patrick C. Wall*

Independent Agricultural Research Consultant, Bahias de Huatulco, Mexico

3.1 Definitions

Sustainability:

- Satisfy human food, feed and fibre needs (and contribute to fuel needs).
- Enhance environmental quality and the resource base.
- Sustain the economic viability of agriculture.
- Enhance the quality of life for farmers, farm workers and society as a whole (NRC, 2010).

Sustainable intensification:

- Sustainable increase in production per unit of land per unit of time.

3.2 Sustainability and Efficiency

There are biophysical, economic, social and political aspects to sustainability, which therefore cannot be attained through a set of agricultural practices alone; rather, the technology needs to be embedded in a comprehensive set of actions that lead to sustainable agriculture. Conservation Agriculture (CA) itself will not lead to sustainable intensification without markets, input supply, knowledge development and sharing, as well as stable and non-prejudicial policies. There have been a number of publications in recent years arguing that CA is only applicable to relatively small groups of farmers (Giller *et al.*, 2009), or exploring the characteristics of farmers for whom

*pcwallmex@gmail.com

CA may be applicable (Corbeels *et al.*, 2013). Undoubtedly there are many impediments to the adoption of CA, but today most accept that, biophysically, CA is functional under most conditions in Africa and more sustainable than current tillage-based systems. However, institutional and market factors limit adoption in many instances (Ndah *et al.*, 2013). This would suggest that far more attention should now be paid to overcoming these institutional bottlenecks: rather than identifying which farmers can benefit from CA, we should be identifying and investing in changing those factors that limit adoption. The Green Revolution in South Asia was based on technology (high-yielding dwarf varieties of wheat and rice, fertilizers, irrigation and pest control) but the Green Revolution took place because there was decided political will and because the institutional aspects necessary for widespread technology adoption (input and output markets, credit, subsidies where deemed necessary, seed production/importation, etc.) were put in place. USAID Director W.S. Gaud, who first used the term 'Green Revolution', stated: 'Good luck – good monsoons – helped bring in the recent record harvests. But hard work, good management, and sound agricultural policies in the developing countries and foreign aid were also very much involved'. Policy in the Green Revolution context included the availability of inputs, credit and markets for outputs, among others (Gaud, 1968).

Growing more for less implies increased efficiency in agricultural systems. It is pertinent to ask here how we shall measure efficiency – and how the farmer will measure efficiency. For sustainable intensification, agriculture needs to produce more (food, feed, fibre, fuel) per unit of land area and do so with as little negative effect on the environment as possible. However, often yield per unit of land (obviously per year or another unit of time) is not the primary objective of farmers, especially smallholder famers. The farmer's priority is normally to produce more income (rather than yield) per hectare but it can also be to produce more income per day worked (Ekboir *et al.*, 2001) (especially where off-farm employment is an important component of the family livelihood strategy), per dollar invested (where, as is often the case, capital is a limiting factor), or even per bag of seed. There is a disparity between different views and aspects of sustainability, depending on who is defining the objectives. Public investment in agriculture, including subsidies, is often used to try to better align society's priorities and farmers' priorities (Mogues *et al.*, 2012).

What measure of efficiency should we use? If we are comparing two agricultural systems, then comparing efficiencies would appear to be relatively simple: define the most limiting factor and whichever system gives more production (of a defined output) for each unit of the most limiting factor is therefore the more efficient – provided, in the context of sustainability, that it also has less impact on the environment. However, comparing efficiencies with respect to yield per hectare between a conventionally tilled system and a CA system – two complex, multi-component systems that often require different equipment and weed control methods, land preparation activities, different planting dates, nutrient amounts and management, plus changes in other management factors – may in fact give erroneous results as to which is the 'best' system. An economic analysis is better able to integrate the different effects and factors than an analysis of yield per se and is therefore arguably far more meaningful for comparing different systems than physical yield. Unfortunately economic analyses are seldom reported in the literature. It should be stressed that of course for this analysis to be meaningful, we should be comparing two locally adapted systems: too often an untried and unadapted CA system imported from another environment has been compared in research trials with a traditional system that has been adapted, practised and fine-tuned by farmers over decades. As stated by Beck *et al.* (1998), 'The problem occurs because an attempt is being made to use research that was designed to test a system component (e.g. tillage) to make judgements about the system as a whole (profitability), and because the comparisons often neglect to optimize cropping strategies for each tillage regime, resulting in agronomic practices that inherently favour one regime' – generally the status quo or conventional tillage. That CA has performed as well as, or often better than, conventional practices in most of the published results from sub-Saharan Africa, especially eastern and southern Africa (Wall *et al.*, 2013), is testimony to the resilience and potential of the system.

Comparisons between systems are academically interesting (though possibly too many resources are dedicated to these comparisons) but far more meaningful is the question: 'How efficient is the CA system?' What is the gap between actual yield and potential yield? Evaluating the efficiency of the system will give an indication of other limiting factors in the system that should be addressed before the system can be considered 'locally adapted'. Fischer *et al.* (2009) differentiated between farmer yield, economically attainable yield and potential yield (set by the environment – temperature, radiation and available water). Interestingly, attainable yield under present market conditions may be very different from attainable yield under efficient market conditions, demonstrating again the effect of policy on farmer decisions and productivity levels. Fischer *et al.* (2009) differentiated between non-water-limited potential yield and the water-limited potential yield of French and Schultz (1984). These authors plotted farm wheat yields against seasonal rainfall and showed that, above a minimum rainfall level that varied somewhat with environment (and probably reflects evaporation and the investment in vegetative growth), the potential yield of wheat in South Australia, and subsequently in other parts of the country, was 20 kg grain/mm of rainfall. If this yield was not obtained, then farmers should investigate the factor(s) that led to the yield reduction – factors that could include poor rainfall distribution in some seasons. The demonstration of the water-limited wheat yield potential by French and Schultz (1984) proved to be a very meaningful measure for South Australian farmers, still used today (and, in fact, amplified to other crops). French and Schultz (1984) also demonstrated that in many cases the yields of wheat in published agronomic research results showing (significant) treatment effects were, in fact, well below the water-limited potential yield, suggesting that there were other factors limiting yield rather than the factors being studied. The utility of the French–Schultz relationship for South Australia stresses the need for a realistic measure of yield potential in any environment so that farmers, and researchers, can measure their crop yields against what they should have been able to achieve. It is possible that yield differences reported in the literature and attributed to a tillage system often reflect the effects of other limiting factors, leading to confusing results, especially in papers based on meta-analysis of a wide range of data from different research, researchers and conditions.

Numerous studies have shown that CA is not a low-input system – system functionality relies on relatively high productivity, not only to produce sufficient crop residues, but presumably also to produce sufficient root mass. Therefore where farmers currently use extremely low-input production strategies, such as in many areas of sub-Saharan Africa, it is doubtful that CA can in fact 'produce more for less'. At the same time these current practices are not sustainable and moving towards more sustainable systems will involve more inputs, whether from renewable on-farm resources, or from off-farm 'imported' non-renewable external inputs. The use of all available resources, including adequate rotations, especially with legume crops, farmyard manure, compost, manual labour, residue transfer and other practices, can help farmers to restrict the need for capital and in many cases these practices may be more sustainable than the use of off-farm inputs. However, inputs, either on-farm (internal) or off-farm (external), are vital and productivity and sustainability cannot be achieved without them.

The use of extractive low-input management practices by the majority of farmers implies that the attainable yield under current market conditions and institutional arrangements is very low and that efforts to improve markets and institutions will have a greater effect on productivity and technology choice than will technology per se: the gap between yield under present market conditions and optimum market conditions is large and so institutional change can result in huge benefits to farmers and society. Overcoming these institutional bottlenecks will no doubt benefit farmers managing conventionally tilled systems as well as those managing CA systems and therefore measures need to be put in place (information, policy, etc.) to encourage the use of more sustainable systems.

Achieving potential yield (or water-limited potential yield) requires optimal levels of nutrients and efficient management to optimize both the aerial and edaphic environments and limit the effects of other organisms (pests, diseases and weeds) on system productivity. Achieving efficient production systems may often require more inputs than smallholder farmers use today, but the key is to use these inputs efficiently – grow more with less wastage – as inefficiency and wastage lead to reduced and/or uneconomic benefits. Efficiency is best measured in terms of the most limiting factor(s): water, nutrients, labour, land, capital investment, etc. If other factors restrict system productivity, efficiency will be reduced. The following paragraphs look at the most common and most important limiting factors to farm productivity in African agriculture and ask whether CA can increase the efficiency of their use.

3.3 CA and Sustainable Intensification

3.3.1 Nutrients, water and risk

Excessive nutrient mining over most of Africa (Stoorvogel *et al.*, 1993) is acute and adequate plant nutrition is often cited as the most limiting factor to crop production in sub-Saharan Africa (Twomlow *et al.*, 2011), while at the same time fertilizer use is very low at less than 10 kg nutrients/ha in sub-Saharan Africa (NRC, 2010) and about 20 kg/ha in eastern and southern Africa in 2009/10 calculated from FAO's FAOSTAT database (Wall *et al.*, 2013). Even lower levels of fertilizer are applied to staple crops; considerably more is applied to cash crops (Groot, 2009). Application of more fertilizer to cash crops, which often have more established markets and input supply chains, clearly shows that the problem is not that farmers do not understand the benefits of fertilizer but rather that they make a conscious decision not to apply fertilizer, or to apply very little, to their staple crops. Fertilizer use by smallholders is not just a function of availability and affordability but also of both production and market risk (Morris *et al.*, 2007). Smallholder farmers, in particular, are averse to risk given their precarious financial situation and their poor access to credit: if fertilizer application to a crop is perceived as risky, it will not be applied (Rockström *et al.*, 2002). At the same time expected profits, together with other market factors, have a large influence on technology adoption (Griliches, 1957), explaining why farmers would tend to use more inputs on cash crops.

One of the major causes of risk in much of Africa is moisture stress, which is often more a function of inefficient use of rainfall than of insufficient or poorly distributed rainfall per se. Between 70% and 85% of rainfall is lost to surface runoff, deep drainage and evaporation rather than being used by crops for productive transpiration in the semi-arid tropics of Africa (Rockström *et al.*, 2002), while in Zimbabwe 30% of rainfall may be lost to runoff alone (Elwell and Stocking, 1988). Therefore, even though total rainfall may be sufficient for optimal crop growth, available water levels may be considerably lower and limit crop productivity. Generally people assume that drought is due to insufficient rainfall or irrigation water but the above data suggests that land and soil management are, in reality, the primary causes of drought in much of sub-Saharan Africa. Soil tillage breaks down soil structure and soil organic matter, reducing water infiltration (and thereby increasing runoff and erosion) and soil water-holding capacity (see a summary of literature in Verhulst *et al.*, 2010), causing droughts even when rainfall should be sufficient for healthy crop yields.

As a result of climate change, increased variability in seasonal rainfall amount and distribution is expected throughout most of Africa coupled with a reduction in rainfall in much of the continent (Lobell *et al.*, 2008) – factors that will aggravate the inefficiencies in rainfall use noted above. CA can reduce the risk of moisture stress by increasing water infiltration and storage (summarized in Wall *et al.*, 2013), reducing compaction impediments to root growth and reducing evaporation (Mrabet, 2008), and therefore remove some of the barriers to smallholder fertilizer use. By improving the crop water balance, CA reduced risk at eight of nine sites in Malawi: yield in the worst seasons was significantly higher under CA than it was under the normal farmer ridged-and-cultivated practice (Wall *et al.*, 2010). Ngwira *et al.* (2013) analysed risk associated with the use of CA across 12 sites in Malawi, including the nine analysed by Wall *et al.* (2010), and found reduced risk at all of the drier lowland sites, but not at the three higher rainfall sites.[1] It is hypothesized that reduced risk will increase the feasibility of farmers using higher levels of fertilizer – once they are convinced of the risk reduction.

Green manure cover crops (GMCC), especially legumes, may be incorporated into the farming system for nitrogen fixation, weed control and other benefits, including the possibility of amelioration of compacted soil (Scopel *et al.*, 2004). However, the use of GMCC still implies costs to the farmer, including seed (often very expensive at the moment) and foregone harvest income, and so farmers often prefer intercropping of, or rotation with, grain legume crops[2] rather than GMCC (Nandwa *et al.*, 2011), especially in areas with unimodal rainfall patterns and short rainy seasons.

CA also markedly reduces soil erosion (generally by over 90%), avoiding not only soil loss but also nutrient losses by erosion. Annual farm losses of soil organic matter through erosion in Zimbabwe were over 850 kg/ha together with approximately 50 kg nitrogen/ha and 8 kg phosphorus/ha (Elwell and Stocking, 1988). Therefore, by reducing erosion, CA reduces nutrient wastage and more crop yield can be produced for every kilogram of fertilizer applied – because it stays where it is applied and, because of the improved crop water balance under CA, the crop has moisture to be able to use the nutrients effectively.

The International Crops Research Institute for the Semi-Arid Tropics (ICRISAT) and the International Maize and Wheat Improvement Centre (CIMMYT) have recommended the use of very low levels of nitrogen fertilizer (micro-dosing) for maize production in the semi-arid areas of Zimbabwe (Twomlow *et al.*, 2011), as has ICRISAT in parts of West Africa (Tabo *et al.*, 2011). Micro-dosing is based on the normal response curve to applied fertilizer and takes advantage of the initial steep slope of fertilizer response. However, it is suggested that micro-dosing is not a feasible technology for CA situations, especially as it is promoted largely for semi-arid situations. In conventional agriculture, nitrogen fertilization focuses on the present season and has little effect on subsequent seasons, whereas under CA, because of the effect on biomass production and residue amounts (especially important in semi-arid situations), N fertilization has a large effect on crop performance not only in the present season but also in subsequent seasons. CA farmers need to fertilize both for current and for future yields, while at the same time maintaining economic efficiency and reducing risk.

3.3.2 CA, labour and fuel use

Labour is frequently the most limiting resource for smallholder farmers and labour savings have been cited in numerous surveys as the principal reason for adoption of CA by smallholders (Ekboir *et al.*, 2001). In case studies in Paraguay, labour demand under CA was 16% lower than under conventional agriculture, but labour productivity (income per day of labour) increased with CA by over 100% because of the combined effects of reduced labour and increased profitability (Sorrenson *et al.*, 1998). However, labour savings depend to a large degree on weed management and the type of CA practised. If herbicides are used, then labour savings from both the lack of tillage and the weed control are large (e.g. in Ghana: Ekboir *et al.*, 2001), whereas if manual weeding is practised, there may be a higher labour requirement in CA than in conventionally tilled fields (Rockström *et al.*, 2001; Djamen *et al.*, 2013), especially in the early years. In Malawi, labour costs were lower in CA systems with chemical weed control than with conventional tillage by between 28% (Ngwira *et al.*, 2012) and 63% (Ito *et al.*, 2007). In these studies, costs of production were higher with CA because of the cost of the herbicides, but the greater yields under CA more than offset the increased costs: net returns were increased by US$130–370/ha, net benefits by 69% and returns to labour 92–100%. Weeds may also be controlled by GMCC: the work of Mariki (2003) in northern Tanzania showed that initially more labour (11%) was used with CA because of the greater weed populations, but after 4 years with a maize–GMCC system (mucuna or lablab alternated with maize in an area with bimodal rainfall) labour use was 45% lower in the CA system than in the conventional system.

The basin system of CA, called Conservation Farming in Zambia and Zimbabwe, also requires more labour than conventional tillage (34 versus 13 person days/ha: Umar *et al.*, 2012). However, the labour requirement for digging basins is in the winter, when competing labour requirements are low, and because of the increased maize yield the returns to labour ($/day worked) in Zambia were five times higher in the basin system than with conventional tillage (Umar *et al.*, 2012). This was more production for more work – but also more production for each day worked. Labour requirements for digging the basins are reduced in subsequent years if the basins are placed in exactly the same place as in previous years; the soil in the previous basins is less compact and more friable. The Zambian Conservation Farming Unit reports a 35–40% reduction in the labour required for digging basins after 4 years of CA (Conservation Farming Unit, 2007).

More efficient machinery use has been one of the drivers of CA adoption on mechanized farms in the Americas. For example, Sorrenson and Montoya (1984, quoted by Derpsch *et al.*, 1991) showed a 60% reduction in tractor hours per hectare and 77% reduction in fuel use in Paraná, Brazil; Landers (1996) showed a 44% reduction, on average, in horsepower/ha across eight farms in the Brazilian Cerrados; and Wall (2002) showed an 80% reduction in fuel use with CA in the lowlands of Bolivia. There are few data on machinery use in CA in sub-Saharan Africa, but on the Agricultural Research Trust (ART) farm near Harare, machinery costs for CA were reduced by 66% compared with conventional tillage (MacRobert *et al.*, 1995), fuel use was reduced and there was considerably more production per litre of fuel used.

3.3.3 Capital

Not only are the returns to investment generally higher under CA than under conventionally tilled fields (Wall *et al.*, 2013) but also the risks of losses are lower (losses are less frequent) under CA (Wall *et al.*, 2010; Ngwira *et al.*, 2013). Cost savings, as noted above, depend to a large degree on the type of CA conducted and weed management methods. Some of the benefits of CA only accrue over time, but to be acceptable to smallholders the CA system must give economic benefits immediately. Because of the effects of CA on moisture saving, these short-term benefits are more likely in drier and non-irrigated environments than they are in wetter or irrigated environments.

3.3.4 Knowledge

One area where growing 'more with less' definitely does not apply is knowledge. CA is more knowledge intensive than traditional low-input systems, partly because it is new but primarily because of the changes in multiple components of the farming system, the need (in most instances) for chemical weed control, the need for good farm and crop management and also the need for farmers to understand the basis of the system and so be able to mould it to their particular conditions. Smallholder farmers are often poorly linked to knowledge systems external to the community (Wall, 2007). Overcoming this barrier and increasing the knowledge base of smallholder farmers is probably the biggest hurdle to be overcome in achieving their widespread adoption of CA. Success will not only depend on enhancing the knowledge of CA and CA systems among researchers, extension (change) agents and policy makers, and the facilitation of farmer-to-farmer knowledge flow, but will also depend on farmer participation in technology development and adaptation, and the development of local innovation systems incorporating agents representing as many as possible of the principal components of the local agricultural value chains, each using their own comparative advantages and information networks to help remove bottlenecks to farm productivity.

3.4 Conclusions

Conservation Agriculture is not a low-input system (although it can be a low external input system), and therefore 'growing more for less' is unlikely, especially in situations, such as those of many of the smallholder farms in sub-Saharan Africa, where farmers currently apply very low levels of inputs. The benefits of CA lie rather in using applied inputs (nutrients, water, labour, fuel) more efficiently than conventionally tilled systems, while at the same time reversing the soil degradation caused by tillage. Risk reduction with CA, especially in water-limited situations, will encourage greater input use, especially nutrients. In the short term, CA generally gives crop yields equal to or greater than yields under conventionally tilled situations, with higher yields more common in situations where moisture stress limits yield in tilled systems. However, economics and labour savings depend, to a large degree, on weed management strategies: if herbicides are used, labour use and production costs are markedly reduced in CA, but if weeds are controlled manually, labour requirements for weeding may offset the labour-saving benefits of reduced tillage. Sustainable intensification of agriculture in Africa will require more than technology alone: institutional change, including adequate policies, input and output markets, may be just as, or often more, important than technology in increasing farmers' economically attainable yields and achieving sustainable intensification.

Notes

[1] It is common for the benefits of CA to be more marked in drier conditions in the short term, due to improved rainfall conservation. However, from personal experience with the trial sites in Malawi, the author believes that rainfall and altitude alone do not explain the differences between sites but that there were specific problems of pests and trial management at the highland sites.

[2] The author believes it is important to differentiate between GMCC and grain legume crops, as much of the nitrogen produced by a legume crop may be exported in the grain. In a GMCC, grain is not harvested and preferably all of the plant material remains on the field. The same species (e.g. *Lablab purpureus* in Tanzania) may be used as a GMCC and a grain legume – when grain is harvested it should be considered a grain legume crop.

References

Beck, D.L., Miller, J.L. and Hagny, M.P. (1998) Successful no-till on the central and northern plains. Presentation at the *American Society of Agronomy Conference*, Baltimore, Maryland. October 1998, 15 pp. Available at: http://www.dakotalakes.com/Publications/asa10_98.pdf (accessed 16 September 2016).

Conservation Farming Unit (2007) *Conservation Farming and Conservation Agriculture. Handbook for Hoe Farmers in Agroecological Regions I and IIa – Flat Culture*. CFU, Lusaka, Zambia, 57 pp.

Corbeels, M., de Graaff, J., Ndah, T.H., Penot, E., Baudron, F., Naudin, K., *et al.* (2013) Understanding the impact and adoption of conservation agriculture in Africa: a multi-scale analysis. *Agriculture, Ecosystems and Environment* 187, 155–170. Available at: http://dx.doi.org/10.1016/j.agee.2013.10.011 (accessed 16 September 2016).

Derpsch, R., Roth, C.H., Sidiras, N. and Köpke, U. (1991) *Controle da erosão no Paraná, Brasil: Sistemas de cobertura do solo, plantio direto e preparo conservacionista do solo.* Sonderpublikation der GTZ, No. 245, Deutsche Gesellschaft für Technische Zusammenarbeit (GTZ), GmbH, Eschborn, Germany, 272 pp.

Djamen Nana, P., Dugué, P., Mkomwa, S., Benoit Da Sanson, J., Essecofy, G., Bougoum, H., *et al.* (2013) Conservation Agriculture in West and Central Africa. In: Jat, R.A., Sahrawat, K.L. and Kassam, A.H. (eds) *Conservation Agriculture: Global Prospects and Challenges*. CAB International, Wallingford, UK, pp. 311–338.

Ekboir, J., Boa, K. and Dankyi, A.A. (2001) The impact of no-till in Ghana. In: García-Torres, L., Benites, J. and Martínez-Vilela, A. (eds) *Conservation Agriculture: A Worldwide Challenge*, Vol. II. ECAF/FAO, Córdoba, Spain, pp. 757–764.

Elwell, H.A. and Stocking, M.A. (1988) Loss of soil nutrients by sheet erosion is a major hidden farming cost. *Zimbabwe Science News* 22(7/8), 79–82.
FAO (Food and Agriculture Organization of the United Nations) FAOSTAT Database. Available at: http://faostat3.fao.org/home/index.html (accessed 16 September 2016).
Fischer, R.A., Byerlee, D. and Edmeades, G.O. (2009) Can technology deliver on the yield challenge to 2050? Paper presented at the *Expert Meeting on How to Feed the World in 2050*, Food and Agriculture Organization of the United Nations, Economics and Social Development Department. 24–26 June 2009. FAO, Rome, 46 pp.
French, R. and Schultz, J. (1984) Water use efficiency of wheat in a Mediterranean-type environment: 1. The relationship between yield, water use and climate. *Australian Journal of Agricultural Research* 35, 743–764.
Gaud, W.S. (1968) The Green Revolution: Accomplishments and apprehensions. Address to the Agency for International Development. Available at: http://www.agbioworld.org/biotech-info/topics/borlaug/borlaug-green.html (accessed 16 September 2016).
Giller, K.E., Witter, E., Corbeels, M. and Tittonell, P. (2009) Conservation agriculture and smallholder farming in Africa: the heretics' view. *Field Crops Research* 114(1), 23–34.
Griliches, Z. (1957) Hybrid corn: an exploration in the economics of technological change. *Econometrica* 25, 501–522.
Groot, J.J.R. (2009) *Update of Fertilizer Supply and Demand – Sub-Saharan Africa*. IFA Africa Forum, IFDC, Cairo, Egypt.
Ito, M., Matsumoto, T. and Quiñones, M. (2007) Conservation tillage in sub-Saharan Africa: the experience of Sasakawa Global 2000. *Crop Protection* 26, 417–423.
Landers, J. (1996) Analisis financeiras. In: *Fasículo de experiencias de plantio direto no Cerrado*. Associaçao de Plantio Direto no Cerrado, Brasilia, Brazil, pp. 217–227.
Lobell, D.B., Burke, M.B., Tebaldi, C., Mastrandrea, M.D., Falcon, W.P. and Naylor, R.L. (2008) Prioritizing climate change adaptation needs for food security in 2030. *Science* 319, 607–610.
MacRobert, J.F., Winkfield, R.A. and Pilbrough, S. (1995) Conservation tillage on the Agricultural Research Trust Farm. In: Vowles, M. (ed.) *Conservation Tillage: A Handbook for Commercial Farmers in Zimbabwe*. LaserPrint, Harare, Zimbabwe, pp. 101–108.
Mariki, W. (2003) The impact of conservation tillage and cover crops on soil fertility and crop production in Karatu and Hanang Districts of northern Tanzania. Tanzania Farmers Service Centre. *TFSC Technical Report 1999–2003*. TFSC/GTZ, Arusha.
Mogues, T., Yu, B., Fan, S. and McBride, L. (2012) *The Impacts of Public Investment in and for Agriculture: Synthesis of the Existing Evidence*. IFPRI Discussion Paper 01217, International Food Policy Research Institute, Washington, DC, 66pp.
Morris, M., Kelly, V.A., Kopicki, R.J. and Byerlee, D. (2007) *Fertilizer Use in African Agriculture: Lessons Learned and Good Practice Guidelines*. The World Bank, Washington, DC, 146 pp.
Mrabet, R. (2008) No-till practices in Morocco. In: Goddard, T., Zoebisch, M., Gan, Y., Ellis, W, Watson, A. and Sombatpanit, S. (eds) *No-Till Farming Systems*. World Association of Soil and Water Conservation (WASWC), Bangkok, pp. 393–412.
Nandwa, S.M., Obanyi, S.N. and Mafongoya, P.L. (2011) Agro-ecological distribution of legumes in farming systems and identification of biophysical niches for legumes growth. In: Bationo, A., Waswa, B., Okeyo, J.M., Maina, F., Kihara, J. and Mokwunye, U. (eds) *Fighting Poverty in Sub-Saharan Africa: The Multiple Role of Legumes in Integrated Soil Fertility Management*. Springer, Dordrecht, pp. 1–26.
Ndah, H.T., Schuler, J., Uthes, S., Zander, P., Traore, K., Gama, M.S., *et al.* (2013) Adoption potential of conservation agriculture in sub-Saharan Africa: results from five case studies. *Environmental Management* 53, 620–635, doi: 10.1007/s00267-013-0215-5.
Ngwira, A., Sleutel, S. and Neve, S. (2012) Soil carbon dynamics as influenced by tillage and crop residue management in loamy sand and sandy loam soils under smallholder farmers' conditions in Malawi. *Nutrient Cycling in Agroecosystems* 92(3), 315–328.

Ngwira, A.R., Thierfelder, C., Eash, N. and Lambert, D.M. (2013) Risk and maize-based cropping systems for smallholder Malawi farmers using Conservation Agriculture Technologies. *Experimental Agriculture* 49(4), 483–503.

NRC (National Research Council) (2010) *Toward Sustainable Agricultural Systems in the 21st Century*. National Academies Press, Washington, DC. Available at: http://www.nap.edu/catalog/12832.html (accessed 16 September 2016).

Rockström, J., Barron, J. and Fox, P. (2001) Water productivity in rain-fed agriculture: challenges and opportunities for smallholder farmers in drought-prone tropical agro-systems. Paper presented at an *IWMI Workshop on Water Productivity in Agriculture*. Colombo, Sri Lanka, 12–14 November 2001.

Rockström, J., Barron, J. and Fox, P. (2002) Rainwater management for increased productivity among small-holder farmers in drought prone environments. *Physics and Chemistry of the Earth* 27, 949–959.

Scopel, E., Triomphe, B., dos Santos Ribeiro, M., Séguy, L., Denardin, J.E. and Kochhann, R.A. (2004) Direct seeding mulch-based cropping systems (DMC) in Latin America. In: Fischer, R.A. (ed.) *New Directions for a Diverse Planet. Proceedings of the 4th International Crop Science Congress*, Brisbane, Australia, 26 September–1 October 2004. Available at: http://regional.org.au/au/asa/2004/symposia/2/2/1406_scopele.htm (accessed 16 September 2016).

Sorrenson, W.J. and Montoya, L.J. (1984) *Implicações econômicas da erosão do solo e das práticas conservacionistas no Paraná, Brasil*. IAPAR/GTZ, Londrina, Brazil, 104 pp.

Sorrenson, W.J., Duarte, C. and Lopez Portillo, J. (1998) *Economics of no-till compared to traditional cultivation on small farms in Paraguay*. Asunción: MAG/GTZ Soil Conservation Project. p. XV.

Stoorvogel, J.J., Smaling, E.M.A. and Jansen, B.H. (1993) Calculating soil nutrient balances in Africa at different scales. 1. Supra-national scale. *Fertilizer Research* 35(3), 227–235.

Tabo, R., Bationo, A., Amadou, B., Marchal, D., Lompo, F., Gandah, M., *et al.* (2011) Fertilizer microdosing and "warrantage" or inventory credit system to improve food security and farmers' income in West Africa. In: Bationo, A., Waswa, B., Okeyo, J.M., Maina, F. and Kihara, J.M. (eds) *Innovations as Key to the Green Revolution in Africa: Exploring the Scientific Facts*. Springer, Dordrecht, pp. 113–122.

Twomlow, S., Rohrbach, D., Dimes, J., Rusike, J., Mupangwa, W., Ncube, B., *et al.* (2011) Microdosing as a pathway to Africa's Green Revolution: evidence from broad-scale on-farm trials. In: Bationo, A., Waswa, B., Okeyo, J.M., Maina, F. and Kihara, J.M. (eds) *Innovations as Key to the Green Revolution in Africa: Exploring the Scientific Facts*. Springer, Dordrecht, pp. 1101–1113.

Umar, B.B., Aune, J.B., Johnsen, F.H. and Lungu, I.O. (2012) Are smallholder Zambian farmers economists? A dual-analysis of farmers' expenditure in conservation and conventional agriculture systems. *Journal of Sustainable Agriculture* 36(8), 908–929.

Verhulst, N., Govaerts, B., Verachtert, E., Castellanos-Navarrete, A., Mezzalama, M., Wall, P., *et al.* (2010) Conservation agriculture, improving soil quality for sustainable production systems? In: Lal, R. and Stewart, B.A. (eds) *Advances in Soil Science: Food Security and Soil Quality*. CRC Press, Boca Raton, Florida, pp. 137–208.

Wall, P.C. (2002) Extending the use of zero tillage agriculture: the case of Bolivia. Paper presented at the *International Workshop on Conservation Agriculture for Sustainable Wheat Production in Rotation with Cotton in Limited Water Resource Areas*, Tashkent, Uzbekistan, October 13–18, 2002.

Wall, P.C. (2007) Tailoring conservation agriculture to the needs of small farmers in developing countries: an analysis of issues. *Journal of Crop Improvement* 19(1/2), 137–155.

Wall, P.C., Thierfelder, C., Govaerts, B. and Verhulst, N. (2010) Conservation Agriculture based systems and their impact on climate change, food security and the poor. Invited paper presented at the *American Society of Agronomy Meetings*, Long Beach, California, USA. October 31–November 3, 2010.

Wall, P.C., Thierfelder, C., Ngwira, A., Govaerts, B., Nyagumbo, I. and Baudron, F. (2013) Conservation Agriculture in Eastern and Southern Africa. In: Jat, R.A., Sahrawat, K.L. and Kassam, A.H. (eds) *Conservation Agriculture: Global Prospects and Challenges*. CAB International, Wallingford, UK, pp. 263–292.

4 Empowering Smallholder Farmers with Profitable and Sustainable Farming Using Conservation Agriculture: The Case of East Africa

SAIDI MKOMWA,[1]* SIMON LUGANDU,[2] PETER KURIA[1] AND WELDONE MUTAI[1]

[1]*African Conservation Tillage Network, Nairobi, Kenya;* [2]*African Conservation Tillage Network, Dar Es Salaam, Tanzania*

4.1 Introduction

Agriculture is the backbone for East Africa (EA) economies and plays a key role in their industrial development and trade. Agriculture accounts for more than 32% of the region's gross domestic product (GDP), employs about 80% of its labour force, accounts for about 65% of foreign exchange earnings and contributes more than 50% of raw materials to the industrial sector (EAC, 2015a). Agriculture is dominated by smallholder farmers, the majority of whom are women, who produce as much as 80% of the food consumed. Most of these farmers have 0.5–2 ha of land, earn less than US$1 a day, and face 3–5 hunger months in a year (Diagana, 2003). Many children under 5 years of age go without a balanced diet. Smallholder agriculture is often seen as a sector in which low incomes, low productivity and vulnerability predominate, with the perception of it being a source of rural poverty and food insecurity, rather than a solution. Some areas in the region are experiencing a reduction in food production by up to 50% due to land degradation, soil erosion, drought and climate change (UNEP, 2009). Malnutrition has devastating impacts on the generation of both today and tomorrow, worsening the health status of populations while reducing their resilience to disease and shock.

*saidi.mkomwa@act-africa.org

According to the World Bank (2008), agricultural development is an essentially pro-poor source of economic growth and is two to four times more effective in raising incomes among the poorest than growth in any other sector. Agricultural growth reduces poverty both directly, by raising farm incomes, and indirectly, through its positive impact on all other sectors of the economy. While agriculture-led growth has played an important role in reducing poverty and transforming the economies of many Asian countries, the strategy is yet to work in Africa.

East Africa is a net importer of food and has a rapidly growing urban population and a middle-income sector. Consequently, producing enough food and other agricultural products is an opportunity for growth and economic transformation. The region has the potential not only to achieve food security but also to become a major player in the global food market. This potential lies in its abundant natural resources and huge population base (145.5 million in 2010; EAC, 2015b), which offer opportunities in terms of a large labour market and an enormous food market. Achieving this potential, however, would require empowerment of the smallholder farmers, who receive very little support, to enable them to move from low-productivity subsistence farming to profitable small family enterprises, which will increase incomes, drive employment and kick-start improvements to the rural economy.

Large areas of sub-Saharan African soils are affected by various types of degradation, including fertility decline. According to the Royal Tropical Institute (RTI, 2011), land degradation and declining soil fertility are a major problem. Land degradation is an important global concern because of its adverse impacts on agricultural production, food security and the environment. Inappropriate land management further increases loss of productivity of resource-poor farmers. This in turn affects their food security and livelihood. Persistent use of conventional farming practices based on intensive tillage, especially when combined with removal or *in situ* burning of crop residues, has magnified soil erosion losses and the soil resource base has been steadily degraded (Montgomery, 2007). Consequently, yields are relatively low despite the high potential for improvement. As the main source of economic activity in EA is agricultural production, declining soil productivity means not only that less food is grown but also that production of cash crops and income are endangered. Thus, rectifying land degradation and enhancing productivity through appropriate soil management and conservation can play a major role in achieving farm household food security and agricultural development.

Weather and climate are important determinants of plant growth, water availability and soil replenishment and affect food security in the region. The region's agriculture is particularly vulnerable to changes in weather patterns, because 93% of agriculture is rainfed (UNDP, 2012). Evidence shows that Africa is warming all year round and faster than the global average and this is likely to continue (AMCEN, 2011). It is predicted that under the lowest warming scenario equatorial EA will experience rainfall increases of 5–20% in December, January and February and decreases of 5–10% in June, July and

August (AMCEN, 2011). Predictions of these effects suggest greater water stress, more frequent droughts and floods, greater variability in rainfall and increases in extreme weather events, with many more negative consequences on agricultural productivity (Christensen *et al.*, 2007). This points to an overall increase of incidences of crop failure due to the adverse weather. Inevitably, effects of climate change will also severely affect the livelihoods of people who are already characterized by high poverty levels, food insecurity, rampant land degradation, weak infrastructure, poor natural resources management and dependence on rainfed agriculture (IFAD, 2011). This eventually poses a greater threat to the survival of the vulnerable population.

There is sufficient evidence that Conservation Agriculture (CA) offers a promising option that, over time, can sustainably increase the productivity of smallholder production systems, as well as their profitability and resilience, without compromising the environment (ACT, 2014). CA is defined as a management system based on three interlinked principles that should be applied in a mutually reinforcing manner: (i) no or minimum mechanical soil disturbance; (ii) permanent soil cover with live or dead plant material; and (iii) crop diversification in space and time. CA systems are not widely used in the EA sub-region, partly because there is limited awareness, capacity, knowledge and policy support for the widespread use of the technologies. The First African Congress on Conservation Agriculture (IACCA) held in March 2014, in an operational context geared to achieve the Comprehensive Africa Agriculture Development Programme (CAADP) goal of 6% growth of the agricultural sector, called for commitment from all national and international stakeholders in the public, private and civil sectors to support the upscaling of CA as a climate-smart agriculture to reach at least 25 million farmers across Africa by 2025 (ACT, 2015a; NEPAD, 2016).

This chapter presents the case for empowering smallholder farmers in EA to increase and sustain agricultural productivity, developing the resilience of agricultural systems to adverse weather events and climate change, and simultaneously to combat degradation of the natural resources base. Rural youth is relatively mobile and migrates to urban areas in search of employment and a better life. The youth has come to associate farming with poverty but they do not hate farming; they only hate the poverty and drudgery of labour associated with it. The sustainable intensification to be promoted based on CA should enable the youth to see a different future in agriculture: not as a struggle to survive but as a business that thrives.

Section 4.2 briefly explains the vulnerability of smallholder farmers in EA to food security and environmental degradation in the context of the major farming systems and practised farmer typologies. Section 4.3 presents an overview of the CA being practised in EA in the context of what type of CA, who does it, how it is done and where it is practised. Section 4.4 presents case studies focusing on smallholder farmers, large-scale commercial farmers and some socio-economic drivers to adoption of CA. Section 4.5 presents the results obtained from the case studies with closer reflections on productivity aspects (yield, labour and drudgery), contributions to climate change resilience and sustainability. The chapter concludes by summarizing the findings and drawing implications of the key roles of CA as a foundation for transforming smallholder agriculture in EA.

4.2 Background

4.2.1 Vulnerability to food security and environmental degradation in East Africa

Vulnerability is a concept that has been around in the poverty literature for a long time and defined in many ways. Chambers (1989) defined the term vulnerability as 'the exposure to contingencies and stress and difficulty in coping with them'. A definition provided by Devereux (2002) is that 'vulnerability denotes both exposure and sensitivity to livelihood shocks'. Vulnerability is therefore a function of people's exposure to risks and of their resilience to those risks. Risk designates the potential of shocks and stresses to affect, in different ways, the state of systems, communities, households or individuals (FAO, 2012). Resilience is the ability of a system and its component parts to anticipate, absorb, accommodate or recover from the effects of a hazardous event in a timely and efficient manner, including through ensuring the preservation, restoration or improvement of its essential basic structures and functions (IPCC, 2012).

Climate change is one of the major causes of reduced agricultural production and productivity in many parts of Africa, including EA. Most crop and livestock farming is rainfed and therefore susceptible to weather fluctuations. The EA region is particularly vulnerable to low agricultural yields and productivity, which lead to severe food shortages, as a result of climate-related impacts such as droughts (EAC, 2011). The East African Community (EAC) shows that the percentage of the national labour force employed in crop cultivation is up to 93% in Burundi, 75% in Kenya, 80% in Rwanda, over 80% in Tanzania and 73% in Uganda. The vulnerability is exacerbated by the high degree of small-scale subsistence farming, where people are dependent on the food crops cultivated as well as the income provided by sales of foodstuff. Any agriculture-related disaster or setback robs farmers of the food as well as the income required to purchase food or gain access to livelihood services. The combination of high rates of poverty and exposure to climate change makes the region highly susceptible to food insecurity. According to Cook and Vizy (2012), large decreases in the number of growing season days (GSDs) are predicted using a regional climate model over most of EA during 2041–2060. Predictions (Cook and Vizy, 2012) show that GSDs face reductions of 10–40% throughout Tanzania and 40–60% over southern Kenya.

The international food security assessment report (2013–2023) for sub-Saharan Africa (SSA), which includes the East African region, is projected to remain the most food-insecure region over the next decade (Meade and Stacey, 2013). The food-insecure population is projected to rise from about 254 million in 2013 to 373 million in 2023. Most economies in the region are based on agriculture and about two-thirds of Africans depend directly on agriculture for their livelihoods.

4.2.2 Scale of land degradation

Land degradation remains a serious threat to livelihoods in EA. Results from satellite-based imagery and remote sensing techniques for sub-Saharan Africa show that about 27% of the land area is subject to degradation processes including soil degradation, overgrazing, or deforestation for intensive crop farming (Kirui and Mirzabaev, 2014). Key hotspot areas include south-western regions in Ethiopia, western Kenya, southern Tanzania and eastern Malawi. These 'hotspot' areas are characterized by a host of factors such as high population pressure, farming activities on steep slopes, frequent famines and intensive crop farming. The causes of land degradation could be proximate or underlying (Lambin and Geist, 2006). Proximate causes are those that have a direct effect on the terrestrial ecosystem. These include biophysical (natural) conditions related to climatic conditions and extreme weather events such as droughts and coastal surges, which may, for example, cause land to become saline. Proximate causes are also related to unsustainable land management practices (anthropogenic) such as over-cultivation, overgrazing and excessive forest conversion. On the other hand, the underlying causes are those factors that indirectly affect proximate causes. Lack of relevant institutions, poverty and insecure land tenure may underlie land degradation by hampering incentives to invest in sustainable land management practices (FAO, 2011). Lack of information and knowledge is considered a major hurdle to reducing land degradation, improving agricultural productivity and facilitating the uptake of sustainable land management among smallholder farmers (Liniger *et al.*, 2011).

4.2.3 Overview of major farming systems and practising farmers in EA

There are many ways to classify farming systems. We shall use the classification of African farming systems suggested by Garrity *et al.* (2012) which is based on: (i) the available natural resource base, including water, land, grazing areas and forest; (ii) climate, particularly length of growing period and altitude; and (iii) the dominant pattern of farm activities and household livelihoods, including field crops, livestock, trees, aquaculture, hunting and gathering, processing and off-farm activities. The farming systems relevant to EA include the following.

- The Maize Mixed Farming System, which is found in sub-humid and humid areas, is dominated by maize with legumes. Livelihoods in this farming system are derived principally from maize, tobacco, cotton, legumes, cassava, cattle, goats, poultry and off-farm work.
- The Agro-Pastoral Farming System is found in semi-arid areas, dominated by sorghum, millet and livestock. Livelihoods are derived from sorghum, some maize, pearl millet, pulses, sesame, cattle, sheep, goats, poultry and off-farm work.

- The Highland Perennial Farming System is mainly found in moist highland areas with good market access above 1400 m above sea level, with a dominant perennial crop, either food or commercial. Livelihoods are derived from diverse activities including tea, coffee, banana, maize, beans, sweet potato, cassava, livestock (including dairy) and off-farm work.
- The Highland Mixed Farming System is in cool highland areas above 1600 m above sea level, with temperate cereals and livestock. Livelihoods are derived from wheat, peas, lentils, broad beans, rape, potatoes, sheep, goats, livestock, poultry and off-farm work.
- The Pastoral Farming System is found in arid areas, dominated by livestock. Livelihood is derived from cattle, camels, sheep, goats, some cereal crops and off-farm work.

On the basis of the agroecological zones in which they operate and the type and composition of their farm portfolio and landholding, smallholder farmers usually cultivate less than 1 ha of land, which may increase to up to 10 ha or more in sparsely populated semi-arid areas, sometimes in combination with livestock of up to ten animals (Dixon *et al.*, 2003). On the basis of farm revenue, smallholder farmers range from those producing crops only for family consumption or subsistence. Most smallholder operations occur in farming systems with the family as the centre of planning, decision making and implementation, operating within a network of relations at the community level. The potential with respect to commercial agriculture is largely untapped and in its infancy (Poulton *et al.*, 2008). Stakeholders are increasingly adopting a value chain approach to promote private sector development in agriculture and are trying to tackle various bottlenecks simultaneously. Previous interventions did not pay adequate attention to the development of market linkages and the role of support entities (Poulton *et al.*, 2008), as they mainly focused on production.

Indigenous CA systems in EA include the perennial banana–coffee-based farming systems in the wet highlands of EA and uniquely the *rweya–kibanja* system in Kagera Region of Tanzania, whereby grass mulch was imported from the grasslands (*rweya*) to be used as mulch in the banana–coffee (*kibanja*) fields. The current CA, however, is largely focused on cereal production and maize in particular. The huge potential for plantation crops (cocoa, coconuts, cashew nuts, palm oil and coffee) is yet to be exploited. In a similar context, CA technologies to support or benefit from livestock and aquaculture integration are not found.

4.3 An Overview of Conservation Agriculture in East Africa

4.3.1 Who are the pioneers?

The introduction and promotion of CA in EA has been by development partners, non-governmental organizations (NGOs) and the host national governments. Sasakawa Global 2000 were the first to introduce CA in EA in the late 1980s. In Tanzania the Selian Agricultural Research Institute (SARI) started working on CA research and extension in 1999 supported by the German Technical Cooperation Agency (GTZ, now GIZ). The Food and Agriculture Organization of the United Nations (FAO) introduced technical cooperation projects that were implemented in partnership with national governments in Kenya, Tanzania and Uganda in the mid-1990s. This was later followed by a more comprehensive but short-lived soil fertility initiative launched by FAO and the World Bank in 1996 in Kenya, Tanzania, Uganda and Rwanda.

Since then more CA projects have been implemented, including the Conservation Agriculture for Sustainable Agricultural and Rural Development (CA for SARD) project funded by the government of the Federal Republic of Germany, executed by FAO and coordinated by the African Conservation Tillage Network (ACT). The project operated in Kenya and Tanzania from 2004 to 2010. Other projects are, or have been: the Regional Land Management Unit (RELMA) project in Kenya and Tanzania (1998–2004); the Legume Research Network Project (LRNP) by Kenya Agricultural Research Institute (KARI, currently KALRO); push–pull technology by the International Centre for Insect Physiology and Ecology (ICIPE, or *icipe*) in collaboration with KARI; the Conservation Agriculture with Trees (CAWT) pilot project in Ghana, Kenya, Tanzania and Zambia supported by the Swedish International Development Agency (SIDA) and implemented by the World Agroforestry Centre (ICRAF) and ACT (December 2010 to May 2012); the Agro-ecology Based Aggradation Conservation Agriculture (ABACO) project in six countries that also included Kenya and Tanzania, funded by the European Union (EU) and implemented by ACT (2010–2014); Farm Mechanization and Conservation Agriculture for Sustainable Intensification (FACASI) funded by the Australian Centre for International Agricultural Research (ACIAR) (2013–2017); the Alliance for a Green Revolution in Africa (AGRA) Soil Health Projects implemented by ACT in Kenya and Tanzania (from 2013); and the Canadian Foodgrains Bank (CFGB) project in Ethiopia, Kenya and Tanzania.

The majority of the early projects were of short-term interventions, running on average from 1.5 to 3 years in duration and focused on CA research and development initiatives meant to assess the performance of different technologies and, overall, the proof of concept. Findings were generally very positive, with some inconsistencies in assessed performance. These arose due to differences of technologies, methodologies and definitions of Conservation Agriculture systems. The next batch of projects (e.g. ABACO, FACASI and AGRA/ACT) dwelt not only on validating the proof of concept but also developing impact pathways, along value chains, to facilitate upscaling of the technologies. With clear barriers to mainstreaming of CA in national programmes, the latest and current projects (e.g. CFGB) have strong emphasis on policy aspects, to ensure that the short-lived project interventions have a lasting impact, not through technology superiority alone, but also policy support.

4.3.2 What type of CA?

CA promotes the three interlinked principles of no or minimum soil disturbance, permanent organic soil cover and crop rotation and/or association. These principles are applied through different approaches and combinations, depending on the farming system and the resource endowments. The bottom line is to ensure simultaneous application of all three principles as the various farm operations are carried out. The main operations are land preparation, planting/seeding and weed management.

Land preparation

CA avoids the inversion tillage that is associated with the problems of hardpans, soil erosion, reduction in the amount of organic matter in the soil and the consequent reductions in soil fertility and crop yield (IIRR and ACT, 2005). Effective implementation of CA calls for adoption of land preparation corrective measures to remove such problems as compacted soil, hardpans, ridges and furrows. Compacted soils and hardpans can be corrected by ripping, subsoiling, planting basins or sowing deep-rooted cover crops such as pigeon pea (*Cajanus cajan*), lablab (*Lablab purpureus*, syn. *Dolichos lablab*) or radish (*Raphanus* spp.) (IIRR and ACT, 2005). The few smallholder farmers with access to tractors have managed to use them effectively to burst and ameliorate plough pans or hard setting of soils at the introductions of CA, resulting in subsequent 50% time and fuel reductions (URT, 2001; CFU, 2016). Where wheeled direct planters are used ridges, bumps and furrows should be removed by performing the last ploughing operation before starting CA. Where irrigation or drainage are practised, permanent broad beds and furrows are recommended (Mkomwa *et al.*, 2015).

Planting/seeding

The choice of which seeding technology to apply depends mostly on the level of mechanization being practised by the targeted communities. A study by Muriuki *et al.* (2012) found that, in Kenya, the hand hoe (66%) was the most widely used implement while the small-scale tractor (17.5%) and ox-plough (14%) came in a distant second and third. The large-scale tractor was used by a mere 2.5% of the farmers. They also reported that about 66% of households in Kenya used the hand hoe to plant in CA-compliant hand-hoe planting basins, 50% used machete, 20% used ox ripper furrows and only 6% of the farmers had adopted jab planting. Farmers in animal-traction farming systems preferred ox rippers (to open planting furrows but also breaking hard/plough pans) and animal-drawn direct seeders. Two-wheeled tractors (2WTs) are a recent but fast-expanding technology in EA. Vastly successful in wetland rice puddling, they are now considered appropriate as they have adequate power for no-till seeding, whereas they fail completely in tillage. Evaluation of different 2WT seeders is ongoing in the FACASI project (Baudron *et al.*, 2015). Tractor-mounted pneumatic seeders, capable of compensating shortcomings in gravity delivery when drilling up and down hillsides, are the preferred choice of the large-scale CA farmers in EA.

Soil cover

Providing adequate soil organic cover is a cornerstone of CA (IIRR and ACT, 2005). For the bimodal or higher rainfall areas, soil cover is usually attained by growing live plants to cover as much of the open spaces as is possible. Mixed cropping with undersown or relayed cover crops or shrubs are some of the strategies. Mono-modal and lower rainfall areas usually rely on production of high biomass, the higher proportion of it being slow-decomposing cereals or shrubs if possible, whose residues would then be used to cover the soil. The choice of cover crops, their agronomy and management with respect to the growth of the main crop are specialized subjects when their utilization with respect to nutrient addition, weed management, feed and food uses are to be optimized. Key contributions on soil cover crops and cropping systems in EA include the work by the Benedictine Fathers of Peramiho, Tanzania, and particularly on *Crotalaria ochroleuca* (Rupper, 1987), the KARI legume research network project (KARI, 2006) and the Southern Highlands of Tanzania screening and promotion of cover crops (including those from Brazil) (ARI Uyole, 2002).

The most commonly used cover crops are pigeon peas (as an intercrop in central and eastern Tanzania) and *Dolichos lablab* (as an intercrop in most of lower-altitude Kenya, northern Tanzania and some parts of Uganda), presumably due to their multi-purpose roles as food, fodder and cash crops. Cowpea (*Vigna unguiculata*), a promising cover crop, has not been widely adopted due to unavailability of desired seed varieties and restricted markets. Desmodium (*Desmodium intortum, D. uncinatum*) and lucerne (*Medicago sativa*) are widely promoted for their preferred forage qualities rather than as soil cover crops. Uptake of other well performing but non-edible soil cover crops such as crotalaria (*Crotalaria ochroleuca, C. paulina*), mucuna (*Mucuna pruriens*) and canavalia (*Canavalia ensiformis*) is limited. Their non-edible attributes in terms of soil fertility improvement as a replacement of inorganic fertilizers or their comparative costs when processed and supplied as livestock feeds need to be quantified to elevate their true/non-facial value. Properly selected cover crops in relay or rotational planting systems such as the oil seeds of canola (*Brassica napus, B. rapa*) and sunflower (*Helianthus annuus* L.) can greatly improve profitability of CA systems as they can contribute highly valuable oils and seed cake. The large-scale farmers in Kenya have installed an oil mill to process such seeds from their CA crop rotations.

Weed management

The most common practices for weed management in EA include minimal mechanical soil disturbance achieved with no-till and manual shallow weeding, ideally chopping and mulching of weeds with no soil disturbance. This way, fewer buried weed seeds are brought to the surface where they can germinate. Maintaining soil surface cover with crop residues or mixed cropping is another popular option to smother weeds and prevent them from growing. The use of desmodium in the push-pull system uses additional allelopathy effects of desmodium to inhibit germination of witchweed (*Striga* spp.) (see Khan *et al.*, Chapter 9, this volume). Herbicides, a critical ingredient of large-scale CA farming, are also used in smallholder CA. Glyphosate is the recommended option and choice of the majority. Its use is highly effective and cost efficient, but training on selection and accurate and safe application is essential.

Integration of CA, livestock and trees

A farming system that successfully combines cropping, livestock and forestry activities stands to gain many benefits that can have a direct impact on whole-farm production, with its ability to supply grains, fibres, livestock products and agro-energy. It will also improve soil fertility with the use of adequate cropping systems and techniques to optimize and intensify its use. Therefore, it allows the diversification of economic activities on-farm and minimizes income risks due to climate and/or market changes. Integrating livestock and crop production in CA means shifting from the traditional systems focused exclusively on livestock or crop to a new approach that sustainably combines both, within the comprehensive framework of the CA approach and core principles.

Livestock feed is at the very interface of positive and negative effects of livestock production (Herrero *et al.*, 2009; Peters *et al.*, 2012) and 7% of the global mitigation potential of agriculture can be reached through feeding improvements in tropical systems alone. Well managed improved forages have a high potential to sequester carbon, decrease nitrous oxide emissions through biological nitrification inhibition and efficient nutrient cycling and reduce methane emissions per unit of livestock produce by improving feed quality and increased productivity (Peters *et al.*, 2012). Increased feed production on agricultural land marginal for cropping, improved land management techniques and recuperation of degraded land also contribute to increased land-use efficiency (Schultze-Kraft and Peters, 1997).

Trees on-farm not only represent a sink for carbon but they also provide shelter, fuel and construction wood, fodder and fruits for self-consumption and the market, and contribute to soil erosion reduction. Trees participate in the capture and recycling of nutrients through exploration of deep soil layers. Likewise, examples of crop water and nitrogen facilitation by *Faidherbia albida* have been documented throughout Africa (Rhoades, 1995). These and many other examples indicate that trees have an important role to play in the design of ecologically intensive agroecosystems to overcome the generally perceived trade-offs between productivity (livelihoods) and ecosystem services. But the management (and abundance) of trees in agricultural landscapes can be challenging when they are owned collectively in a rural community, which calls for complex approaches to natural resource management.

Recent perspectives (according to Tittonell, 2013) argue that the model of intensification per unit area or per animal, which is deeply rooted in the mind of scientists from the Green Revolution generation, is now obsolete. Furthermore, integrated crop–livestock systems (IC–LS), implying a diverse range of integrated ecological, biophysical and socio-economic conditions, have been a foundation of agriculture for hundreds of years and have a significant role to play in sustainable development. In recent decades, there have been practical innovations in integrated production systems based on CA that harness synergies between the production sectors of crops, livestock and agroforestry that ensure economic and ecological sustainability while providing ecosystem services. Such IC–LS increase environmental resilience through increased biological diversity, increased water infiltration and runoff/erosion control, effective/efficient nutrient cycling/recycling and improved soil health; they also provide ecosystem services, enhance forest and watershed preservation and contribute to adaptation and mitigation of climate change (FAO, 2010). The objective to balance livestock production, livelihoods and environmental protection with sustainable intensification of crop–livestock systems seems to be a feasible pathway to reverse degradation, alleviate poverty and improve food and income security of resource-poor farmers (Delgado *et al.*, 1999; Herrero *et al.*, 2009). However, many challenges exist in the design of CA-based crop–livestock–tree systems that are adapted to local conditions and circumstances and that are widely acceptable to smallholder farmers in EA.

The initiatives addressing CA-based agroforestry in EA include the SIDA-supported RELMA, the Land Management and Environment Programme (LAMP) and Soil Conservation and Agroforestry Project in Arusha (SCAPA) projects in northern Tanzania. Others include the SIDA-financed ICRAF–ACT-implemented CAWT project. Muriuki *et al.* (2012) report from a CAWT project survey that 60% and 90% of households in Tanzania and Kenya, respectively, practised livestock keeping within their farms. Livestock keeping can therefore encourage adoption of CA due to supply of manure to crop fields, efficient and value-added utilization of surplus produce, supply of farm power and efficient utilization of household labour for managing livestock during crop-production off-seasons. Unplanned livestock keeping, however, can discourage adoption because of competing usage of biomass as mulch or fodder. Competition could also be for space between cover crops and fodder, especially if the cover crops are not edible by animals.

4.3.3 Who are the CA farmers and where are they?

The majority of the CA farmers in EA are smallholders managing 2 ha on average. They constitute 99% of the estimated total of 36,870 farmers in the sub-region. In terms of areas under CA, smallholders contribute some 22,748 ha, or 36.6% of the nearly 62,000 ha in EA, as summarized in Table 4.1. Table 4.2 lists the 20 large-scale CA farmers in the sub-region. These have three to four decades of experience of CA, unlike their counterpart smallholders who have been engaged in CA only in the last 10–15 years. The large-scale farmers have transited from conventional inversion tillage driven by economic benefits. The major crops grown by smallholders are maize, sorghum and pulses and the priorities for the large-scale farmers are barley, sorghum, wheat, maize, beans, canola and safflower.

Table 4.1. Distribution and characteristics of the Conservation Agriculture (CA) farmers in East Africa (EA).

	Number of farmers			Area under CA (ha)				Area under small-holders (%)
Country	Small-holders[a]	Large-scale commercial[b]	Total	Small-holder	Large-scale commercial	Total	Small-holder farmers (%)	
Kenya	11,100	10	11,110	6,210	15,415	21,625	99.9	28.7
Tanzania	14,450	10	14,460	10,678	21,940	32,618	99.9	32.7
Uganda	11,000	**20**	11,000	5,800	2,000	7,800	100.0	74.4
Rwanda	200		200	40	0	40	100.0	100.0
Burundi	100		100	20	0	20	100.0	100.0
Total	**36,850**		**36,870**	**22,748**	**39,355**	**62,103**	**99.9**	**36.6**

[a]The smallholder farmers manage on average 0.5 ha under CA; they use hand tools (hand hoe, jab planter and dibble stick) and animal traction implements.
[b]Large-scale commercial farmers manage 400–1600 ha; they use high-capacity GPS-guided tractor-powered seeders.

Table 4.2. The large-scale CA farmers in EA.

Country	County/Region	Name of farm	CA area (ha)
Kenya	Laikipia	Lengetia	1,800
	Nakuru (Rongai)	Madrugada	1,200
	Nakuru (Njoro)	Bruce Nightingale	240
	Nakuru (Rongai)	Luke Nightingale	400
	Meru	Kisima	2,400
	Narok	Hugo Wood	800
	Timau	Wangu Embori	1,225
	Machakos – Athi river	Jacobs	350
	Uasin Ngishu-Moiben	Kruger	2,000
		Others	5,000
			15,415
Tanzania	Iringa	Rutuba farm	500
	Iringa, Kilolo	Clinton Foundation	1,800
	Arusha, Manyara	Quality Foods	16,000
	Arusha, Manyara	Field master	2,000
	Arusha	Naberera	600
	Kilimanjaro	West Kilimanjaro	200
	Manyara	HADS Babati	120
	Arusha	Oljoro	200
	Arusha	Monduli	120
	Dodoma	Kibaigwa Flour Supply (Service Provider – ripping)	400
			21,940
Uganda			2,000
Rwanda			–
Burundi			–
Total large-scale			**39,355**

A study conducted by ICRAF and ACT in Kenya and Tanzania (Muriuki *et al*., 2012; Lugandu, 2013) concluded that CA (all three principles simultaneously) has been adopted by 5% of sampled smallholder farmers. Adoption of the individual practices contributing to CA was 19%, 21% and 61% for no-till, soil cover and crop rotations, respectively, in Tanzania. This is a good achievement for a set of technologies involving a dimensional opposite paradigm shift, in a time frame of less than a decade, with the key interventions introduced on project mode. Adoption is, however, low compared with the huge potential benefits to be reaped with full adoption. The factors influencing adoption of CA in the target countries include the age of the household head, household size, access to training resources, knowledge dissemination and farmers' perception of CA as potentially mitigating climate change.

In Kenya, adoption of CA practices is different in different parts of the country, largely because of different driving factors. Table 4.3 provides a summary of CA practices, as adopted in Kenya, in the context of the hotspot districts and responsible institutions. They also present the opportunities and highlight the constraints to upscaling of CA.

4.4 Case Studies

4.4.1 Introduction of CA to smallholder farmers in Laikipia County, Kenya

Introduction

This case study is based on the work of the EU-funded agroecology-based aggradation CA (ABACO) project in Laikipia County, Kenya. The ABACO project was implemented by the African Conservation Tillage Network (ACT) and operated also in East Africa (Tanzania), West Africa (Burkina Faso) and Southern Africa (Zimbabwe, Mozambique, Madagascar) from 2011 to 2015. The CA being promoted is a farming approach that encourages the application of three principles: (i) minimum soil disturbance; (ii) permanent soil cover; and (iii) crop rotations/associations. CA has been promoted concurrently with good agronomic practices (improved seeds, timely planting, optimal plant spacing and effective weed control). Inorganic fertilizers were promoted and used by all farmers in all of the treatments. Some 300 smallholder farmers (40% women) in ten villages used ten farmer field schools (FFS) to introduce and adapt innovative CA technologies (ACT, 2015b). The typical 0.08 ha FFS validation plot was subdivided into five treatments that addressed the three CA principles and aligned to provide research findings that address the farmers' key production constraints. Treatments of one FFS form a replicate of another similar FFS.

Experimental design

The experiments followed a balanced randomized complete block design (RCBD) with five treatments replicated seven times. The experiments were implemented by seven FFS. The treatments were: farmers' practice (FP); no-till with no cover crops (NNC); no-till with *Dolichos lablab* (NDL); no-till with pigeon peas (NPP); and no-till with dolichos and pigeon peas (NDP).

Table 4.3. Conservation Agriculture technologies situation analysis for Kenya.

Technologies	Practices	Adoption: No Low	Adoption: Medium	Adoption: High	Location (Districts)	Opportunities to scaling up	Constraints	Institutions/ Partners
Seeding/ Planting	Hand hoe	√			Machakos, Mbeere, Laikipia, Nakuru, Bungoma, Siaya	Medium: low cost; possible timeliness of planting	High labour demand	CA SARD (FAO, ACT, KARI, KENDAT, MOA, FAPEAGRO)
	Jab planter	√			Nakuru, Bungoma, Siaya, Laikipia	High: timeliness, affordable	Quality not assured in local supply	CA SARD; importers (BrazAfrica, Farm Eng); local w/shop producers (Ekima, Femo works, LoTech); EU-ABACO
	Use of oxen and donkeys for ripping and subsoiling		√		Machakos, Mbeere, Laikipia, Nakuru, Bungoma, Siaya	Medium: livestock and implements locally available; local knowhow available	Poor access to credit; low produce prices; increasing costs to maintain oxen	CA SARD; local producers; EU-ABACO; AGRA-ACT-KARLO
	Use of oxen and donkeys for direct seeding	√			Machakos, Mbeere, Laikipia, Nakuru, Bungoma, Siaya	High: through hire services	High acquisition cost; high cost for maintenance of oxen	CA SARD; local producers; importers; EU-ABACO
	Use of 2-wheel tractor seeders	√			Laikipia	High: through hire services	High acquisition cost	CA SARD; FACASI
	Use of tractor rippers	√			Laikipia, Nakuru	High: through hire services	High acquisition cost	CA SARD; local producers
	Crop residues	√			Machakos, Mbeere, Laikipia, Nakuru, Bungoma, Siaya	Medium: high benefits /returns to soil cover	Free range livestock; low awareness on benefits	CA SARD; AGRA-ACT-KARLO

Soil fertility management with N-fixing crops, shrubs	Cover crops such as pigeon peas, mucuna and lablab			√	Machakos, Mbeere, Laikipia, Nakuru, Bungoma, Siaya and other districts	High: used as food; good market value; soil fertility improvement	Low awareness; roaming livestock; difficult supply of disease free seeds; inedible, no market (e.g. Mucuna, Canavalia)	CA SARD; KARI legume research network; ILRI; NGOs; MDG villages; ICRISAT; ICIPE; (desmodium push–pull); AGRA-ACT-KARLO
Weed management	Herbicides (weed wiper, knapsack, pedestal, animal pulled sprayers)	√			Nakuru. Bungoma, Siaya	High: high efficacy (when properly used); comparatively lower cost	Mixed messages to environmental degradation; improper use	Private sector agro dealers; Syngenta; ICIPE; CA SARD; Kenya Farmers Association
	Cover cropping, use of crop residues		√		Machakos, Mbeere, Laikipia, Nakuru, Bungoma, Siaya	High: common practice among farmers esp. in banana coffee systems; highly beneficial to moisture utilization	Free-range grazing livestock; other competing uses of residues (thatching, fuel wood, hand crafts)	CA SARD; EU-ABACO; AGRA-ACT-KARLO
Labour and cost management	Zero tillage, ripping			√	Machakos, Mbeere, Laikipia, Nakuru, Bungoma, Siaya	High: elderly farmers and unreliable rainfall are pushing for labour-saving technologies	Limited: investments and promotion of the state-of-the-art knowhow and awareness of the benefits	CA SARD, MOA, GTZ, CIRAD

Continued

Table 4.3. Continued.

Technologies	Practices	Adoption: No Low	Adoption: Medium	Adoption: High	Location (Districts)	Opportunities to scaling up	Constraints	Institutions/ Partners
Integration of livestock as sources of power				√	Machakos, Mbeere, Laikipia, Nakuru, Bungoma, Siaya	Medium: about 25% of cultivated area tilled by work animals	Fodder shortage due to decreasing farm sizes and grazing lands; frequent droughts; livestock diseases and outbreaks (CBPP, trypanosomiasis)	CA SARD, MOA Extensionists, CA equipment producers and suppliers, paravets
Crop rotations and cropping systems to manage weeds			√		Machakos, Mbeere, Laikipia, Nakuru, Bungoma, Siaya	Medium	Limited awareness; more weight put on yield rather than maintenance of the soil fertility status	KARI, universities, CA SARD; EU-ABACO; AGRA-ACT-KARLO; ACT-NORAD

CA SARD: Conservation Agriculture for Sustainable Agriculture and Rural Development; EU-ABACO: European Union-funded Agroecology-Based Aggradation Conservation Agriculture; AGRA: Alliance for a Green Revolution in Africa; KARI: Kenya Agricultural Research Institute; KALRO: Kenya Agricultural and Livestock Research Organization; NORAD: Norwegian Agency for Development Cooperation; KENDAT: Kenya Network for Dissemination of Animal Traction Technologies; FAPEAGRO: Foundation for Research Support and Development of Agribusiness; FACASI: Farm Mechanization and Conservation Agriculture for Sustainable Intensification.

The experiments were conducted at Laikipia County, Kenya, and ran for three consecutive years (2012, 2013 and 2014). A detailed description of the study area, treatments structure and trial implementation is given in ACT (2015b).

Management of the selected treatments is further described below:

1. Farmers' practice (FP): hand hoe tillage, use of machete or hand hoe for opening planting holes, and hand hoe weed control (twice).
2. No-till with no cover crops (NNC) (Fig. 4.1): ox-drawn ripper (for opening planting furrows and to remove hardpans) (Fig. 4.2), shallow weeder (Fig. 4.3) and/or herbicide (glyphosate) for weed control, seeding with jab planter (Fig. 4.4) or animal-drawn direct seeder (Fig. 4.5), no cover crop grown.
3. No-till with dolichos lablab (NDL): ox-drawn ripper, shallow weeder and/or glyphosate herbicide for weed control, seeding with jab planter or animal-drawn seeder, dolichos lablab undersown in the maize 3 weeks after planting the maize.
4. No-till with pigeon peas (NPP): ox-drawn ripper, shallow weeder and/or glyphosate herbicide for weed control, seeding with jab planter or animal-drawn seeder, pigeon peas planted at same time as the maize.
5. No-till with dolichos and pigeon peas (NDP): ox-drawn ripper, shallow weeder and/or glyphosate herbicide for weed control, seeding with jab planter or animal-drawn seeder, a combination (simultaneous) of pigeon peas and dolichos lablab cover crops.

In addition to the live soil cover crops, soil cover was also achieved by retention of crop residues (mainly maize stover). An estimated 50% of the biomass produced from the plots remained on the plots, while the rest was fed to livestock or other uses.

Fig. 4.1. No-till with no cover crops (NCC) treatment.

Fig. 4.2. Animal-drawn ripper. The ox ripper is used to break up surface crusts or impermeable sub-surface soil layers and opens a narrow furrow in the soil, about 10 cm wide and 15 cm deep. It can remove compaction or plough pans and produces better results in heavy soils. Ripping is done on dry soils before the rains. It helps to increase water capture and subsequent infiltration and reduces soil erosion, especially on bare land. Ripper attachments are fitted to the traditional plough frame, thus lowering the cost of acquisition. Seed and fertilizer are placed manually in the furrow and covered. Only 15–20% of the land is disturbed by the ripper, compared with ploughing.

Fig. 4.3. Shallow weeder. The shallow weeder is a hand-pulled tool for scraping off weeds with minimal soil disturbance and without displacement of soil cover. It has a hardened cutting blade 15–25 cm wide and a metal bracket to screw or push-mount wooden handle. Typical weight 0.3–0.4 kg. The rake angle allows only for scraping and not for digging.

Data collection and analysis

Maize and stover were harvested at maturity; the cobs in each plot were separated from the stover and fresh weight was determined. Maize stover was cut at ground level and total above-ground fresh weight determined. Grain yield was measured in kg/ha. Rainfall measurements were collected from Kalalu weather station located within 10 km radius from each group.

Fig. 4.4. Jab planter. The jab planter is a manually operated planting tool with the ability to penetrate surface residues/mulch in untilled fields and simultaneously dispense seeds and fertilizer at the required depth. It is a precision hand tool in the metering of seeds and fertilizers and their placement at desired depth and optimal seed–fertilizer spacing. It allows farmers to plant from an upright standing position. The jab planter is of light weight (1.5–3 kg), easy to maintain and operate, and generally of low cost. However, training is required on correct use and maintenance. It can also be used for intercrop seeding as well as in hilly and stony soil conditions.

Maize and stover yield data were subjected to analysis of variance using SAS 9.2 software (SAS Institute, 2004) to test levels of significance due to treatments. Significantly different means were separated using least significance difference test (LSD) at p=0.05.

Cost–benefit analysis considered the cost of farm inputs: seeds, fertilizer, insecticide and herbicide purchased, labour costs and returns from sale of produce. Time taken to perform every activity was carefully recorded and labour valued at local wage of Kenyan Shillings (KSh) 200 per 8 h day (1 US$ = 101.8 KSh, March 2016) (Table 4.4). Returns were based on prevailing market price for maize grain. Cost of farm inputs (seeds, diammonium phosphate (DAP), calcium ammonium nitrate (CAN), insecticides and herbicides) were based on retail prices as per agro-input stockists. An assumption made during cost–benefit analysis was that differences between

Fig. 4.5. Animal-drawn seeder. The animal-drawn direct seeder is designed to seed into surface mulch in untilled soil. The implement has separate seed and fertilizer hoppers and a cutting disc (coulter). The coulter cuts through the residues; a ripper tine opens a furrow; the precision metered seed and fertilizer are placed in the furrow; and the drive wheel covers the seeds – all in a single operation. The seeder can be pulled by a single ox or a pair of oxen or donkeys. Seeding with the animal traction direct seeder is fast and efficient and it disturbs little soil (slit 5 cm wide).

Table 4.4. Parameters used in the economic analysis of the implemented treatments.

	Actual values		
Parameter	2012	2013	2014
Price of maize seed (KSh/kg)	135	150	160
Labour cost (KSh/day)	200	200	200
Price of DAP fertilizer (KSh/50 kg)	2500	2500	3000
Price of CAN fertilizer (KSh/50 kg)	1500	1500	2000
Price of insecticide (KSh/l)	1800	2000	2000
Price of herbicides (KSh/l)	1400	1400	1500

KSh = Kenyan Shilling; DAP = diammonium phosphate; CAN = calcium ammonium nitrate.

costs of production for a single tilled plot are due to adoption of a tillage strategy and cost of other agronomic practices (e.g. fertilization and crop protection). Economic analysis was performed on cumulated costs and benefits over 3 years in Laikipia.

Results and discussion

GRAIN YIELD Maize grain yield was not significantly influenced by the treatments (p=0.409) during Year 2012 (Table 4.5). On the other hand, maize grain yields were significantly influenced by the treatments (p=0.02) during Year 2013. Additionally, grain yields were highest during Year 2013 compared with Year 2012 and Year 2014. This could be as a result of increased total rainfall amounts (Fig. 4.6) as compared with Year 2012 and Year 2014. The grain yields increased by 68% and 67% under no-till with pigeon peas (NPP) and no-till with dolichos lablab (NDL) treatments, respectively, compared

Table 4.5. Maize grain yield (kg/ha) during Year 2012, Year 2013 and Year 2014 in Laikipia.

	Grain yield (kg/ha)		
Treatment	Year 2012	Year 2013	Year 2014
FP	2189.4[a]	1516.8[b]	979.4[c]
NNC	1758.3[a]	2059.5[ab]	1239.0[ab]
NDL	2220.5[a]	2538.8[a]	1475.3[a]
NPP	1781.9[a]	2549.5[a]	1413.7[a]
NDP	1814.0[a]	2164.7[ab]	1152.1[b]
p	0.409	0.02	0.0017

FP= farmers' practice; NNC= no-till with no cover crops; NDL= no-till with dolichos lablab; NPP= no-till with pigeon peas; NDP= no-till with dolichos and pigeon peas. Same superscript letters in the same column denote no significant difference between the treatments.

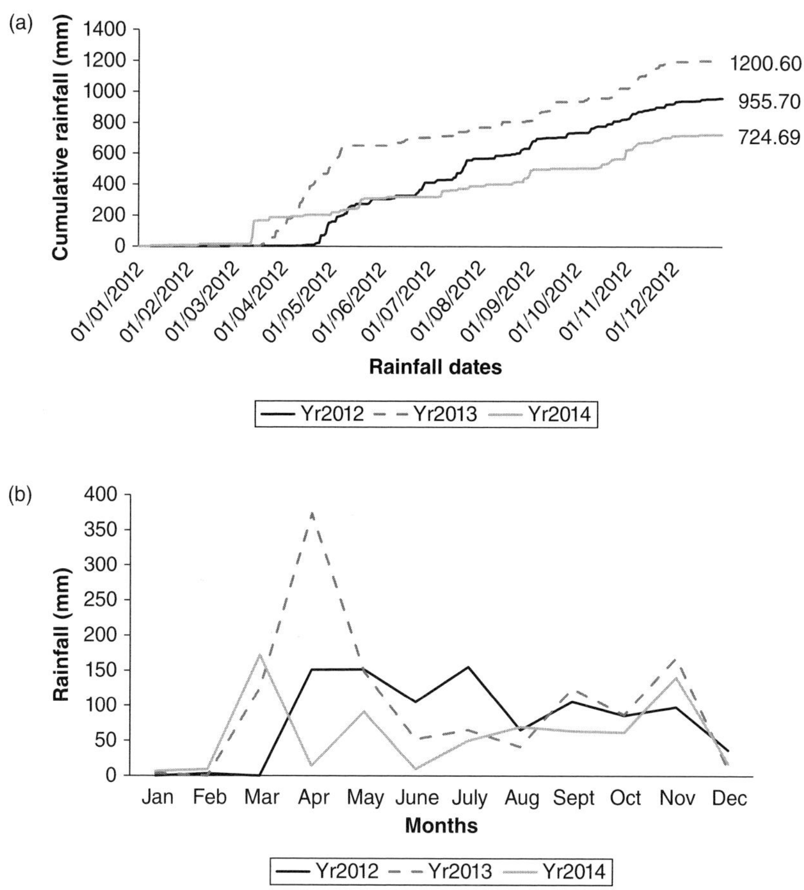

Fig. 4.6. (a) Cumulative rainfall amounts and (b) total rainfall amounts at Laikipia, Kenya, during Years 2012, 2013 and 2014, measured at Kalalu meteorological station.

with the FP. Under no-till with dolichos and pigeon peas (NDP) and no-till with no cover crops (NNC) treatments, yields increased by 42% and 35%, respectively. In Year 2012, low yields could have been as a result of waterlogging experienced during the cropping season, which poses a severe limitation to maize growth and leaching of plant nutrients, particularly nitrogen that is vital for maize crop.

In Year 2014, there was a decline in total rainfall amounts (Fig. 4.6) and in quantities of maize grain yields (Table 4.5). However, in comparison with the FP, the yields still increased by 50% and 44% under NDL and NPP, respectively. Moreover, NNC and NDP treatments increased the yields by 26% and 17%, respectively. The high increase in maize yields under NPP and NDL treatments could be attributed to the ability of intercrop to cover the soil surface quickly and reduce soil erosion, suppress weeds, enhanced biological N_2-fixation and larger quantities of better quality organic matter inputs produced in cereal–legume intercrop. Although there were legumes under NDP treatment, the increase in grain yields was still lower as compared with NPP and NDL. This could be attributed to increased soil moisture competition among the maize and the two legumes.

From the findings, NDL was the best technique in increasing maize yields in Laikipia. In comparison with FP, NDL treatment increased maize yields in Laikipia by 1.4%, 67% and 50% during Years 2012, 2013 and 2014, respectively (Table 4.5 and Fig. 4.7). NPP and NDP treatments, respectively, followed closely.

The NDL treatment produced the highest yield in year one (2220 kg/ha) followed closely by the conventional hand-hoe tilled FP (2189 kg/ha). The hand-hoe tilled FP produced the lowest and significantly different grain yield in the subsequent years declining by 30.7% in year 2 (2013) and by 55.2% in year 3 (2014). The trendsetter maize with dolichos lablab treatment maintained the highest and significantly different maize grain yields (only comparable to NPP) (Fig. 4.7).

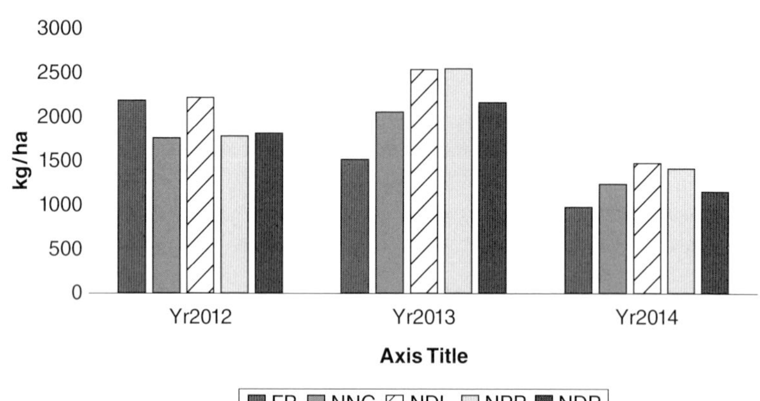

Fig. 4.7. Crop yield variations with Conservation Agriculture (CA). FP, farmers' practice; NNC, no-till with no cover crops; NDL, no-till with dolichos lablab; NPP, no-till with pigeon peas; NDP, no-till with dolichos and pigeon peas.

STOVER YIELD The treatments did not significantly influence stover yield during Year 2012 (p=0.439) and Year 2014 (p=0.356). Conversely, stover yields were significantly high in Year 2013 (p=0.012) in comparison with the FP. NDP treatment increased stover yield by 118%. This is attributed to the high total rainfall amounts received during the year.

COSTS–BENEFIT ANALYSIS During Year 2012 NNC treatment had the lowest net benefit of KSh 8305 in comparison with FP treatment. NDL treatment had the highest net benefit in Year 2012, followed by the FP (Table 4.6). Additionally the two treatments (NDL and FP) also had the highest maize yields in Year 2012 (Table 4.5). The net benefits significantly (p=0.007) increased in Year 2013 under NPP, NDL and NDP by 153%, 153% and 130%, respectively, compared with FP. There was no significant difference between FP and NNC treatments in Year 2013. The net benefits were highest in Year 2013 (Table 4.6) and similarly the same year experienced the highest amount of rainfall (Fig. 4.6) and highest maize yields (Table 4.5). Cost–benefit analysis is a useful tool that influences the decision of farmers to adopt or not to adopt CA. Farmers are more likely to adopt practices that have higher net benefits.

During Year 2014, the net benefits significantly (p<0.0001) increased under NDL, NPP and NDP treatments by 211%, 168% and 130%, respectively, compared with FP treatment that resulted in a loss. There were losses under both FP and NNC treatments in Year 2014 (Table 4.6). The net benefits decreased in Year 2014 as compared with Years 2012 and 2013 (Table 4.6). This could be attributed to the decrease in maize yield in Year 2014 (Table 4.5), which could have been as a result of decrease in rainfall amount in the same year (Fig. 4.6). Treatments that resulted in higher maize grain yields (Table 4.5) also recorded the highest net benefits.

The benefit–cost ratio (BCR) significantly increased under NDP and NDL treatments by 68% and 31%, respectively, compared with FP in Year 2012. NDP, NPP and NDL treatments recorded the highest BCR, respectively, in Year 2013. There was no significant difference between NNC and FP treatments in Year 2013 (Table 4.7).

Table 4.6. Net benefit for Year 2012, Year 2013 and Year 2014 in Laikipia.

	Net benefit/ha (KSh)		
Treatment	Year 2012	Year 2013	Year 2014
FP	43165[a]	26569[b]	–4215[b]
NNC	8305[b]	27236[b]	–21751[c]
NDL	46583[a]	67301[a]	13145[a]
NPP	32245[a]	67431[a]	11304[a]
NDP	41185[a]	61165[a]	9734[a]
p	0.009	0.0007	<0.0001

KSh = Kenyan Shilling; FP = farmers' practice; NNC = no-till with no cover crops; NDL = no-till with dolichos lablab; NPP = no-till with pigeon peas; NDP = no-till with dolichos and pigeon peas. Same superscript letters in the same column denote no significant difference between the treatments.

During Year 2014, NDP treatment increased the BCR by 400% compared with FP, which recorded a loss. Both NDL and NPP practices increased BCR by 300% in comparison with the FP. Both FP and NNC treatments recorded losses during Year 2014 (Table 4.7).

CLIMATE RESILIENCE OF CA Taking 2012 as the base year, yields in Year 2013 increased for all of the CA treatments but decreased by 31% for the FP when rainfall increased by 244.9 mm from 955.7 mm to 1200.6 mm (Fig. 4.8). Yields decreased below the base for all treatments in Year 2014 when rainfall decreased to 724 mm. The decrease was, however, higher (55%) for the FP, while it varied from 21% to 36% for the CA treatments.

4.4.2 Influence of large-scale CA farmers on smallholder CA farmers

According to Kassam *et al.* (2015), CA in 2013 was practised on nearly 157 million hectares of global annual cropland. It is estimated that 90% of the farmers involved in CA are smallholders in Africa, Asia and Latin America; and close to 90% of the global area under CA is large-scale

Table 4.7. Benefit–cost ratio for Year 2012, Year 2013 and Year 2014 in Laikipia.

Treatment	Year 2012	Year 2013	Year 2014
FP	1.6^{ab}	1.1^{b}	-0.1^{cb}
NNC	0.2^{c}	0.7^{b}	-0.4^{c}
NDL	2.1^{ab}	2.9^{a}	0.4^{ab}
NPP	1.4^{cb}	3.0^{a}	0.4^{ab}
NDP	2.7^{a}	4.1^{a}	0.5^{a}
p	0.01	0.0012	0.017

FP= farmers' practice; NNC= no-till with no cover crops; NDL= no-till with dolichos lablab; NPP= no-till with pigeon peas; NDP= no-till with dolichos and pigeon peas. Same superscript letters in the same column denote no significant difference between the treatments.

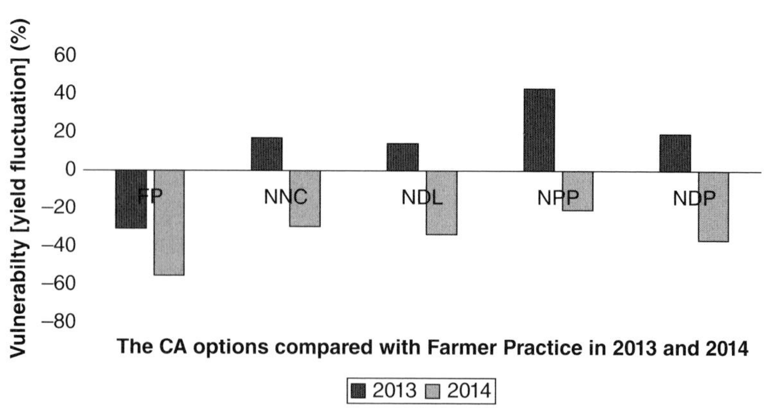

Fig. 4.8. Cushioning climate-change yield fluctuations with Conservation Agriculture (CA).

commercial farming in the Americas and Australia. For EA about 99% of the 17,500 farmers are smallholders managing on average 0.5 ha and they farm 15.8% of the area under CA (Table 4.1). The large-scale farmers manage anywhere from 500 ha to 2500 ha. There is a strange nothing-in-between gap for medium-scale farmers.

In EA, the major large-scale farmers who have adopted CA and the areas they manage are listed in Table 4.1. These farms have used CA in their wheat and barley farms for two to four decades, and there are newcomers all the time. The main technologies practised are no-tillage crop rotations and fallow systems. These farms have invested in no-till agricultural machinery and they share crop residue between mulch and livestock (Kaumbutho and Kienzle, 2007). The adoption of CA by this category of farmers is mainly because of a high land-to-labour ratio that calls for mechanization and their net income on-farm is maximized for every unit reduction in operational costs particularly saved on land preparation in terms of fuel, production costs and increased profitability of their farms. Apart from the direct benefits associated with reduced farming costs, reduced annual precipitation is probably one of the most important factors that forced these large-scale farmers into the adoption of soil and water conservation technologies like Conservation Agriculture. Accumulation of crop residue over time provided permanent soil cover, which harvested *in situ* most of the rainfall received and prevented moisture from evaporating. Increasing costs of production with the upsurge in oil prices and declining produce prices are other important factors that influenced adoption of CA for them to remain competitive.

The coexistence of large-scale commercial CA farms and smallholder CA farmers presents unique opportunities to influence adoption of CA by smallholder and medium-scale farmers. Some of these opportunities are discussed below.

Learning from the innovators

Smallholders in the vicinity of the large-scale commercial farmers learn by observation how the large-scale farmers practise CA. While a small proportion of these neighbours give up as they get intimidated by the unaffordably high costs of machinery investment, or even dubbing them 'witches' when they fail to internalize the outstanding performance of CA in years of dry spells, the majority will extract useful lessons that they proceed to implement on their own farms. Such lessons have triggered smallholders to search for advice or alternative hand-tool, animal traction or tractor CA technologies or services.

Large-scale farmers have also at times organized for CA outreach services to their smallholder neighbours. Such was the case with Mr and Mrs Laurie Sessions of Lengetia Farm in Laikipia, Kenya, with their open desire to influence small-scale farmers to adopt CA, which they believed was the only viable option available for crop farmers in the region. They have had to employ an extension officer to provide advice on smallholder CA to ten farmer groups around Lengetia Farm. The smallholders were supported to establish demonstration plots to facilitate learning about CA.

The eight (longstanding) large-scale farmers in Kenya are working in unison as a cooperative in what they call the G8. Through this arrangement, they have acquired a joint oil-milling machine for canola and sunflower. They also hire direct seeding machinery services from each other as appropriate to offset high equipment ownership costs. Smallholders are learning about the benefits and the need to cooperate.

Inputs and services support

Quality assurance and affordability on some CA inputs, particularly herbicides, is a challenge for smallholders. Smaller packages to match smaller plots are not on the market and are usually more expensive compared with bulk packages. It is also difficult to verify herbicide efficacy for a multitude of miniature packages compared with bulk consignments. Large-scale farmers, such as Lengetia Farm, have used their economies of scale and ability to buy in bulk to secure price discounts and assured quality products. They have been willing to pass some of their savings to the smallholder farmers in their vicinity. They have alternatively supported local entrepreneurs to manufacture shallow weeders, which they stock and sell to smallholders.

In the desire to support smallholders, Lengetia Farm has also acquired two sets of two-wheel tractors and seeders for rental at subsidized fees to the smallholders. The power in 2WTs is not adequate for tillage, but sufficient for no-tillage seeding and herbicide spraying.

Field Masters Ltd (Mick Dennis) of Arusha, Tanzania, have incorporated contract zero-tillage crop planting and optional herbicide spraying as part of their business plan. They cover several regions in eastern Tanzania to spread the planting window. Their services are available to farms averaging 200 ha.

Contract farming opportunities

Quality Food Products (Ekko Oosterhuis, http://www.qfp.co.tz) of Arusha, Tanzania, have incorporated contract zero-tillage crop planting to smallholders starting from 20 ha as part of their business plan. Their contract business includes some farmers who hire them just to plant on a cash basis, but most of their customers also contract with them to market their produce. Thus the two sides of their business complement each other well. Farmers who qualify for their contract farming do also, in addition to produce market and no-tillage services, qualify for a guarantee on inputs loan.

4.4.3 Socio-economic drivers to CA adoption – survey results

Methodology

A socio-economic survey to determine adoption and drivers to adoption of CA was conducted in the villages that had the ten FFS that participated in the EU-funded ABACO project from 2011 to 2014 in Laikipia County. A total of 63 households were randomly selected for the study from eight communities of Mazingira, Hurukia Mutungi, Umande scheme, Kalalu promotion, Nyumba, Ruaimiri, Mutirithia and Kiruma Muramati. A questionnaire containing both open-ended and closed-ended questions was developed and administered to the households. The questionnaires were filled and entered in Statistical Package for Social Sciences (SPSS). Frequencies, percentages, means and cross-tabulation were used to analyse the data.

Results and discussion

PROFILES OF THE RESPONDENTS The majority of the respondents from the study area were females (55%). The majority (83%) of the respondents were married, with 13% being widowed and 4% being separated. A majority of the respondents (52%) had attained primary education, 38% had attained secondary school education, 5% of the respondents had attained tertiary education while a similar number (5%) had not attained any formal education. The average age of the head of household from the study area was 57 years. The average household size in the study area was five members, with Ruaimiri having the highest household size of ten members while Hurukia Mutungi had the lowest household size at three members.

ADOPTION OF CONSERVATION AGRICULTURE Some 30% of the respondents in the study area had taken up CA within 3 years of its introduction. The Mazingira group had the highest number of households (71%) who had taken up CA, followed by Kalalu group (66%). None of the respondents in Nyumba, Umande Scheme, Kiruma Muramati, Ruaimiri and Mutirithia had adopted CA. The amount of land under CA as compared with the total land cultivated shows that the majority of the respondents (51%) had less than 25% of their land under CA, 25% had fair levels of CA application with 26–50% of their cultivated land under CA, 3% had good levels of CA adoption of 51–75%, 14% had excellent levels of adoption with over 75% of their cultivated land under CA. The average amount of land under CA in the study area was 0.37 ha, with Umande scheme having the largest (0.52 ha) average piece of land under CA.

KNOWLEDGE ON CONSERVATION AGRICULTURE AND LAND USE A majority (46%) of the respondents in the study area were stated to have had a fair level of knowledge on CA, 43% had good levels of knowledge, 9% had poor levels of knowledge while only 2% of the respondents had excellent levels of knowledge on CA. The average total land size under cultivation in comparison with knowledge on CA was 0.94 ha in the study area, with those in possession of excellent knowledge having the largest piece of land under cultivation (1.2 ha) while those with poor knowledge levels had the least amount of cultivated land (0.78 ha). Those with excellent levels of knowledge on CA had higher amounts of land under CA as compared with those with poor levels of knowledge on CA.

COSTS OF LABOUR The average cost of family labour per month in the conventional hand-hoe tilled plots was KSh 4830 while in CA plots it was KSh 1497. When comparing hired labour per month, labour cost for conventional plots was KSh 5079 while for CA plots it was KSh 1227. These results indicate that CA is cost effective when compared with conventional farming.

RELATIONSHIP BETWEEN HOUSEHOLD SOCIOECONOMIC ATTRIBUTES AND ADOPTION OF CA There was a positive significant relationship between gender of the household head and the level of adoption of CA (p=0.006). This implies that adoption of CA was higher in the male-headed households and lower in the female-headed households. A positive significant relationship (p=0.015) existed between education levels of the household heads with an increase in the level of education leading to an increase in adoption of CA in the study area.

A negative relationship existed between adoption of CA and marital status, implying that adoption was likely to be higher among the married FFS members as compared with unmarried members (Table 4.8).

STAKEHOLDER REFLECTION MEETING ON CA UPSCALING PATHWAYS Reflection meetings with key stakeholders in Laikipia County, mainly from the government, NGOs, smallholder and large-scale farmers and the private sector, identified a list of 22 constraints to adoption of CA. A pairwise ranking of the constraints is in Table 4.9.

These findings shine a light on the prominent role of social issues/incentives to unlock adoption, as opposed to the technical and technology push approaches in isolation. The approach of using groups of famers, reaching a minority proportion, is clearly being challenged by the second priority constraint of not having a common goal in the village. The need to integrate and exploit synergies between CA and livestock production and of developing unique CA technologies that support targeted livestock (and vice versa) is highlighted by constraints 3 and 4. It can be concluded from these findings that adoption of CA is a complex decision and is determined by a number of factors, including social, cultural and economic perspectives. The situation has also been extensively discussed by FAO (2010).

4.5 Conclusions and Recommendations

Conservation Agriculture offers a potential solution to improving farming productivity among smallholder farmers in EA. There are wide and varied entry points in the transition from conventional inversion tillage to CA based on the choice of cropping systems, mechanization level and farmer typologies. The choice to be made is usually not obvious and the subsequent effectiveness and profitability of the adoption depends on these initial steps. It is necessary to empower smallholder farmers in this transition. Crop yields increase when no-till equipment and cover crops (e.g. dolichos lablab or pigeon peas) are employed by smallholder farmers. The no-till with dolichos and pigeon peas treatment which recorded the highest BCR throughout the 3 years reflect that CA is offering labour-saving properties and a potential solution to farm power shortages suitable for households under labour stress. The selection and correct use of CA implements, cover crops and cropping systems is knowledge intensive. Adequate capacity building of extensionists and farmers is a crucial aspect to enhance adaptation and adoption.

Table 4.8. Relationship between household socio-economic attributes and adoption of CA.

Adoption of CA	Gender	Age	Marital status	HH size	Education level	Household income	Cows	Goats	Chicken
Pearson correlation	0.374**	−0.203	−0.076	0.254	0.338*	0.096	0.146	0.028	0.016
Sig. (2-tailed)	0.006	0.144	0.587	0.066	0.015	0.453	0.298	0.845	0.909
N	53	53	53	53	51	53	53	53	53

*, significantly different; **, highly significantly different.

Table 4.9. Pairwise ranking of CA constraints to adoption (Laikipia East CA stakeholders; October 2012).

	1	2	3	4	5	6	7	8	9	10	Rank
Choice of crop enterprise (CCE)	x	BA	WM	CAI	SC	KN	CG	ACC	CCE	CCE	**8**
Beliefs & attitudes (BA)	x	x	BA	BA	BA	BA	BA	BA	BA	BA	**1**
Weed mgt. (WM)	x	x	x	WM	SC	KN	CG	ACC	WM	WM	**6**
CA inputs (CAI)	x	x	x	x	SC	KN	CG	ACC	CAI	CAI	**7**
Soil cover (SC)	x	x	x	x	x	SC	CG	ACC	SC	SC	**4**
CA knowledge (KN)	x	x	x	x	x	x	CG	ACC	KN	KN	**5**
Common goal in the village (CG)	x	x	x	x	x	x	x	CG	CG	CG	**2**
Animal/crop conflict (ACC)	x	x	x	x	x	x	x	x	ACC	ACC	**3**
Herbicide effects (HE)	x	x	x	x	x	x	x	x	x	CC	**10**
Climate change (CC)	x	x	x	x	x	x	x	x	x	x	**9**

Climate change is already having a profound negative impact and threatens agricultural production. Apart from CA performing better than conventional farming practices with respect to yields, costs and labour, it also enhances resilience of the production system resulting from climate-change variations. During poor rainfall periods, yields from conventional farming practices decreased at a higher rate as compared with the yields resulting from CA. When more soil cover is provided and soil organic matter builds up, resilience to drought increases.

Through social surveys and working with the community groups it is concluded that promotion and adoption of CA is a knowledge-intensive process. Smallholder farmers have limited awareness of CA and its potential benefits, hence causing only a small proportion of the group to adopt CA practices. It can be added that CA is not fully mainstreamed into respective rural development policies, programmes and strategies or implemented, due to the lack of awareness among policy and decision makers. Farmers who were exposed to CA knowledge increased the land under CA. This emphasizes the strategic role of training and general capacity building. It also underscores the need for systemic capacity building, to enable agricultural universities, research institutions and government agricultural extension in general to streamline CA into university curricula, research programmes and extension services.

We recommend that smallholders should be exposed to CA cropping systems and linked to CA mechanization options in order to empower them to achieve profitable and sustainable farming. It is recommended that the level of engagement of national stakeholders in the public, private and civil sectors to support the upscaling of CA as a climate-smart technology should be increased and an enabling environment created to facilitate access to CA technology. The enabling policy environment will allow investment financing; and technological development including private sector involvement in CA-related value chains is necessary. To enhance adoption, CA knowledge dissemination is required. It is also recommended that access to CA entrepreneurial service provision by producer farmers should be enhanced for increased adoption of market-led CA. Small-scale farmers largely do not own CA tools, equipment, draught animals and machinery for different farm operations such as land preparation, planting, spraying, threshing, shelling and transportation. Availability of CA services is crucial to influence the choices on how individuals exploit their natural resources.

Acknowledgements

The authors acknowledge funding from the European Union through the lead organization, African Conservation Tillage Network (ACT) of Nairobi, Kenya, grant No. Europe Aid/138500/C/ACT/Multi for the Agroecology-Based Aggradation – Conservation Agriculture: Tailoring Innovations to Combat Food Insecurity in Semi-Arid Africa (ABACO). We also thank Anne Kimaita and Githonga J.J. of Laikipia County Agriculture Office for their skilled guidance and assistance with the field activities respectively.

References

ACT (2014) Conservation agriculture: building entrepreneurship and resilient farming systems. In: *Book of Condensed Papers of the First Africa Congress on Conservation Agriculture, 18–21 March, 2014, Lusaka*. African Conservation Tillage Network, Nairobi. Available at: http://www.africacacongress.org/sites/default/files/attachments/book_of_condensed_papers.pdf (accessed April 2016).

ACT (2015a) Declaration of the first Africa congress on conservation agriculture. In: *Proceedings of the First Africa Congress on Conservation Agriculture*. African Conservation Tillage Network, Nairobi. pp. 30–31.

ACT (2015b) *Agro-ecology Based Aggradation Conservation Agriculture (ABACO). End of Project Report*. African Conservation Tillage Network, Nairobi.

AMCEN (2011) Addressing climate change challenges in Africa; A practical guide towards sustainable development. *African Ministerial Conference on Environment*. United National Environment Programme (UNEP) Regional Office of Africa, Nairobi, Kenya

ARI Uyole (2002) *Evaluation of Cover Crops in the Southern Highlands of Tanzania. Annual Research Progress Report*. Uyole Agricultural Research Institute, Mbeya, Tanzania.

Baudron, F., Sims, B., Justice, S., Kahan, D.G., Rose, R., Mkomwa, S., *et al.* (2015) Re-examining appropriate mechanization in Africa: two-wheel tractors, conservation agriculture, and private sector involvement. *Food Security* 7(3), 889–904.

CFU (2016) Private mechanised minimum-till service provision for small and medium scale farmers. Available at: http://conservationagriculture.org/uploads/pdf/MECHANISED-MIN-TILL-PROGRESS.pdf (accessed 16 September 2016).

Chambers, R. (1989) Editorial introduction: vulnerability, coping and policy. *IDS Bulletin* 20, 1–7. doi: 10.1111/j.1759-5436.1989.mp20002001.x.
Christensen, J.H., Hewitson, B., Busuioc, A., Chen, A., Gao, X., Held, I., *et al.* (2007) Regional climate projections. In: Solomon, S., Qin, D., Manning, M., Chen, Z., Marquis, M., Averyt, K.B., Tignor, M. and Miller, H.L. (eds) *Climate Change 2007: The Physical Science Basis. Contribution of Working Group I to the Fourth Assessment Report of the Intergovernmental Panel on Climate Change*. Cambridge University Press, Cambridge, United Kingdom and New York. pp. 847–941.
Cook, K. and Vizy, E. (2012) Impact of climate change on mid-twenty-first century growing seasons in Africa. *Climate Dynamics* 39, 2937–2955. doi: http://dx.doi.org/10.1175/JCLI-D-12-00455.1.
Delgado, C., Rosegrant, M., Steinfeld, H., Ehui, S. and Courbois, C. (1999) *Livestock to 2020: The Next Food Revolution*. IFPRI Food, Agriculture, and the Environment Discussion Paper 28. Washington, D.C.: IFPRI.
Devereux, S. (2002) *State of Disaster: Causes, Consequences, and Policy Lessons from Malawi*. June 2002. ActionAid, Lilongwe, Malawi.
Diagana, B. (2003) *Land Degradation in Sub-Saharan Africa: What Explains the Widespread Adoption of Unsustainable Farming Practices?* Department of Agricultural Economics and Economics, Montana State University, June 2003.
Dixon, J., Tanyeri-Abur, A. and Wattenbach, H. (2003) Context and framework for approaches to assessing the impact of globalization on smallholders. In: Dixon, J., Taniguchi, K. and Wattenbach, H. (eds) *Approaches to Assessing the Impact of Globalization on African Smallholders: Household and Village Economy Modeling, Proceedings of Working Session Globalization and the African Smallholder Study*. FAO and World Bank, Food and Agriculture Organization of the United Nations, Rome.
EAC (2011) *The East African Community Climate Change Master Plan (EACCCMP)*. EAC, Arusha, Tanzania.
EAC (2015a) *Regional Meeting of Experts on the Development of a Standard Climate Smart Agriculture (CSA)*. Research and Training Curriculum, 25–26 June 2015, Nairobi, Kenya, Background Paper. Ref: EAC/BG/2015.
EAC (2015b) East African Community Statistical Portal. Available at: http://www.eac.int/statistics/ (accessed 16 September 2016).
FAO (2010) An international consultation on integrated crop-livestock systems for development. *Integrated Crop Management* 13, 2010.
FAO (2011) *Sustainable Land Management in Practice. Guidelines and Best Practices for Sub-Saharan Africa*. FAO, Rome.
FAO (2012) Building resilience for adaptation to climate change in the agriculture sector. *Proceedings of a Joint FAO/OECD Workshop,* 23–24 April 2012.
Garrity, D., Dixon, J. and Boffa, J. (2012) *Understanding African Farming Systems: Science and Policy Implications*. Australian Centre for International Agricultural Research and Australian International Food Security Centre.
Herrero, M., Thornton, P.K., Gerber, P. and Reid, R.S. (2009) Livestock, livelihoods and the environment: understanding the trade-offs. *Current Opinion in Environmental Sustainability* 1(2), 111–120.
IFAD (2011) *Addressing climate change in East and Southern Africa*. Available at: https://www.ifad.org/documents/10180/eee9750c-bcd5-4ffc-a7ea-56b1209e0049 (accessed 15 November 2016).
IIRR and ACT (2005) *Conservation Agriculture: A Manual for Farmers and Extension Workers in Africa*. International Institute of Rural Re-construction, Nairobi; African Conservation Tillage Network, Nairobi. ISBN 9966-9705-9-2.
IPCC (2012) *Managing the Risks of Extreme Events and Disasters to Advance Climate Change Adaptation (SREX)*. Special Report of the Intergovernmental Panel on Climate Change (IPCC). IPCC Secretariat, Geneva, Switzerland.

Kassam, A.H., Friedrich, T., Derpsch, R. and Kienzle, J. (2015) Overview of the worldwide spread of Conservation Agriculture. *Field Actions Science Reports*, 8. Available at: http://factsreports.revues.org/3966.

KARI (2006) *Enhancing Agricultural Productivity in East Africa: Development and Up-scaling of Green Manure Legume Technologies in Kenya*. Legume Research Network Project. KARI, Nairobi.

Kaumbutho, P. and Kienzle, J. (eds) (2007) *Conservation Agriculture as Practised in Kenya: Two Case Studies*. African Conservation Tillage Network, Centre de Coopération Internationale de Recherche Agronomique pour le Développement, Food and Agriculture Organization of the United Nations (FAO), Nairobi.

Kirui, O. and Mirzabaev, A. (2014) *Economics of Land Degradation in Eastern Africa*. ZEF Policy Brief No. 10. ZEF, Center for Development Research, Walter-Flex-Strasse 3, Bonn, Germany.

Lambin, E.F. and Geist, H.J. (2006) *Land-Use and Land-Cover Change. Local Processes and Global Impacts*. Springer, Berlin, Germany (WP 6319).

Liniger, H.P., Studer, R.M., Hauert, C. and Gurtner, M. (2011) *Sustainable Land Management in Practice – Guidelines and Best Practices for Sub-Saharan Africa*. TerrAfrica, World Overview of Conservation Approaches and Technologies (WOCAT) and FAO. FAO, Rome.

Lugandu, S. (2013) *Factors Influencing the Adoption of Conservation Agriculture by Smallholder Farmers in Karatu and Kongwa Districts of Tanzania*. REPOA Research Report 13/1.

Meade, B. and Stacey, R. (2013) *International Food Security Assessment, 2013–2023*. GFA-24, U.S. Department of Agriculture, Economic Research Service, Washington, DC.

Mkomwa, S., Kaumbutho, P. and Makungu, P. (2015) Farm machinery for Conservation Agriculture. In: Farooq, M. and Siddique, K.H.M. (eds) *Conservation Agriculture*. Springer International Publishing, Cham, Switzerland. pp. 109–131. doi 10.1007/978-3-319-11620-4_1.

Montgomery, D.R. (2007) Soil erosion and agricultural sustainability. *Proceedings of the National Academy of Sciences of the USA* 104, 13268–13272.

Muriuki, J., Dulla, H., Mkomwa, S. and Mowo, J. (2012) *Conservation Agriculture with Trees (CAWT) Project: Scaling-Up the Science and Practice of Conservation Agriculture in Sub-Saharan Africa*. SIDA/ICRAF/ACT Conservation Agriculture with Trees Final Project report. ICRAF, Nairobi.

NEPAD (2016) *Africa Climate Smart Agriculture Alliance launched*. Available at: http://www.nepad.org/content/africa-climate-smart-agriculture-alliance-launched (accessed 14 November 2016).

Peters, M., Rao, I., Fisher, M., Subbarao, G., Martens, S., Herrero, M., *et al.* (2012) Tropical forage-based systems to mitigate greenhouse gas emissions. In: Hershey, C.H. and Neate, P. (eds) *Eco-efficiency: From Vision to Reality* (Issues in Tropical Agriculture series). Centro Internacional de Agricultura Tropical (CIAT), Cali, CO. 2013 pp. 1–20.

Poulton, C., Tyler, G., Hazel, P., Dorward, A., Kydd, J. and Stockbridge, M. (2008) *All-Africa Review of Experiences with Commercial Agriculture: Lessons from Success and Failures*. FAO, Rome.

Rhoades, C. (1995) Seasonal pattern of nitrogen mineralization and soil moisture beneath *Faidherbia albida* (syn *Acacia albida*) in central Malawi. *Agroforestry Systems* 29, 133–145.

RTI (2011) *Towards Integrated Soil Fertility Management in Tanzania: Developing Farmers' Options and Responsive Policies in the Context of Prevailing Agro-ecological, Socio-economic and Institutional Conditions*. Royal Tropical Institute, KIT Publishers.

Rupper, G. (1987) Cultivation of marejea (*Crotalaria ochroleuca*). Experience of Peramiho. In: *Proceedings of a Writers' Workshop*, Peramiho, Songea, Tanzania. Benedictine Publications, Ndanda Peramibo, pp. 9–12.

SAS Institute Inc. (2004) *SAS/STAT 9.1 User's Guide*. SAS Institute Inc., Cary, North Carolina.

Schultze-Kraft, R. and Peters, M. (1997) Tropical legumes in agricultural production and resource management: an overview. *Giessener Beitraege zur Entwicklungsforschung* 24, 1–17.

Tittonell, P.A. (2013) *Inaugural lecture upon taking up the position of Chair in Farming Systems Ecology at Wageningen University on 16 May 2013*. Available at: https://www.wageningenur.nl/upload_mm/8/3/e/8b4f46f7-4656-4f68-bb11-905534c6946c_Inaugural%20lecture%20Pablo%20Tittonell.pdf (accessed 16 September 2016).

UNDP (2012) *Africa Human Development Report 2012*. United Nations Development Programme. Regional Bureau for Africa (RBA), New York.
UNEP (2009) *UNEP Year Book 2009*. Division of Early Warning and Assessment (DEWA), United Nations Environment Programme, Nairobi.
URT (2001) Development of conservation and no tillage based systems for sustainable use of the natural resource base. Project proposal in support of Soil Fertility Initiative. United Republic of Tanzania, Ministry of Agriculture and Food Security.
World Bank (2008) *The Agenda for Agriculture Based Countries of Sub-Saharan Africa. Agriculture for Development Policy Brief*. Policy Brief extracted from the World Bank's 2008 World Development Report, Agriculture for Development.

5 Implementing the Principles of Conservation Agriculture in Malawi: Crop Yields and Factors Affecting Adoption

W. TRENT BUNDERSON,[1]* ZWIDE D. JERE,[1] CHRISTIAN THIERFELDER,[2] MPHATSO GAMA,[3] BLESSINGS M. MWALE,[1] SPENCER W.D. NG'OMA,[1] RICHARD M. MUSEKA,[1] JOHN M. PAUL,[1] BRAND MBALE,[1] OBEDI MKANDAWIRE[1] AND PHILLIP TEMBO[1]

[1]*Total LandCare, Lilongwe, Malawi;* [2]*CIMMYT, International Maize and Wheat Improvement Center, Zimbabwe;* [3]*Machinga Agricultural Development Division, Malawi*

5.1 Introduction

Malawi faces complex social, economic and environmental problems that threaten a steepening dependency on foreign aid. The critical issues have been well documented (UNICEF, 1993; Bunderson and Hayes, 1995; World Bank, 1995; Bunderson *et al.*, 2002; Ellis *et al.*, 2003; GoM, 2007a,b; UNDP, 2007; Denning *et al.*, 2009; Thierfelder and Wall, 2011; Thierfelder *et al.*, 2013a; Wall *et al.*, 2013). The heart of the crisis is the nation's high and growing population, which is placing increased pressure on agricultural land, the country's most important natural resource. One result is that land holdings are shrinking and becoming more fragmented. Marginal areas have been brought under cultivation and fallowing has been replaced by continuous cropping under the destructive and labour-intensive tillage practice of ridging. Recurrent drought, reduced export earnings and declining terms of trade have magnified these problems.

The challenge today is to find sustainable non-destructive ways to increase the productivity of smallholder agriculture at a rate faster than that of the population. However, despite enormous efforts to promote production-increasing technologies among smallholder households, nearly 60% live below the poverty line. Another 20% are only marginally better. The

*trentbunderson@yahoo.com

situation facing smallholder farmers in Malawi, and women in particular, is well documented (UNICEF, 1993; World Bank , 1995; GoM, 2007a,b; UNDP, 2007; Takane, 2008; Tchale, 2009; Carr and Hartl, 2010). Households lack basic needs such as adequate food, water, fuel, health, shelter and education. Outlooks for reversing this predicament are bleak. Family labour, so vital to smallholder households in Malawi, is decreasing from the necessity to engage in off-farm employment. This often coincides with peak periods of labour demand, forcing households into a vicious cycle of increasing poverty and food deficits. Particularly vulnerable are female-headed households who are burdened by the added tasks of collecting fuel-wood and water, cooking, child rearing and other domestic chores. In their struggle to survive, smallholder farmers are unable to meet their immediate short-term needs while simultaneously managing their natural resources in a sustainable manner.

Although Malawi's farmer input subsidy programme has helped to increase smallholder crop yields (Denning *et al.*, 2009), its long-term sustainability, cost effectiveness and impact on productivity are major concerns, especially with projected reductions in the scale of the programme (Holden and Lunduka, 2010, 2013; Lunduka *et al.*, 2013). Without the subsidy, access to fertilizer and improved seed is beyond the reach of most farmers. At the same time, it should be emphasized that achieving long-term food security is not just about hybrid seed and fertilizer. Ultimately, the aim is to transform smallholder farmers from basic subsistence to sustainable self-sufficiency and profitability with reduced vulnerability to climate change and lower dependence on external assistance (UNDP, 2007; Bunderson *et al.*, 2015). In this context, Conservation Agriculture (CA) is attracting attention based on claims that it offers opportunities for farmers to mitigate the impacts of climate change on farm productivity (Thierfelder and Wall, 2010a), to reduce the loss of valuable rainfall and topsoil from runoff (Munyati, 1997; Thierfelder and Wall, 2009), to adapt more effectively to adverse weather conditions (Bunderson *et al.*, 2015), to improve the health of their soils (Thierfelder and Wall, 2010b, 2011) and to reduce labour and input costs (Ngwira *et al.*, 2013b).

The focus of this chapter is threefold: (i) to describe the CA system used by Total LandCare (TLC) with smallholder farmers in Malawi; (ii) to compare maize and groundnut yields under CA with conventional ridge tillage in different parts of Malawi; and (iii) to evaluate smallholder adoption of CA over time with recommended actions to address key challenges.

5.2 CA System Promoted by TLC

The CA system promoted by TLC is based on over 10 years of practical experience with the International Maize and Wheat Improvement Centre (CIMMYT) to maximize its impacts and benefits under different farm circumstances (Ito *et al.*, 2007). It incorporates the three core principles of minimum soil disturbance, good soil cover and crop rotations/associations but with some flexibility to allow farmers to start with minimum soil disturbance as the base for building a robust CA system. The core principles of TLC's system are described below and in Fig. 5.1.

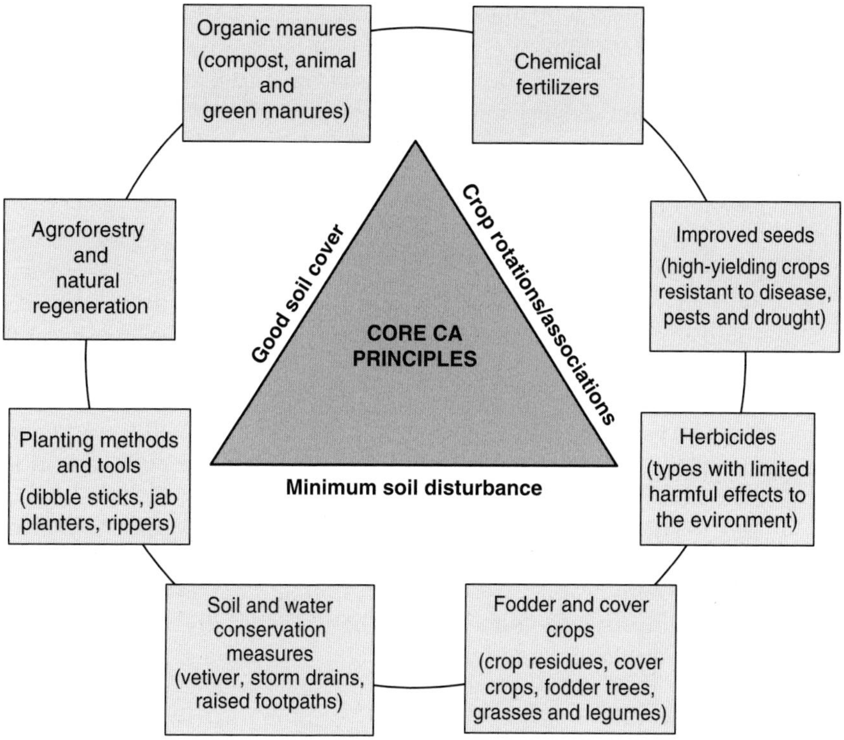

Fig. 5.1. TLC's system of Conservation Agriculture (CA) complemented by other good practices depending on local farm conditions. Minimum soil disturbance is the focal starting point, supported by the other two principles and complementary practices depending on the local farm situation.

The message on CA is simple: 'Make small planting holes with a dibble stick or hoe (mimics the age-old practice of planting across Africa before the introduction of ridging), retain crop residues produced *in situ* and diversify crops with rotations and/or intercrops'.

5.2.1 Basic principles of CA

The three basic principles of CA are:

- minimal soil disturbance by manual or mechanical means;
- good soil cover with biomass; and
- diversification of cropping.

Minimal soil disturbance by manual or mechanical means

- No ploughing, ridging, tillage or heavy weeding by manual or mechanical means. Old ridges flatten out or disappear within one or two seasons.
- Direct seeding into: (i) small planting holes on the flat or tops of old ridges using a dibble stick, hoe, or other suitable tool; or (ii) rip lines using rippers drawn by animals or tractors.

Good soil cover with biomass

The aim is to retain crop residues and other biomass on the soil surface with no burning or importation of biomass from adjacent fields. Ideally, permanent ground cover with organic biomass is desirable but this is not always possible during initial years of transition to CA, due to factors affecting the volume of biomass production and retention, including low crop productivity, type of crops grown, burning and grazing by livestock.

Use of crop residues should begin in Year 1 if available (i.e. not burned, grazed or removed). Since this is not always possible, farmers are given flexibility to organize the protection and use of crop residues by Year 2. This may entail working with community leaders to establish by-laws to protect the residues and other biomass from being burned or removed (e.g. by mice hunters, or by others who want to sabotage the practice out of spite or jealousy).

The benefits of this practice are to protect the soil from the elements, to maximize the capture of rainfall, to minimize water runoff and loss of topsoil, to improve soil structure and organic matter, to help suppress weeds and to increase beneficial activities of soil biota, such as microorganisms, termites and earthworms. These factors improve soil properties to increase water-holding capacity, to reduce runoff and erosion, to increase soil fertility and to improve the effectiveness of fertilizers by reducing nutrient losses from volatilization and leaching.

Diversification of cropping through rotations, associations (intercropping) or sequences (relay cropping) with emphasis on legumes

The purposes of diversification are to improve soil health, to reduce weeds, pests and diseases, to improve soil cover, to offer a more nutritious diet for the farm household and to increase income from sales of surplus produce. Legume intercrops are simply planted at the recommended inter-plant spacing between the rows of the main crop. This may be done at the time of planting the main crop, as for pigeon peas, or 4–6 weeks later for a relay crop such as cow peas. The system of choice depends on the interests, needs and resources of the farmer. For example, intercropping may be the only choice for farmers with limited land unless yields are high.

5.2.2 Complementary practices

Options to integrate other good agronomic practices depend on the local agroenvironment and farmer priorities and resources to increase effectiveness, productivity and profitability:

- Improved seed of different crops, including varieties resistant to drought and diseases.
- Cost-effective use of chemical and organic fertilizers at rates to optimize returns, e.g. 45–70 kg N/ha and 20 kg P_2O_5/ha.
- Judicious use of herbicides such as glyphosate and Stella Star with proper knapsack sprayers and nozzles (good quality sprayers conforming to

international standards with jet or flat yellow deflector nozzles or other 'wick type' applicators) helps to reduce labour costs for weed control with minimal soil disturbance, and may be complemented with other practices (e.g. surface residues, intercropping, spot hand-weeding, and scraping weeds off the soil surface without deep cultivation, etc.).

- Agroforestry:
  - Integration with *Faidherbia albida* in several forms: (i) with existing natural trees; (ii) managing regenerating seedlings and root-stock to minimize damage and mortality from weeding, burning and browsing; and (iii) planting 5–6-week-old air-pruned seedlings spaced 10 m × 10 m with later thinning (Bunderson *et al.*, 2002).
  - Other agroforestry practices include interplanting leguminous shrubs such as *Sesbania* and *Tephrosia* 1 m apart between crop rows, or *Gliricidia* 1 m apart with a row spacing of 4–5 m (Bunderson *et al.*, 2002).
- Integration with other conservation measures, e.g. dense hedges of vetiver grass on the contour at intervals of 5–15 m, depending on the slope. Details for planting vetiver grass with CA are provided in Bunderson *et al.* (2015).

5.3 Long-term Assessments of CA versus Conventional Ridge Tillage (CRT)

5.3.1 Yield assessment

Assessments of crop yields involved a joint programme between TLC, CIMMYT and the Ministry of Agriculture, Irrigation and Water Development (MoAIWD) to establish and evaluate CA and CRT under typical farm conditions. The results have been well documented in multiple publications that have received national and international recognition (Thierfelder *et al.*, 2012, 2013a, 2015a,b; Ngwira *et al.*, 2013a,b, 2014a,b; Bunderson *et al.*, 2014, 2015).

The numbers of on-farm trials and sites increased with time and each was monitored annually from the date established. All trials were managed by farmers with technical support from TLC and MoAIWD. Each trial included three plots of 0.1 ha each as follows: (i) maize under CRT with no residues; (ii) maize under CA with residues; and (iii) maize and a legume intercrop under CA with residues. In later years, groundnuts were evaluated in rotation with maize by splitting the original plots in half.

After land preparation for each practice, all plots were treated in the same manner in terms of planting time, crop variety, plant spacing and fertilizer application (type, rate and timing). All treatments were kept weed free. This involved manual hoe-weeding in the CRT plots, while herbicides and spot hand-weeding were used in the CA plots as follows. A systemic herbicide (glyphosate) was applied to the sole maize plot at the time of planting, followed by a selective pre-emergence herbicide (Bullet®), which was replaced later by Harness®. The CA maize–legume plot received glyphosate only as the initial measure for weed control.

Results on yields show the superiority of CA relative to CRT (Fig. 5.2). Maize yields under the two CA treatments were higher than the CRT, ranging from 11% to 70% for all sites.

Yield increases under CA were more significant in years of low rainfall such as 2009/10 and 2011/12 (Fig. 5.3).

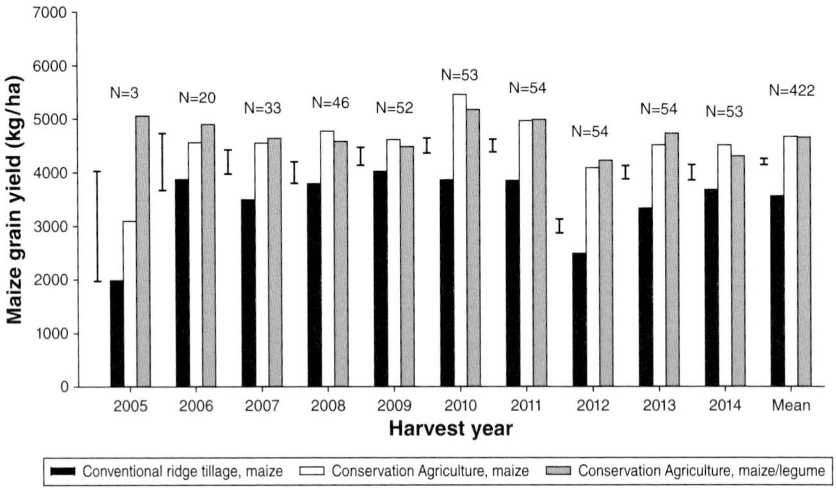

Fig. 5.2. Mean maize yields on farmer fields under Conservation Agriculture (CA) versus conventional ridge tillage (CRT) from 2004/05 to 2013/14 (*p* is significant between CA and CRT except in 2005/06) (from TLC, Machinga ADD, CIMMYT On-Farm Trials with CA versus Conventional Ridge Tillage). The error bars represent the standard error of the difference (SED) between the means at $p<0.05$.

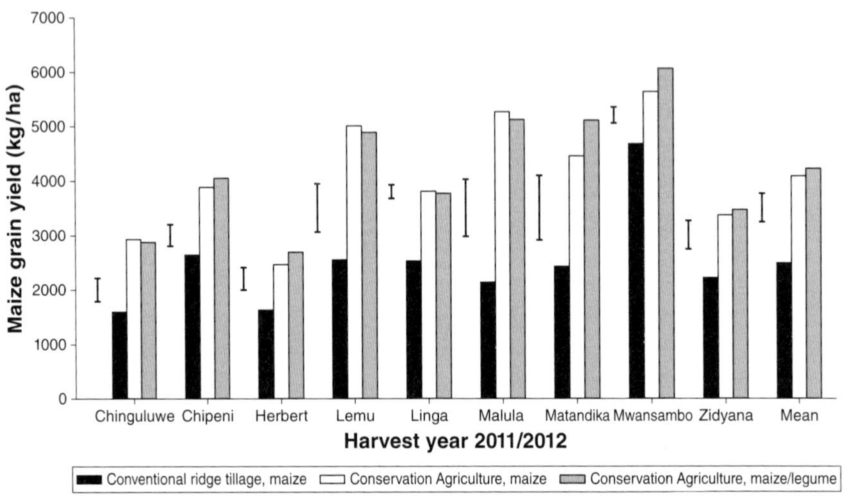

Fig. 5.3. Mean maize yields on 54 farmer fields under Conservation Agriculture (CA) versus conventional ridge tillage (CRT) in a year of low rainfall, 2011/12 (six farmers per site) (from TLC, Machinga ADD, CIMMYT On-Farm Trials with CA versus Conventional Ridge Tillage). The error bars represent the standard error of the difference (SED) between the means at $p<0.05$.

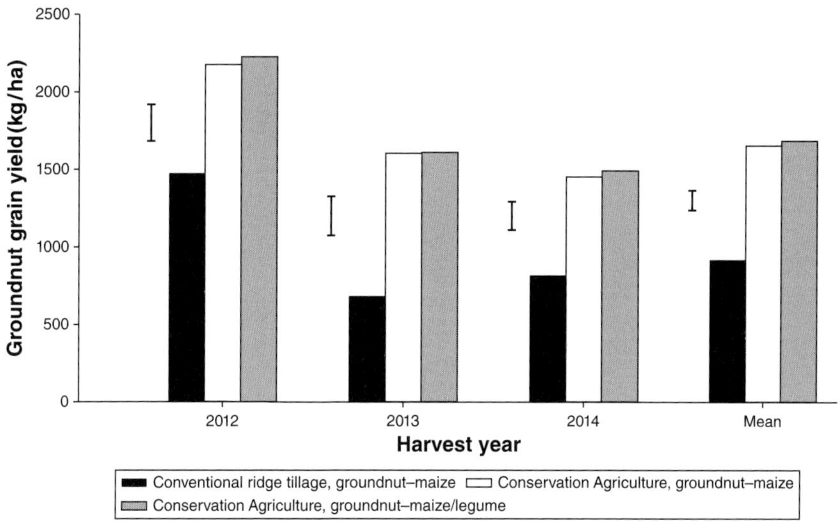

Fig. 5.4. Mean groundnut yields on farmer fields (6 replicates per site) after Conservation Agriculture (CA) and conventional ridge tillage (CRT) (from TLC, Machinga ADD, CIMMYT On-Farm Trials with CA versus Conventional Ridge Tillage). The error bars represent the standard error of the difference (SED) between the means at $p<0.05$.

Farmers also realized significant benefits from growing groundnuts under CA in rotation with maize. The key advantage is the ability to halve the row spacing, which is not possible with ridging or planting basins because there is no space to double the number of ridges or basins, not to mention the increased labour required. Figure 5.4 shows increased yields of groundnuts under CA after maize, ranging from 45–50% in 2011/12 to 93–99% in 2012/13, with a mean increase of 57–60% across the 3 years. Ground cover was also doubled, which helps to reduce the risks of water runoff and rosette disease.

Ultimately, cereal and legume crops grown under CA increased benefits to households in terms of greater productivity, food security and incomes. There were also better prospects to improve household nutrition by increasing the consumption of groundnuts.

5.3.2 Gross margins and labour costs of CA versus CRT

Tables 5.1 and 5.2 show gross margins and labour costs for all farm operations under CA versus CRT based on data collected from the on-farm trials managed by farmers combined with information from the MoAIWD on labour costs for common farm operations.

In terms of labour, the results show savings of 47% and 33% for sole maize and intercropped maize, respectively, under CA versus CRT, which is significant considering that most of the labour saved is for the physically demanding tasks of manual ridging and weeding with hoes. The lower

Table 5.1. Labour inputs of two CA systems versus CRT. (From TLC, CIMMYT, MoAIWD on-farm trials.)

Labour input (6 h days)	CRT maize	CA maize	CA maize + legume
Land preparation/clearing	7.50	0.00	0.58
Ridging	36.00	0.00	0.00
Distributing crop residues on the ground	0.00	6.80	7.15
Planting maize	9.44	10.08	10.08
Planting legume intercrop	0.00	0.00	13.50
Basal dressing	12.28	12.60	13.56
1st weeding	24.63	3.85	3.25
Top dressing (CAN)	11.43	12.00	12.60
Drawing water (herbicide use)	0.00	2.40	1.20
Roundup application	0.00	4.17	4.17
Harness application	0.00	4.17	0.00
2nd weeding/banking	23.29	4.25	2.20
Harvesting maize (stooking/collecting cobs)	12.69	12.69	12.69
Harvesting legume (uprooting plants/collecting pods)	0.00	0.00	11.42
Total labour inputs	**137.26**	**73.01**	**92.40**
Labour savings %	**0%**	**47%**	**33%**

savings for intercropping are due to the extra labour for planting and harvesting the legume crop but this is offset by the yield of the legume.

Gross margins are based on valuing all labour costs, whether labour was hired or family was used. Crop yields under the two CA systems and CRT were averaged over 10 years from 2004/05 to 2013/14 and involved up to 12 different sites across Malawi, each with six farmers per site.

The results show gross margins of US$806 and US$1034 for sole and intercropped maize, respectively, under CA, versus US$468 under CRT. The intercropping system clearly shows higher returns to land, labour and capital.

5.3.3 Planting basins and direct seeding

Some advocates in southern Africa promote planting basins dug by hand (Aagaard, 2011; Johansen *et al.*, 2012). This method of planting was originally developed by the Zimbabwean farmer Brian Oldrieve (Oldrieve, 1989, 1993) and later expanded massively by the Conservation Farming Unit of Zambia (Haggblade and Tembo, 2003).

Construction of basins is a one-off operation with the primary aim to conserve water by harvesting rainwater and to help break shallow hard pans in areas where they have formed over time. Crop yields may also be improved within the basins, relative to conventional tillage systems, due to increased efficiency of nutrient uptake from the concentration of chemical

Table 5.2. Costs, revenues and gross margins of CA versus conventional ridge tillage (CRT) (all figures in units/ha). From TLC, CIMMYT, Machinga ADD on-farm trials from 2004/05 to 2013/14.

	Crop yields (kg/ha) across sites and years							
	Crop		CRT maize		CA maize		CA maize/legume	
	Maize		3424		4463		4675	
	Legume intercrop[a]		0		0		468	
		US$	CRT maize		CA maize		CA maize/legume	
	Unit	/unit	Qty	Total US$	Qty	Total US$	Qty	Total US$
REVENUE FROM HARVEST								
Maize	kg	0.27	3424	929	4463	1211	4675	1269
Cowpea /pigeon pea intercrop (av.)	kg	0.57	–	–	–	–	468	267
TOTAL REVENUE (with weighted legume av.)	kg	0.30	3424	929	4463	1211	5143	1536
VARIABLE COSTS								
Material inputs:								
Maize seed	kg	2.29	25	57.25	25	57.25	25	57.25
Cowpea /pigeon pea seed	kg	2.19	0	–	0	–	40	87.56
23:21:0:4	50kg bag	45.43	2.5	113.58	2.5	113.58	2.5	113.58
Urea	50kg bag	48.29	2.7	130.38	2.7	130.38	2.7	130.38
Roundup	litre	8.00	0	–	2.5	20.00	2.5	20.00
Harness	litre	8.57	0	–	1.0	8.57	–	–
Total cost of material inputs				**301.21**		**329.78**		**408.77**
Labour costs (6 h day/person):								
Land clearing weeds/crop residues	day	1.00	7.50	7.50	–	–	0.58	0.58
Ridging	day	1.25	36.00	45.00	–	–	–	–
Laying down stover	day	1.00	–	–	6.80	6.80	7.15	7.15
Planting maize	day	1.00	9.44	9.44	10.08	10.08	10.08	10.08
Planting legume intercrop	day	1.00	–	–	–	–	13.50	13.50
Basal dressing	day	1.00	12.28	12.28	12.60	12.60	13.56	13.56
1st weeding	day	1.25	24.63	30.79	3.85	4.81	3.25	4.06

Continued

Table 5.2. Continued.

			Crop yields (kg/ha) across sites and years					
	Crop		CRT maize		CA maize		CA maize/legume	
	Maize		3424		4463		4675	
	Legume intercrop[a]		0		0		468	
		US$	CRT maize		CA maize		CA maize/legume	
	Unit	/unit	Qty	Total US$	Qty	Total US$	Qty	Total US$
Top dressing	day	1.00	11.43	11.43	12.00	12.00	12.60	12.60
Drawing water (herbicide use)	day	1.00	–	–	2.40	2.40	1.20	1.20
Roundup application	day	1.00	–	–	4.17	4.17	4.17	4.17
Harness application	day	1.00	–	–	4.17	4.17	–	–
2nd weeding/banking	day	1.35	23.29	31.44	4.25	4.25	2.20	2.97
Harvesting maize	day	1.00	12.69	12.69	12.69	12.69	12.69	12.69
Harvesting legume	day	1.00	–	–	–	–	11.42	11.42
Total labour costs	days		**137.3**	**160.57**	**73.01**	**75.46**	**92.40**	**93.98**
Sprayer costs (shared with 10 farmers):								
Depreciation (based on total days used)	500 days	7.71	0	–	4.17	0.06	4.17	0.06
Maintenance (50% of depreciation)	500 days	7.71	0	–	2.08	0.03	2.08	0.03
Total sprayer costs						**0.10**		**0.10**
TOTAL COSTS				**461.78**				
Gross margin per hectare				**468**		**806**		**1034**
Benefit–cost ratio				2.01		2.99		3.06
Break-even maize yield @ current average price (kg/ha)				1701		1493		1686
Break-even price @ current yield (US$/kg)				0.13		0.09		0.10
Total labour required (days and US$)			137.3	160.57	73.0	75.46	92.4	93.98
Gross margin return to labour (US$/day)				3.91		5.49		7.02
Gross margin if maize yield or price drops by 30%				189		443		573

[a]Yield is the average yield of cowpeas and pigeon peas.

fertilizers, organic manures and/or lime within the basins (Aagaard, 2011). However, access to animal manure, compost and lime is limited in Malawi. For example, the average number of cattle per farm household is 0.5, but 10–15 cattle are needed to secure sufficient manure from animal enclosures to apply 10 t/ha, the amount generally needed for a cost-effective return on yield. Although lime is locally produced in Malawi, transport costs from the source to the field are prohibitive and hence lime is rarely used.

In Malawi, where rainfall is generally more reliable than in many parts of East and Southern Africa, the added value of water conservation in basins has not been established against the high labour costs of digging basins. This is particularly relevant when compared with no-till with small planting holes and retention of crop residues which help to maximize the capture of rainfall and minimize runoff (Figs 5.5 and 5.6). That said, there are certain niches with low rainfall in Malawi where the advantages of basins have applicability.

An assessment of the micro-basin planting method reveals several drawbacks, as follows.

Labour costs

Results show that digging basins in Malawi to the specifications recommended by the Conservation Farming Unit (CFU), i.e. 35 cm long × 15 cm wide × 20 cm deep, is five times more labour intensive and time consuming than

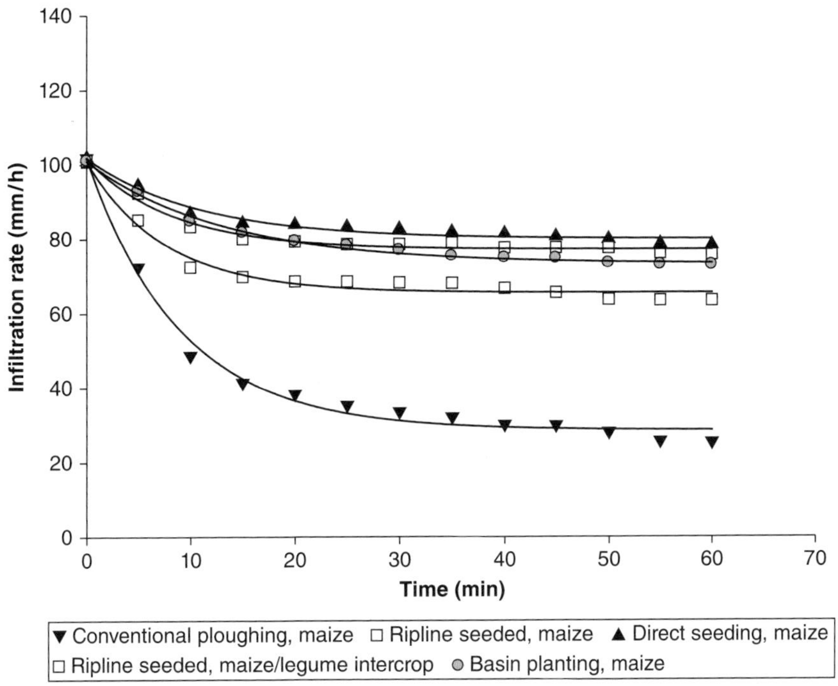

Fig. 5.5. Effects of conventional ploughing and CA systems on water infiltration in Zimbabwe (from Thierfelder and Wall, 2012). Note that all CA systems retained crop residues.

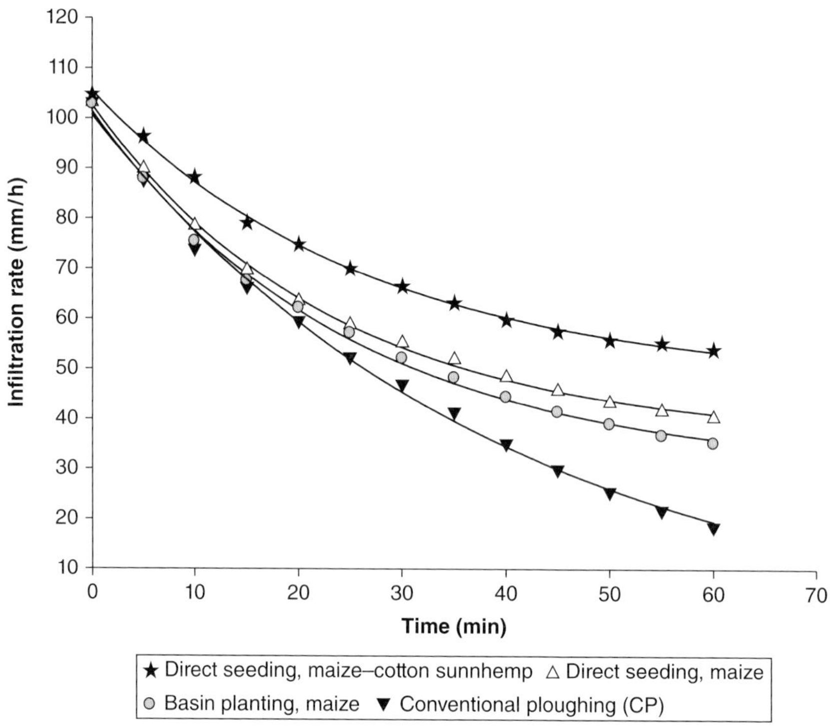

Fig. 5.6. Effects of conventional ploughing and CA systems on water infiltration in Zambia (from Thierfelder and Wall, 2010b). Note that all CA systems retained crop residues.

conventional ridging. Even though digging basins is intended as a one-off operation, the high labour costs involved make it prohibitive to undertake, which is reflected by low adoption rates and the small areas under basins relative to no-till using a dibble stick or hoe. This is especially true with the CFU recommendation to flatten the ridges before digging basins (Aagaard, 2011).

Tables 5.3, 5.4 and 5.5 show estimated labour costs for constructing ridges, levelling ridges and digging basins from the studies reported in this paper. The figures are based on the current recommended ridge spacing of 75 cm, which equates to 13,333 m of ridges/ha, and the CFU spacing for basins of 75 cm between rows and 70 cm within the row, which equates to 19,047 basins/ha. The recommendation to also flatten the ridge is more labour demanding than ridging, because the soil from each ridge must be evenly distributed across the width of the furrow. This adds time and effort to dig the basins, which need to be dug early in the dry season before spreading the crop residues, to avoid interference with digging. It should be stressed that the labour costs for each task relate to the small plots used in this study. If the pace is maintained at 6 h/day, the labour costs are 20 person-days for ridging, 22 for levelling ridges and 102 for digging basins. However, it is impossible to maintain this rate of work over an extended period of time. Consequently, actual labour costs/ha were estimated by doubling the figures recorded in the study to better reflect reality based on reported costs for ridging of about 40 person-days/ha, which is roughly double the rate recorded for the task in this study.

The above labour estimates are shown in Tables 5.3, 5.4 and 5.5 but farmer interviews suggest they are conservative for digging basins.

Soil disturbance

For ridges spaced 75 cm apart, farmers construct 13.3 km of ridges/ha by hand every year. The soil moved, based on the studies reported here, equates to 720 t/ha based on a measured average of 54 kg of soil/m in a standard size ridge. In contrast, the amount of soil moved for digging basins is 285–305 t/ha based on the CFU recommended basin spacing of 70 cm × 75 cm and a basin size of 35 cm long × 15 cm wide × 20 cm deep which contains 15–16 kg of soil. Adding in the 720 t of soil moved for levelling the ridges, the total amount of soil moved for digging basins is over 1000 t/ha.

Table 5.3. Labour costs for constructing ridges on common loamy clay soils in Malawi.

			Projected labour cost of ridging per ha							
			Labour cost/ha (at same pace as ridging 10 m)				Estimated actual labour cost/ha			
Farmer	Time to ridge 10 m (mins)	Time to ridge 1 m (mins)	Hours/ha	Days /ha @ 6 h /day	Labour cost /ha @ $1/ day	Metres /person /day	Hours /ha	Days /ha @ 6 h / day	Labour cost /ha @ $1 /day	Metres /person /day
Owen	5.13	0.51	114	19	19	702	228	38	38	351
Chikobudzo	5.30	0.53	118	20	20	679	236	39	39	340
Lofat	6.21	0.62	138	23	23	580	276	46	46	290
Jalek	4.85	0.49	108	18	18	742	216	36	36	371
Average	**5.37**	**0.54**	**119**	**20**	**20**	**676**	**239**	**40**	**40**	**338**

Table 5.4. Labour costs for levelling ridges before digging basins on common loamy clay soils in Malawi.

			Projected labour cost of levelling ridges per ha							
			Labour cost/ha (at same pace as 14 m of ridge or 20 basins)				Estimated actual labour cost/ha			
Farmer	Time to level 14 m of ridge (20 basins) (mins)	Time to level 1 m of ridge (mins)	Hours /ha	Days /ha @ 6 h /day	Labour Cost /ha @ $1 /day	Metres /person /day	Hours /ha	Days /ha @ 6 h /day	Labour cost /ha @ $1 /day	Metres /person /day
Owen	9.34	0.67	148	25	25	540	297	49	49	270
Chikobudzo	7.63	0.55	121	20	20	661	242	40	40	330
Lofat	8.45	0.60	134	22	22	596	268	45	45	298
Jalek	8.27	0.59	131	22	22	609	263	44	44	305
Average	**8.42**	**0.60**	**134**	**22**	**22**	**602**	**267**	**45**	**45**	**301**

Table 5.5. Labour costs for digging basins on common loamy clay soils in Malawi.

			Projected labour cost for digging basins (without back-filling) per ha							
			Labour cost/ha (at same pace as digging 10 basins)				Estimated actual labour cost/ha			
Farmer	Time to dig 10 basins (mins)	Time to dig 1 basin (mins)	Hours /ha	Days /ha @ 6 h / day	Labour cost /ha @ $1/day	Metres /person /day	Hours /ha	Days /ha @ 6 h /day	Labour cost /ha @ $1 /day	Metres /person /day
Owen	16.8	1.68	533	89	89	214	1067	178	178	107
Chikobudzo	19.3	1.93	613	102	102	187	1225	204	204	93
Lofat	23.5	2.35	745	124	124	153	1490	248	248	77
Jalek	17.5	1.75	556	93	93	206	1111	185	185	103
Average	**19.27**	**1.93**	**612**	**102**	**102**	**190**	**1223**	**204**	**204**	**95**

The Zambian *chaka* hoe is recommended to ease the onerous task of digging in hard soil as it has a narrow heavier blade (Aagaard, 2011). However, the Malawi hoe is better suited for levelling ridges, because its lighter and wider blade moves more soil with each stroke. In this study, all farmers preferred using the local hoe for ridging and levelling ridges. Promoting the *chaka* hoe means adding another implement for 3 million smallholders to buy for digging basins in Malawi.

Inconsistency in basin size, spacing and back-filling

Observations of basins constructed by hundreds of farmers across Malawi show tremendous variability in their size and spacing, which greatly compromises their purpose and function. It is rare to find basins constructed by farmers to the specified size and spacing recommended by CFU. Clearly, the problem lies with poor understanding and delivery of the extension message. This problem is often compounded by failure to return the soil to the basin. This means that crops are planted at the bottom of the basin, which can lead to other problems such as water-logging, fungal diseases and cut-worms. In many cases, the basins are actually pits 50 cm deep or more. The high levels of labour and soil disturbance involved in digging pits of this size means that they do not qualify as CA. The risk of distorting basin size adds another complication for getting the right message across to undertake basin planting properly.

Risk of missing the planting rains

The time and effort to flatten ridges and dig basins, even in soil softened by rain, creates the high risk of missing the first planting rains, with serious implications on yields.

Limited flexibility

The fixed position of basins at 75 cm × 70 cm is not compatible with the recommended spacing of different crops in Malawi such as tobacco, cassava, cotton and grain legumes. Use of basins also raises a question about its value for intercropping because the main crop is planted in basins while the intercrop is planted on the flat between basins as there is no possibility to dig more rows of basins.

Uniform plant spacing
Concentrating several seeds in the confined space of basins compromises crop growth due to competition for light, moisture and nutrients between the plants from seeds in the same planting station versus direct seeding with one seed per hole, which ensures uniform plant spacing for optimal growth and yields with minimal intra-competition.

Dibble sticks versus planting basins
In Malawi, the preferred and widely accepted method of planting by farmers is the no-till system of making small planting holes with a dibble stick or hoe, a practice that is aligned with and endorsed by the Agricultural Technology Clearing Committee (ATCC) under the Department of Agricultural Research Services (DARS) in the MoAIWD (Ligowe *et al.*, 2013). The true value of this system is that it mimics the age-old method of planting before the introduction of ridging during the colonial period. It has proved effective with different crops across a wide range of agroecological zones. A key attribute favoured by farmers is flexibility of no-till planting to accommodate the spatial requirements of many crops to optimize yields, minimize inter-plant competition and increase ground cover. Equally important is that it is fast and easy, which makes it well suited to a wide range of farmers and farm workers, from the young, women and elderly to those who are ill or physically challenged.

The arguments above are supported by the results of long-term trials under the Sustainable Intensification of Maize–Legume Systems in Eastern and Southern Africa (SIMLESA) project in the lowlands of Malawi, where rainfall is generally lower than the rest of the country (Nyagumbo *et al.*, 2015). These trials showed significant increases in maize yields under CA using a dibble stick versus planting basins and conventional ridge tillage (Figs 5.7 and 5.8). Note that both systems of CA retained crop residues. The message is that the payoff does not justify the labour costs of flattening ridges and digging basins.

5.4 Adoption of CA and Related Challenges

The unique synergies created between TLC and CIMMYT have generated strong scientific evidence that CA practices in Malawi are more profitable, viable and gender sensitive than conventional ridge tillage systems (Ngwira *et al.*, 2012, 2013a,b, 2014a,b; Thierfelder *et al.*, 2012, 2013a,b, 2015a,b; Bunderson *et al.*, 2014). The results have provided the basis to improve, promote and adapt CA practices with farmers in different parts of Malawi. This has enabled TLC to attract funding from multiple donors to scale up CA using an innovative farmer-led participatory approach. Donors of special note include the International Fund for Agricultural Development (IFAD), Royal Norwegian Embassy in Malawi and Zambia, British Aid through

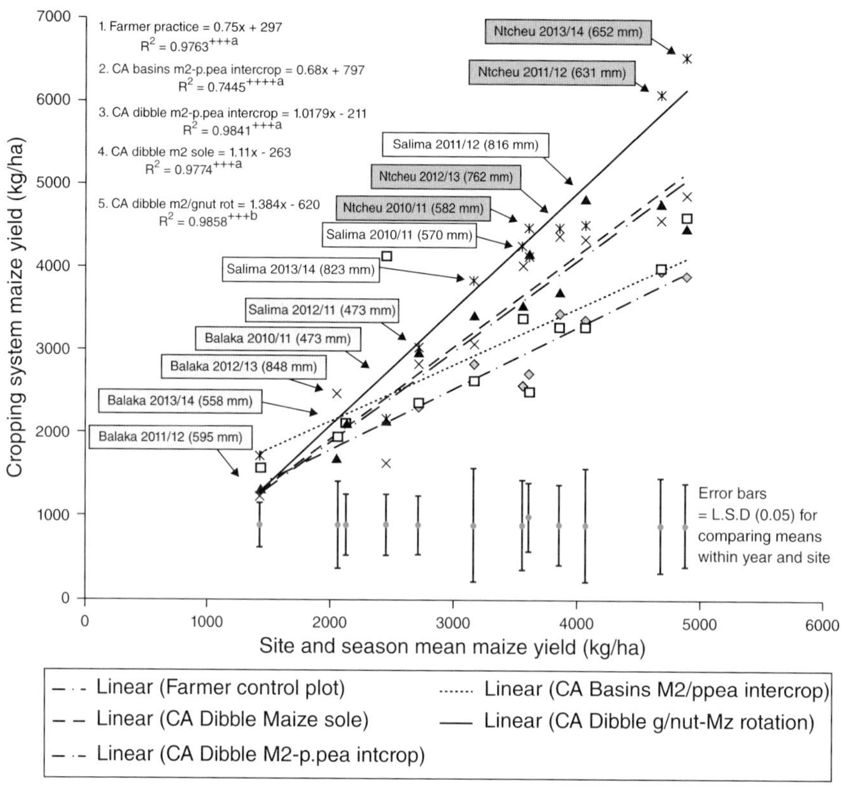

Fig. 5.7. Linear regressions of maize yields under different cropping systems against site and season means in Malawi lowland sites from 2010/11 to 2013/14 (from Nyagumbo *et al.*, 2015). Error bars denote least significant difference (LSD) (0.05) to separate means from each site and season. Labels for each site show season and total rainfall (mm). Treatment regression equations followed by the same superscript letter are not significantly different at $p<0.05$. gnut, groundnut; mz, maize; p.pea, pigeon pea; rot, rotation.

the Division for International Development (DFID) and the United States Agency for International Development (USAID).

The outcome has led to a steady increase in the area and number of farmers practising CA from 46 farmers on 14 ha in 2005/06 to 63,928 farmers on 17,710 ha in 2014/15 (Fig. 5.9). Of the 180,000 households supported by TLC in Malawi, about 35% are practising CA.

Despite excellent progress with developing good practices for CA and demonstrating its multiple benefits, adoption levels have been lower than expected. This situation prompted TLC to conduct three surveys across different parts of the country to identify the underlying drivers and barriers to adoption from the perspective of farmers (Mwale *et al.*, 2013, 2014a,b). The first involved 785 households across three TLC projects funded by different donors. The second survey involved 505 households in areas supported under the CFU/COMESA Conservation Agriculture Regional Project (CARP) and the third involved 855 households in areas supported by the

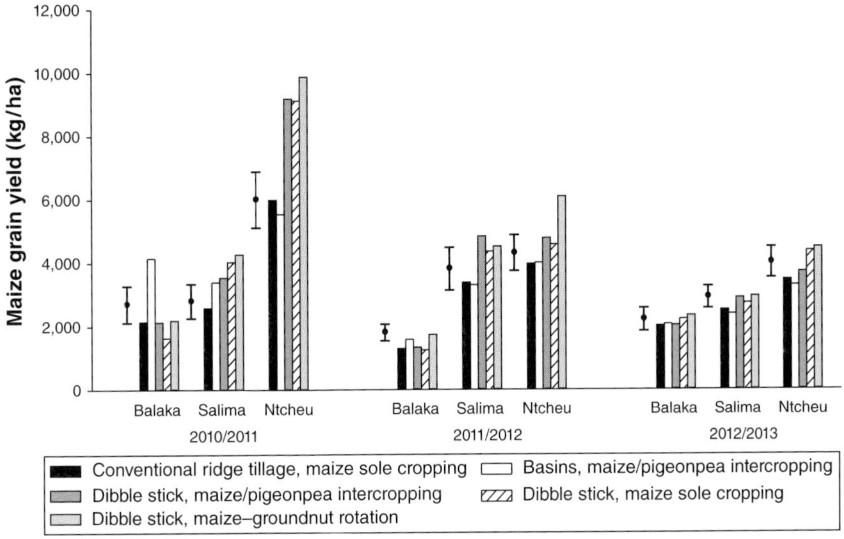

Fig. 5.8. Maize yields under different CA systems and conventional ridge tillage in three lowland sites over 3 years (from Nyagumbo *et al.*, 2015). The error bars represent the standard error of the difference (SED) between the means at $p<0.05$.

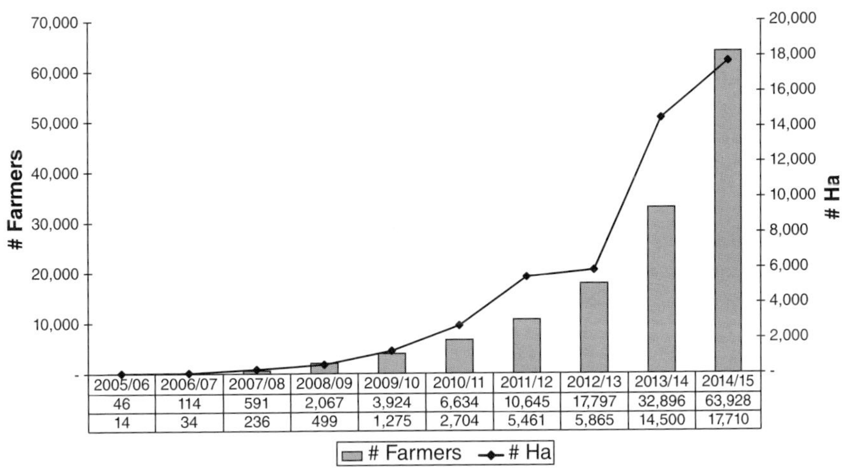

2005/06	2006/07	2007/08	2008/09	2009/10	2010/11	2011/12	2012/13	2013/14	2014/15
46	114	591	2,067	3,924	6,634	10,645	17,797	32,896	63,928
14	34	236	499	1,275	2,704	5,461	5,865	14,500	17,710

Fig. 5.9. Farmers practising CA across TLC programmes, 2005/06 to 2014/15. (Note that these are annual not cumulative figures across years and exclude results from a new project – Sustainable Agricultural Production Program funded by IFAD with the MoAIWD.)

DFID project for Building Resilience to Climate Change (BRCC). The results of all three surveys were consolidated for two groups of farmers as follows: (i) farmers who had never attempted CA (828 farmers); and (ii) farmers who had practised CA for two or more years (998 farmers). The results are presented in Table 5.6 with explanations of the challenges below.

Table 5.6. Key factors driving and impeding the adoption of CA.

Most important factor impeding adoption[a]	Number of responses	% of total respondents
Lack of information / poor understanding of CA and/ or conflicting messages and approaches, including promotion of ridges and basins	404	49%
Limited access to inputs including residues/biomass and/or labour costs for collecting biomass	226	27%
Resistance to change (culture of ridging and clean fields)	161	19%
Other, e.g. concerns about control of weeds, pests and diseases; fear that increased termite and earthworm activity under CA is harmful	37	4%
Total number of non-adopters	828	100%

Factors driving adoption[b]	Number of responses	% of total respondents
Increased food security and crop yields[c]	784	79%
Reduced labour costs	625	63%
Reduced effects of dry spells on crop yields (moisture conservation)	392	39%
Soil improvement	379	38%
Increased incomes (from reduced costs or sales of surplus production)	217	22%
Total number of respondents practising CA for 2 or more years	998	100%

[a]Responses on barriers were limited to the most important factor identified by each farmer.
[b]Responses on drivers of adoption included multiple factors identified by each farmer.
[c]Recent interviews separate from these surveys reveal great interest in CA with legume crops because yields can be doubled simply by halving the row spacing, which is not possible with ridges or fixed basins. Another attraction rarely reported is that the incidence of *Striga* with maize is greatly reduced under CA, which can significantly affect yields.

5.4.1 Challenges

Lack of information and poor understanding of CA

Many farmers have not been exposed to or received proper training on CA and how to implement it in practice. This lack of knowledge has been compounded by inconsistent and often conflicting extension messages and advice on the principles and practices of CA in Malawi. Several examples are noteworthy.

- The ATCC of the MoAIWD has officially endorsed CA practices using dibble sticks, retention of crop residues *in situ*, and crop associations, intercropping and/or rotations (see DARS Extension Circular by Ligowe *et al.*, 2013).

However, the MoAIWD has also maintained the antiquated policy of contour ridging, which directly contradicts the basic premise of CA.

- The extension messages conveyed to farmers by extension staff from many organizations often insist that all three principles of CA must be undertaken at the same time. Most farmers do not have the ability to do this. A common example is that crop residues and weeds are often burned by mice hunters to force mice into holes where they can be easily trapped and caught. The message should be that if all three principles cannot be undertaken, farmers can begin CA with no-tillage to be followed later with innovative ways to integrate the other two practices as soon as possible (see Fig. 5.1).
- The basin system of planting recommended by the CFU in Zambia has added complications and confusion among farmers (as well as extension staff) over the best practice to adopt. This concept has been compromised further by distorting the message on the recommended basin size and spacing. The results have led to a wide range of basins from very shallow holes to deep pits that do not qualify as CA due to the heavy work involved and amount of soil disturbance.
- The labour-intensive task of digging, levelling ridges and distributing the soil across the furrow has been an added barrier to adoption. For those who have used basins, the area has generally not expanded, due to the labour costs involved and the fact that the fixed position of basins does not fit the recommended spacing of other crops (see points above on basins). The bottom line is that the basin system is not well accepted by Malawi farmers because of the high labour costs involved for little or no gain in yield relative to minimum tillage on the flat or old ridges.
- Another misguided perception, created mainly by field staff, is the need to secure or even import large quantities of crop residues or other biomass for distribution over the field. This is a common factor that has limited the rate of uptake and area coverage of CA on individual farms, not only because of the scarcity of biomass but also because of the high labour costs to collect, transport and distribute the biomass on a different plot of land.
- In attempts to document success with CA, some organizations promote and report adoption if any one of the three principles is undertaken. For example, if farmers practise any form of rotation or intercropping, it is considered (incorrectly) to qualify as CA and is reported as such. If this were the case, almost every farmer in Malawi would qualify as practising CA. The bottom line is that no-tillage is the critical factor for a practice to be considered as CA.
- The poor understanding of CA and contradictions over what it means reflect the absence of a standard training programme on CA for extension officers and students of agriculture that has been certified and endorsed by qualified practitioners.

Misconceptions about inputs and tools

Many farmers believe that CA cannot be undertaken without specific inputs and tools, which is often the message conveyed to them by extension staff (e.g. hybrid seed, fertilizers, herbicides, knapsack sprayers, jab planters and even *chaka* hoes, the CFU tool used in Zambia for making planting basins). These beliefs have limited the uptake of CA and its scale on farms. While quality inputs and tools are important to maximize the benefits of CA, they are not a prerequisite and farmers have flexibility to undertake the practice without them as they would with conventional ridge tillage.

Resistance to change

Farmers are reluctant to break the deep-rooted culture of ridging and clean fields in favour of CA unless, through experience and innovation, they find it to be a better practice with clear benefits that are compatible with their crops and system of farming. The reluctance to try new practices is exacerbated by the fear of ridicule from the community for undertaking a radically different system of planting, which has sometimes led to acts of sabotage out of spite and jealousy.

Other factors

Many farmers are concerned about the difficulties in controlling weeds, pests and diseases under CA. A common fear is that the increased abundance of termites and earthworms is harmful, whereas in fact they have beneficial effects on the soil and crops.

5.4.2 Drivers

The surveys also identified four consistent drivers for adopting CA:

- increased food security from higher and more stable crop yields;
- significant savings in labour costs, with many farmers alluding to the time available to expand and diversify farming or to engage in other productive activities;
- reduced impacts of dry spells or low rainfall due to higher moisture levels in the soil; and
- increased incomes from reduced labour and input costs combined with sales of surplus production, especially with grain legumes, as sole crops and as relay or intercrops.

Collectively, the factors driving adoption reflect higher returns to land, labour and capital which provide the basis for improving farm productivity and profitability with enhanced resilience to climate change.

5.5 Changing Attitudes and Actions to Scale Up Adoption of CA

Although there has been diverse interest and investment in promoting CA in Malawi, which has been reinforced by positive results with farmers in many parts of the country, adoption rates have been much lower than expected. This has raised concerns that there are problems with the extension delivery of CA on the ground. In response, TLC has made extensive efforts to investigate and publicize the true nature of the problem with the aim to galvanize collective action to address the challenges in a collaborative and systematic manner. The results are influencing the mind-set among major stakeholders and implementers in Malawi to change the focus and direction for promoting CA.

Key developments include the following:

- Acknowledgement that extension staff and farmers have been confused by conflicting extension messages, which has affected adoption rates and upscaling efforts across projects.
- Increasing appreciation of the importance of keeping the basic principles of CA as simple as possible for widespread application while recognizing the need for farmer flexibility to adapt CA to use with different crops, agroecological zones and farmer-specific circumstances.
- Increasing pressure for the MoAIWD to reverse its support of contour ridging, because it is adding more confusion and is counter-productive to efforts at scaling up CA.
- Growing recognition that conventional linear extension models have had poor success due to limited involvement of farmers, which has perpetuated the poor understanding of and attraction for CA.

TLC extension services involve a dynamic participatory approach with community leaders, farmers, researchers and extension staff to jointly identify, plan and evaluate best interventions to address priority farmer needs and interests. This system is similar to the innovation approaches described by Ekboir (2002) and Thierfelder and Wall (2011). The concept is simple: adoption of a technology depends on the interests of farmers; therefore, active farmer participation in evaluating and adapting a technology to their specific needs and circumstances is critical to attract interest in adoption. A good starting point is to engage truly good farmers who have influence in the community and who are willing to experiment on their own farms. Agreed action plans include farmer-led training, demonstrations, field days and exchange visits to showcase best practices and their benefits for upscaling targeted interventions. The premise for success is that it is farmer driven with a focus on developing a robust interactive model of extension and research based on local experiences, knowledge and innovation. The results facilitate sharing outcomes and challenges among farmers, researchers and extension agents on what works well, what does not and how to improve, refine and adapt innovations to meet the demands of different situations.

This approach runs against traditional linear extension models where technologies developed on station are passed on to extension service providers and then to farmers with the expectation of immediate adoption. Interactive participatory approaches, with farmers in the forefront, provide opportunities to understand and respond to the innovations, interests, needs and resources of the farmer to create conditions favourable for adopting and upscaling within the context of the farmer.

To accelerate and scale up adoption of CA with more farmers on larger areas of their farms, there is a critical need to provide evidence of its performance with different crops and farmers in diverse agroecologies, to harmonize extension messages among implementing organizations (or at least to minimize conflicting messages), to strengthen the knowledge base of CA among farmers and staff and to improve access to agricultural inputs and tools by farmers. Ultimately, the success for scaling up is that it must be farmer driven, it cannot be forced upon them from the outside. The challenge is to understand the context to make this a reality. Subsequent actions led by farmers can then be formulated to create the necessary preconditions to facilitate and accelerate faster adoption of CA at scale.

Key actions underway to transform the upscaling of CA in Malawi are as follows.

- Agreement for greater collaboration among donors, government departments, research institutions, universities, non-governmental organizations (NGOs), projects and the private sector to harmonize basic extension messages on CA. The aim is to dispel myths and misconceptions about: (i) the true meaning of CA and its benefits; (ii) how to implement CA in practice, even in the absence of inputs and tools; (iii) use of crop residues *in situ* versus importing them from other fields; and (iv) how to manage weeds, pests, diseases, burning and livestock effectively in an integrated manner.
- Facilitate the widespread diffusion of CA by fostering innovative farmer-based approaches that have led to sustained adoption of CA technologies.
- Demands for the National CA Task Force (NCATF) to take a more proactive role in coordinating extension efforts among all implementers by sharing lessons and experiences through regular meetings, publications, workshops, field trips and symposia.
- Nomination of TLC and CIMMYT to provide leadership on the technical and extension components of CA, based on their long track record of success with farmers on the ground.
- Commissioning of TLC by the NCATF to develop national guidelines for implementing CA in Malawi with flexibility for adaptations to location and farmer-specific circumstances.
- Recommendations for the DARS and Lilongwe University to develop a curriculum for training extension staff based on farmer experiences and innovations on what has worked well, where and why.
- Production of quality extension and training materials by a central provider.
- Dissemination of the existing wealth of knowledge and experience-based results from the past 10 years about the value and impact of CA in Malawi through a wide range of media to better inform and educate key stakeholders, including the MoAIWD, NGOs, donor agencies, projects, policy makers and the private sector. A major aim is to translate successful examples of farmer-led out-scaling into streamlined activities by all agencies involved which will support more sustained adoption.

The overall focus is to adapt CA to diverse farm conditions and agroecologies while avoiding conflicts from promoting distinctly different systems of CA that only add more confusion to extension staff and farmers.

5.6 Conclusions

TLC's experience over the past 10 years shows that CA has tremendous potential to increase the productivity and profitability of crops grown by smallholder farmers at lower cost with enhanced resilience to climate change due to higher and less variable yields under CA relative to CRT during years of low rainfall. The challenge going forward is to coordinate and foster a farmer-driven programme among a broad spectrum of implementers and stakeholders to scale up CA massively across different landscapes. Achieving this aim will require greater collaboration among all stakeholders, including government, donors, NGOs, projects and the private sector, to strengthen the understanding of CA and to harmonize extension messages to reduce confusion and conflict with farmers and extension staff.

References

Aagaard, P. (2011) *The Practice of Conventional and Conservation Agriculture in East and Southern Africa*. Conservation Farming Unit Zambia, Lusaka, Zambia.

Bunderson, W.T. and Hayes, I.M. (1995) Agricultural and environmental sustainability in Malawi. In: *Proceedings of the African Training for Leadership and Advanced Skills (ATLAS) Field Services Conference on Sustainable Agriculture for Africa: Issues for Policy-Makers*. Abidjan, Côte d'Ivoire, July 17–21, 1995. African Academy of Sciences, Nairobi, Kenya.

Bunderson, W.T., Jere, Z.D., Hayes, I.M. and Phombeya, H.S.K. (2002) *Land Care Practices in Malawi*. Malawi Agroforestry Extension Project Publication No. 42. Total LandCare, Lilongwe, Malawi.

Bunderson, W.T., Thierfelder, C., Jere, Z.D., Gama, M., Museka, R., Ng'oma, S.W.D., *et al.* (2014) Building resilience to climate change in Malawi: trends in crop yields under conservation agriculture and factors affecting adoption. Paper presented at the *1st African Congress of Conservation Agriculture*. ACT, Lusaka, Zambia.

Bunderson, W.T., Jere, Z.D., Museka, R.M., Mbale, B., N'goma, S.W.D., Paul, J.M., *et al.* (2015) *Conservation Agriculture in East and Southern Africa: Evidence Based Results and Guidelines for Implementation*. Total LandCare Booklet Publication No. 4, June 2015. Total LandCare, Lilongwe, Malawi.

Carr, M. and Hartl, M. (2010) *Lightening the Load: Labour-Saving Technologies and Practices for Rural Women*. International Fund for Agricultural Development (IFAD), Rome.

Denning, G., Kabambe, P., Sanchez, P., Malik, A., Flor, R., Harawa, R., *et al.* (2009) Input subsidies to improve smallholder maize productivity in Malawi: toward an African Green Revolution. *PLoS Biology* 7(1), e1000023.

Ekboir, J. (2002) *CIMMYT 2000–2001 World Wheat Overview and Outlook: Developing No-Till Packages for Small-Scale Farmers*. CIMMYT, El Batán, Mexico.

Ellis, F., Kutengule, M. and Nyasulu, A. (2003) Livelihoods and rural poverty reduction in Malawi. *World Development* 31, 1495–1510.

GoM (2007a) *Economic Report*. Ministry of Economic Planning and Development, Government of Malawi, Lilongwe, Malawi.

GoM (2007b) *Malawi: Poverty and Vulnerability Assessment – Investing in Our Future*. GoM/World Bank, Lilongwe, Malawi.

Haggblade, S. and Tembo, G. (2003) *Conservation Farming in Zambia*. EPTD Discussion Paper No. 108. IFPRI, Washington, DC.

Holden, S.T. and Lunduka, R.W. (2010) *Impacts of the Fertilizer Subsidy Programme in Malawi: Targeting, Household Perceptions and Preferences*. Noragric Report No. 54. Norwegian University of Life Sciences, Ås, Norway.

Holden, S.T. and Lunduka, R.W. (2013) Who benefit from Malawi's targeted farm input subsidy program? *Forum for Development Studies* 40, 1–25.

Ito, M., Matsumoto, T. and Quinones, M.A. (2007) Conservation tillage practices in sub-Saharan Africa: the experience of Sassaka Global 2000. *Crop Protection* 26, 417–423.

Johansen, C., Haque, M., Bell, R., Thierfelder, C. and Esdaile, R. (2012) Conservation agriculture for small holder rainfed farming: opportunities and constraints of new mechanized seeding systems. *Field Crops Research* 132, 18–32.

Ligowe, I.S., Ngwira, A. and Kamalongo, D. (2013) *Conservation Agriculture (CA) – A Way to Sustainable Agriculture*. Extension Circular, Department of Agricultural Research Services, Lilongwe, Malawi.

Lunduka, R.W., Ricker-Gilbert, J. and Fisher, M. (2013) What are the farm-level impacts of Malawi's farm input subsidy program? A critical review. *Agricultural Economics* 44(6), 563–579.

Munyati, M. (1997) Conservation tillage for sustainable crop production systems: results and experiences from on-station and on-farm research (1988–1996). *The Zimbabwe Science News* 31(2), 27–33.

Mwale, B.M., Ng'oma, S.W.D., Paul, J.M., Tembo, P., Sawasawa, H., Museka, R.M., *et al.* (2013) *Impacts of Conservation Agriculture on Farm Households in Malawi: Survey Results of 4 Projects Implemented by Total LandCare*. Total LandCare, Lilongwe, Malawi.

Mwale, B.M., Ng'oma, S.W.D., Paul, J.M., Tembo, P., Sawasawa, H., Museka, R.M., *et al.* (2014a) *Survey Results to Assess the Impacts of Conservation Agriculture on Farm Households in Malawi. DFID Project on Building Resilience to Climate Change*. Total LandCare, Lilongwe, Malawi.

Mwale, B.M., Ng'oma, S.W.D., Paul, J.M., Tembo, P., Museka, R.M., Mbale, B., *et al.* (2014b) *Assessing the Impacts of Conservation Agriculture on Farm Households in Malawi. CFU/COMESA Conservation Agriculture Regional Program*. Total LandCare, Lilongwe, Malawi.

Ngwira, A.R., Aune, J.B. and Mkwinda, S. (2012) On-farm evaluation of yield and economic benefit of short term maize legume intercropping systems under conservation agriculture in Malawi. *Field Crops Research* 132, 149–157.

Ngwira, A., Thierfelder, C., Eash, N. and Lambert, D. (2013a) Risk and maize-based cropping systems for small holder Malawi farmers using conservation agriculture technologies. *Experimental Agriculture* 49(04), 483–503.

Ngwira, A.R., Thierfelder, C. and Lambert, D.M. (2013b) Conservation agriculture systems for Malawian smallholder farmers: long-term effects on crop productivity, profitability and soil quality. *Renewable Agriculture and Food Systems* 28(04), 350–363.

Ngwira, A., Aune, J.B. and Thierfelder, C. (2014a) On-farm evaluation of the effect of the principles and components of conservation agriculture on maize yield and weed biomass in Malawi. *Experimental Agriculture* 50(04), 591–610.

Ngwira, A., Johnsen, F.H., Aune, J.B., Mekuria, M. and Thierfelder, C. (2014b) Adoption and extent of conservation agriculture practices among smallholder farmers in Malawi. *Journal of Soil and Water Conservation* 69(2), 107–119.

Nyagumbo, I., Mkuhlani, S., Pisa, C., Kamalongo, D., Dias, D. and Mekuria, M. (2015) Maize yield effects of conservation agriculture based maize-legume cropping systems in contrasting agro-ecologies of Malawi and Mozambique. *Nutrient Cycling in Agroecosystems* 105(3), 275–290 (July 2016); first available online 20 September 2015.

Oldrieve, B. (1989) Conservation tillage in action. In: Vowles, M. (ed.) *Conservation Tillage: A Handbook for Commercial Farmers in Zimbabwe*. Commercial Grain Producers Association, Harare, Zimbabwe.

Oldrieve, B. (1993) *Conservation Farming for Communal, Small Scale, Resettlement and Co-operative Farmers of Zimbabwe: A Farm Management Handbook*. Prestige Business Services (Pvt) Ltd, Harare, Zimbabwe.

Takane, T. (2008) Labour use in smallholder agriculture in Malawi: six village case studies. *African Study Monographs* 29(4), 183–200.

Tchale, H. (2009) The efficiency of smallholder agriculture in Malawi. *African Journal of Agriculture and Resource Economics* 3(2), 101–121.

Thierfelder, C. and Wall, P.C. (2009) Effects of conservation agriculture techniques on infiltration and soil water content in Zambia and Zimbabwe. *Soil and Tillage Research* 105(2), 217–227.

Thierfelder, C. and Wall, P.C. (2010a) Investigating Conservation Agriculture (CA) systems in Zambia and Zimbabwe to mitigate future effects of climate change. *Journal of Crop Improvement* 24(2), 113–121.

Thierfelder, C. and Wall, P.C. (2010b) Rotations in conservation agriculture systems of Zambia: effects on soil quality and water relations. *Experimental Agriculture* 46(03), 309–325.

Thierfelder, C. and Wall, P.C. (2011) Reducing the risk of crop failure for smallholder farmers in Africa through the adoption of conservation agriculture. In: Bationo, A., Waswa, B., Okeyo, J.M.M., Maina, F. and Kihara, J.M. (eds) *Innovations as Key to the Green Revolution in Africa*. Springer, Dordrecht, Netherlands, pp. 1269–1277.

Thierfelder, C. and Wall, P.C. (2012) Effects of conservation agriculture on soil quality and productivity in contrasting agro-ecological environments of Zimbabwe. *Soil Use and Management* 28(2), 209–220.

Thierfelder, C., Cheesman, S. and Rusinamhodzi, L. (2012) A comparative analysis of conservation agriculture systems: benefits and challenges of rotations and intercropping in Zimbabwe. *Field Crops* Research 137, 237–250.

Thierfelder, C., Chisui, J.L., Gama, M., Cheesman, S., Jere, Z.D., Bunderson, W.T., *et al.* (2013a) Maize-based conservation agriculture systems in Malawi: long-term trends in productivity. *Field Crops Research* 142, 47–57.

Thierfelder, C., Mombeyarara, T., Mango, N. and Rusinamhodzi, L. (2013b) Integration of conservation agriculture in smallholder farming systems of southern Africa: identification of key entry points. *International Journal of Agricultural Sustainability* 11(4), 317 330.

Thierfelder, C., Matemba-Mutasa, R. and Rusinamhodzi, L. (2015a) Yield response of maize (*Zea mays* L.) to conservation agriculture cropping system in southern Africa. *Soil and Tillage Research* 146, 230–242.

Thierfelder, C., Rusinamhodzi, L., Ngwira, A.R., Mupangwa, W., Nyagumbo, I., Kassie, G.T. and Cairns, J.E. (2015b) Conservation agriculture in southern Africa: advances in knowledge. *Renewable Agriculture and Food Systems* 30, 328–348.

UNDP (2007) *Human Development Report 2007/08: Fighting Climate Change in a Divided World*. United Nations Development Program, New York.

UNICEF (1993) *Situation Analysis of Poverty in Malawi*. UNICEF, Lilongwe, Malawi.

Wall, P.C., Thierfelder, C., Ngwira, A.R., Govaerts, B., Nyagumbo, I. and Baudron, F. (2013) Conservation agriculture in eastern and southern Africa. In: Jat, R.A., Sahrawat, K.L. and Kassam, A.H. (eds) *Conservation Agriculture: Global Prospects and Challenges*. CAB International, Wallingford, UK.

World Bank (1995) *Malawi Agricultural Sector Memorandum: Volumes I & II*. Lilongwe, Malawi.

6 Innovation Systems and Farmer Perceptions Regarding Conservation Agriculture in Cabo Delgado, Mozambique

Baqir Lalani,[1]* Peter Dorward,[1] Amir H. Kassam[1,2] and Jose Dambiro[3]

[1]*University of Reading, UK;* [2]*Food and Agriculture Organization of the United Nations, Rome, Italy;* [3]*Aga Khan Foundation (Mozambique), Pemba, Mozambique*

6.1 Introduction

There has been some debate regarding the ability of Conservation Agriculture (CA) to provide benefits to smallholder farmers in sub-Saharan Africa. Key areas of contention have surrounded yield, labour, soil quality and weeding, with particular focus on the suitability of CA to benefit the poorest, especially where external inputs are out of reach. Moreover, CA research and promotion in Southern Africa has also been criticized by some researchers for being top-down and inflexible (Andersson and Giller, 2012; Grabowski and Kerr, 2014).

Recent research and development efforts within East and Southern Africa, as a case in point, have clearly shown that constraints to adoption of CA can be overcome but stress the need for local innovation systems that involve farmers exchanging among themselves and use of participatory methods that help to adapt CA principles to local conditions (Wall *et al.*, 2013).

This chapter explores one such innovation system using a district in Cabo Delgado, Mozambique, as a case study. The aims of the chapter are twofold: firstly, to describe the process of construction of the innovation system, and by way of actors' perceptions of the system, to better understand key components central to its formation and functioning; and secondly, to explore farmers' perceptions of CA (including motivation to comply with certain actors in the innovation system) and the effectiveness of the current innovation system in reaching its target beneficiary – the very poor.

*b.lalani@pgr.reading.ac.uk; belalani@aol.com

Section 6.1 provides an overview of the evolution in approaches to agricultural development, putting into context the emergence of innovation systems thinking. Section 6.2 provides details of the methodological approach taken and includes background on the district under study. Section 6.3 provides a timeline of the key events that have formed the innovation system and explores perceptions/interactions and types of partnerships that exist among stakeholders within the innovation system. Section 6.4 provides details of the farmers' perceptions of CA in the study district. Finally, Section 6.5 concludes and considers implications for other regions in sub-Saharan Africa.

6.1.1 Evolution of approaches to agricultural development

Agricultural development for part of the 20th century mainly involved the so-called top-down or reductionist approach. This has largely been seen as a supply-led modus operandi, whereby research institutions developed high-yielding varieties (HYVs), namely wheat, rice and maize, alongside improved application rates of synthetic fertilizer, pesticide and irrigation in order to maximize yield. Typically, transfer to farmers has been via the transfer of technology (TOT) model, which assesses the new technology under controlled conditions to be then passed on to extension agents for dissemination to farmers upon completion (Pimbert, 1994) .

Notwithstanding the well documented impact such technologies and models have had on increasing agricultural productivity and household well-being per se (Mendola, 2007), it has been argued that these have generally been successful in more favourable environments (Wiggins and Cromwell, 1995). Thus critics have shown wide distain for the neglected role of the context-specific needs of beneficiaries and likely sustainability of much of the new technology being produced (Pimbert, 1994; Sumberg, 2005). This includes the impact on the natural resource base (Pingali, 2012). In recent years, this has prompted a gradual shift to more demand-led participatory research, which has involved farmers as key decision makers in the process of technology development.

The gradual shift from the 1960s to date has included the Adoption and Diffusion model (transfer of technology model, TOT, as outlined above), the Farming Systems Research (FSR) model that emerged in the 1970s, the Agricultural Knowledge and Information System (AKIS) in the 1990s and more recently the Agricultural Innovation Systems (AIS).

FSR involved diagnosing constraints and needs of farmers relevant to their farming system. It focused on a multidisciplinary approach with partnerships between farmers, social scientists, technical specialists and more recently extension personnel and policy makers, together with attempts to increase efficiency through the provision of packages of interventions (Norman, 2002; Klerkx *et al.*, 2012). The modus operandi involved on-farm testing and modification of technologies. This approach has been criticized for its lack of focus on resource-poor farmers and often poor communication between researchers and farmers, which also inhibited the communication of the knowledge gathered (Chambers and Jiggins, 1987). FSR has further been criticized for focusing largely on farm-level issues and neglecting the broader system in which farmers are rooted (Bingen and Gibbon, 2012).

This led to the emergence of AKIS, which was less linear in its approach. AKIS, in contrast to FSR, has focused on strengthening systems that assist in the generation and dissemination of knowledge (Röling, 1994). The system was heavily criticized for the emphasis put on the role of the agricultural research system as the centre of innovation rather than the concept of multiple knowledge bases and the role of different kinds of actors involved in agricultural innovation (Hall *et al.*, 2001).

In response to these drawbacks, the AIS has emerged in the 2000s with a primary focus on improving the capacity of farmers to innovate. This perspective largely recognizes the role of numerous actors (beyond the agricultural research system) that are able to contribute to agricultural innovation through diverse ways, including promoting better knowledge flows and the development, transfer and further adaptation of technologies (Temel *et al.*, 2002). It therefore seeks to increase the capacities of actors that include smallholder farmers, allowing them to learn, innovate and change (Hall *et al.*, 2006).

It is clear that the concept of an innovation has thus altered over time from being one centred around merely a technological advancement, such as a new seed variety, to one that includes and hinges on new forms of social and organizational norms (Leeuwis, 2011). More specifically, innovation systems thinking at its core presupposes that it is these other nodes (social, organizational, etc.) that bring about innovation (Leeuwis, 2011). Moreover, actors that have usually been excluded from the top-down approach to innovations development, such as input dealers and non-governmental organizations (NGOs), have become more prominent in the innovation process (Spielman, 2005).

The following explains some of the components involved in the AIS approach to innovation systems thinking from a sub-Saharan Africa (SSA) perspective and includes some of the current criticisms. It is through this lens that this study aims to explore the construction of agricultural innovation and its effectiveness in reaching the poorest farmers. An innovation here is defined as 'any new knowledge introduced into and utilized in an economic or social process' (OECD, 1999).

6.1.2 What role of the market and the state? The emergence of civil society actors

With respect to SSA in particular, there have been a number of constraints cited that have hindered the development of innovations. These include, for example, weak demand for innovations from farmers and the highly bureaucratic and hierarchical nature of relationships that exist between research, extension and farmers (Sumberg, 2005). This is compounded by a number of market and government failures that have often given rise to civil society organizations attempting to fill the vacuum. This has been the case, for instance, where NGOs have been involved in the dissemination of improved seed varieties or agricultural service delivery (such as agricultural extension services) to smallholders, given little private interest and waning governmental structures (Wiggins and Cromwell, 1995).

In relation to CA it has been suggested that, in specific circumstances, CA use requires the aid of subsidized inputs such as fertilizer from NGOs (Nkala *et al.*, 2011; Nkala, 2012; Grabowski and Kerr, 2014). Furthermore, it has been argued (Nkala, 2012) that such subsidized inputs are imperative in helping the poorest farmers to benefit from CA. However, Ngwira *et al.* (2014) questioned the long-term sustainability of financial incentives provided for CA use, as it can omit the cognitive aspect and instead farmers may be using CA because incentives exist.

6.1.3 Community-based approaches

To fill the void often left by the market and the state, community-based approaches to common property management, service delivery or as conduits for market access in the form of associations/cooperatives have been widespread. NGOs and development practitioners have also been instrumental in engendering the need for collective action, often through group formation.

Feder *et al.* (2010) have shown that, as with markets and states, community-based approaches can fail too. Giving examples of a number of community-based extension (CBE) initiatives, which include farmer field schools (FFS), they highlight how they often have fallen prey to mismanagement of funds and 'elite' capture involving often wealthier participants accessing resources and social exclusion of the vulnerable and marginalized (women and poorer households, for instance). Thus it has been argued that CBEs are not always suitable and different systems may be needed in certain circumstances (Hayami, 2009).

There has therefore been a move to develop partnerships between diverse actors in order to provide a stimulus for innovation. As Leeuwis (2011) quite aptly noted, critical to innovation development and design is the role of communication, and thereby many facilitating actors can play roles as diverse as knowledge brokerage, mediation and matching supply with demand.

6.1.4 Multi-sector partnerships and innovation platforms

More recently, innovation platforms (IPs) have been promoted that involve complementary action taken by diverse actors. For example, Rubyogo *et al.* (2010) placed great importance on developing partnerships between stakeholders. For instance, actors including the National Agricultural Research System (NARS), private companies and farmers' organizations have enabled 3.8 million households across SSA to access new bean varieties. Often the seed sector has been plagued by difficulties, including very low multiplication of certain varieties that would be useful for marginal areas. This initiative was thought largely successful as it decentralized seed multiplication of favourable varieties that were identified by farmers to particular agroecological zones, often through locally based producers supported by extension services. Farmers were also aided with better information by private companies providing improved seed labelling and packaging (Rubyogo *et al.* 2010).

Although there exists a number of examples of successful partnerships, concern has been raised about the modalities of partnerships and the number of partnerships (Hoffman and Christinck, 2007), including the difficulty in managing and maintaining such partnerships given differing goals, organizational cultures and personalities involved (Sanginga *et al.*, 2007). There is also the problem of high turnover of field staff in local NGOs and NARS. A further limitation is that the use of platforms and partnerships could be seen as another 'blueprint' for success (Sumberg, 2005). Klerkx *et al.* (2009) thus put forward another notion of 'innovation brokers' that can help to mediate between partners. For example, in the case of no-tillage uptake in South America, innovative brokers were instrumental in the development of the innovation. Many different networks tried various forms and approaches to organize innovation until it was suitably adapted to local conditions (World Bank, 2011). Hounkonnou *et al.* (2012) highlighted the importance of such platforms, showing that communities with IPs have achieved higher poverty reduction than those without.

6.1.5 Is this innovation system reaching the poorest?

Given the diversity of stakeholders often involved in innovation systems, the challenge is to develop innovations that are likely to generate a wide impact on poverty alleviation (Ortiz *et al.*, 2103). How does one analyse the impact of an innovation, or to what extent the system is helping the poor? EIARD (2006) suggested that assessments should look at the perceptions of stakeholders, which also include opinions of the community, i.e. individuals and groups.

Biggs and Matsaert (2004) also asserted that gathering views from different stakeholders involved in the innovation system is important to understanding particular needs and solutions. Furthermore, Ekboir *et al.* (2009) suggested incorporating social network analysis (SNA) to understand information flows and the size, efficiency and connectedness within a particular network. This has been used widely by other authors examining agricultural innovation systems (Ortiz *et al.*, 2013). Hall *et al.* (2006b) also advocated the use of sector timelines and partnership linkages in evaluating an innovation system.

Household-level formal surveys are useful in looking at adoption and impact, though they have been criticized for looking at adoption in a static state and failing to see adoption as a 'process', or the fact that adoption could be ambiguous, e.g. differences within household ownership and use (RIU, 2010). The following describes the methodological approach taken in this study.

6.2 Methodological Approach

A mixed-methods approach that includes both qualitative and quantitative data was used. Thirteen key informant interviews and three focus groups were conducted during August 2014 with various actors involved in the innovation system of CA in Metuge district of Cabo Delgado (see Section 6.2.2). To explore the first aim of understanding the construction and functioning of the present innovation system, an actor matrix using the Biggs and Matsaert (2004) framework was constructed and each actor was asked to provide a score for the other actors in the system reflecting their perceived role in the overall system. Using principles of social network analysis (Borgatti *et al.*, 2009), each actor was also asked whether they had any informal or formal ties to the actors with respect to either information on CA or specific goods or services related to CA. The typology of partnerships and learning presented by Hall *et al.* (2006b), a schematic of the partnerships and interactions within the innovation system, is also presented.

In addition, this chapter draws on results from a household survey administered in September 2014 in a total of six communities in Metuge District in order to gather perceptions of farmers using CA and those not using CA. It also includes exploration of farmers' perceptions on motivation to comply with social referents regarding information on CA, which sheds light on whom they consider important actors within the innovation system. Farmers were also disaggregated by wealth to compare perceptions among poorer and better-off farmers in order to examine whether the innovation system is effective in reaching the poorest farmers (see Section 6.2.1). From a list of users and non-users of CA, 250 households were randomly selected (using probability proportional to population size). A total of 197 were interviewed (145 users of CA and 52 non-users).

6.2.1 Data analysis

In order to compare poorer and better-off farmers, principal component analysis (PCA) was conducted to establish a wealth index. As is common in a number of poverty studies, the first principal component (PC1), which explained the majority of variance in the data, was used as the index (Edirisinghe, 2015). Households were then ranked into terciles with respect to the level of wealth. Given the ordinal (Likert-type data) used, tests of statistical significance were done using the Mann–Whitney U test.

6.2.2 Background to case study

Cabo Delgado is the northernmost province in Mozambique, located within the coastal plain. The majority of inhabitants in the province rely on subsistence (mainly rainfed) agriculture, which, given its geographical location, is compounded by poor market access and limited infrastructure, including roads. Recent research has ranked Cabo Delgado as one of the poorest provinces in Mozambique (Fox *et al.*, 2005) with the highest prevalence of stunting in the country (FAO, 2010).

Its climate is sub-humid (also termed moist savannah) characterized by a long dry season (May–November) and rainy season (December–April). Annual rainfall in the province is 800–1000 mm though the intensity of rainfall sometimes results in heavy flooding throughout the province. The main crops incorporated into the cropping system used in Metuge district (the case study district) are maize, cowpea, sesame and pigeon pea. Lablab and mucuna are also grown. External inputs have seldom been used in the district and wider province and this includes compost or animal manure.

Conservation Agriculture has been defined by three linked principles: (i) continuous no or minimum mechanical soil disturbance; (ii) permanent soil mulch cover with organic matter; and (iii) diversification of crop species grown in rotations, associations or sequences (FAO, 2015).

Usage of CA is defined in this chapter by a farming household simultaneously applying all three principles of CA, which are:

- No-tillage with or without the use of micro-pits in the first few seasons. There are three manual seeding systems commonly used in the district: (i) dibble stick; (ii) planting micro-pits as an entry point into CA; and (iii) jab planters (Box 6.1 and Fig. 6.1). The use of dibble sticks and jab planters is still limited and micro-pits have been more commonly used in the district.
- Soil cover, i.e. mulching (covering at least 30% of the cultivated soil surface).
- Cropping system diversity using rotation/association/sequence involving at least three different crops during the season. The majority (90%) of the farmers surveyed used three or more crops in the cropping system. Only a small number (10%) of CA users from those surveyed used two crops. As these farmers were following all the other principles, and were engaged in the CA adoption process, their responses have been incorporated into the farmers' perceptions of CA presented in Section 6.4.

Box 6.1. Manual Conservation Agriculture no-till seeding practices. Adapted from Johansen *et al.* (2012) and notes from author.

A dibble stick is a pointed stick that is used to open small holes (in crop residue) for planting seeds in no-till soil.

The second practice is the use of micro-pits in the initial few seasons to break soil compaction. In Cabo Delgado (Mozambique), AKF-CRSP has promoted the use of micro-pits (about 20 cm long × 10 cm wide × 15 cm deep).

The third practice used is the jab planter. This was first imported into Zimbabwe and Mozambique in the early 2000s and there have been attempts to make jab planters locally. Two compartments are mounted on the planter: one for fertilizer (animal manure and ash can be used as a substitute for chemical fertilizer) and one for seed. Below the compartments are two tips, which are pushed into the soil (making small holes) and then released, allowing the fertilizer and seed to drop into the planting holes. The jab planter can pierce through mulch-covered soil with relative ease.

Fig. 6.1. (a) Micro-pit without mulch. (b) Micro-pit with organic mulch cover, i.e. crop residues, grass and other biomass. (c) Jab planter. (d) Dibble stick being used by a farmer.

Non-CA users or conventional farmers are defined here as farmers practising conventional tillage agriculture with the use of hand-hoes. They may, however, be practising intercropping and/or rotation, and growing three or more crops during the season. Only a handful of non-CA users were found to be practising mulching.

CA adoption has steadily increased in Cabo Delgado in recent years. This has been supported by the institutional presence of the Aga Khan Foundation's Coastal Rural Support Programme (AKF-CRSP), which began promoting CA in the province in 2008. AKF has taken a different approach to other NGOs elsewhere in Mozambique and SSA as it has not provided or subsidized the use of external inputs such as fertilizers and herbicides. Instead the focus has very much been on improvements in soil fertility through use of legumes (as green manure) and perennials. Moreover, this has also focused on using different sources of mulch, including the retention of crop residues but also vegetation biomass such as grass or other dead plant material either produced on-farm or brought in from surrounding areas, i.e. bush areas. Compost projects have also been initiated, though compost is not widely used among CA users in this case study district or the wider province. Only one farming household was found to be using a pesticide or compost in the household survey undertaken. No fertilizers or herbicides were found to be used (Lalani, 2016).

The household survey revealed that the majority of CA farmers were using a three-crop sequence during the growing season, with maize–cowpea and cassava or maize–cowpea and sesame being the most common. These were the most common three-way sequences for non-CA farmers as well. Many non-CA farmers also cultivated only two crops such as maize and cassava in the growing season. Among the CA farmers, 19% were using four-way crop mixes, e.g. maize/cowpea/pigeon pea/cassava or maize/cowpea/cassava/sesame. Only two households among the non-CA cohort were using four crops.

The former crop mix seemed the most popular among poorer households with smaller plots and the latter more common among wealthier households with larger land holdings (Lalani, 2016).

6.3 Conservation Agriculture Innovation System – Metuge District

The timeline of key activities that occurred in the set-up of the innovation system is presented in Table 6.1.

6.3.1 Actors' innovation matrix and social network analysis

Table 6.2 shows the actor matrix using the Biggs and Matsaert (2004) framework. Each actor provides a score for the other actors in the system reflecting the strength of their perceived role in the overall system. Specific ties that may exist between actors, be these informal or formal, related to exchange of information or specific goods and services with respect to CA, are explored in Table 6.3.

The key findings from the actor matrix and social network analysis are summarized as follows.

- The actor innovation matrix (Table 6.2) signals a strong role for a number of actors, including the AKF facilitator, district coordinator and seed multiplication group. In contrast, the government role through the agriculture directorate (District Services of Economic Affairs, or *Serviços Distritais das Actividades Económicas*, SDAE) is seen as weaker.
- Table 6.3 shows the results of the social network analysis. It follows a directed graph format where existence of a tie/interaction between actors, related to information (informal/formal) or goods/services with respect to CA, is signalled by one (1) and absence of a tie by zero (0). In a directed graph format, A may have a tie with B but this may not be reciprocal, e.g. B might not seek information or goods/services from A (Scott, 2000). Overall, there is a dense network in this innovation system signalled by the high network density of approximately 0.69 overall. Network density measures the extent to which the nodes in the network are tied to

Table 6.1. Timeline of CA innovation system set-up.

Activities	Responsible	Year	Role
MOU signed between Aga Khan Foundation (AKF) and Ministry of Agriculture	AKF higher-level staff/ Agricultural directorate (SDAE)	2007	Work on CA part of strategic plan on agrarian reform with government
AKF staff spent 3 weeks in 2008 training facilitators on CA principles in order to create demonstration plots in district		2008	Facilitators help to set up demonstration fields in each village with select number of farmers
Establishment of farmer field schools (FFS) and seed multiplication groups	AKF District level and AKF Village facilitators	2010	
Setting up benchmark farmers	FFS and AKF facilitators	2012	Benchmark farmers are those that have used CA for at least 3 years and chosen to exhibit the full CA system in use to other FFS members
Linkage with associations and setting up of sales commission for a variety of crops including legumes – 'connecting' role with rural shops	AKF and FFS	2013	
Assisting village development organizations (VDOs) in CA training and to identify community promoters. Establishment of CA clubs and CA forum (regional meeting of all CA club members and invitation to agricultural directorate and other NGOs working in region to participate). Use of theatre to improve awareness of CA and land degradation, e.g. burning of crop residues	AKF	2013/2014	CA clubs consist of a select few from the FFS in each village or community that are experienced CA farmers and are chosen to go to other villages without FFS to engage with communities on CA

Table 6.2. Actor innovation matrix.

Type of actor	CA cl	CA cp	BF	AKF vf	AKF dc	AKF ad	SDAE	SMG	SC	RS	F F S	VDO	OF	WWF	Med
CA club	–	2	2	3	3	2	1	–	3	2	2	2	2	1	1
CA community promoter	–	–	–	–	3	1	2	2	3	2	3	2	2	1	1
Benchmark farmers	1	2	–	3	2	3	2	2	1	1	2	2	2	1	1
AKF village facilitator	1	3	1	–	–	–	1	3	3	2	3	1	1	1	1
AKF district coordinator	1	2	1d	3	–	–	1e	3	1	1	3	1	–	1a	1c
AKF agriculture director	3	3b	3	3	3	–	3	3	–	3	3	1	2	2a	1c
Member of Agr. directorate SDAE	–	2	3	3	3	3	–	2	2	1	3	2	2	1	1
Seed multiplication	2	3	3	3	2	3	2	–	2	3	2	3	3	1	1
Sales commission/association	–	2	3	3	3	3	1	2	–	2	2	2	–	1	1
Rural shops	–	2	–	3	3	3	3	–	1	–	–	1	1	1	1
FFS female (four members)	1	2	2	3	3	1	1	1	1	1	–	3	2	1	1
FFS male(four members)	1	1	3	3	2	2	2	2	2	1	–	2	2	1	3
VDO	2	3	2	3	3	3	3	3	2	2	2	1	1	–	–
Other farmers	2	3	3	2	2	1	1	3	2	2	–	2	–	1	1
World Wildlife Fund	–	2	1	3	3	3	1	1	1	2	1	2	1	–	1
Media[a]	–	–	–	–	–	–	–	–	–	–	–	–	–	–	–
Average score per actor	1.6	2.3	2.1	2.9	2.7	2.3	1.7	2.3	1.9	1.8	1.9	1.8	1.8	1.1	1.1

[a]Unable to meet media representatives (radio, television) due to political sensitivities at the time.
The numbers in the matrix indicate strong (3) and weaker linkages (2 and 1). Blank (–) means does not know or not applicable. Reads from left column to row cell, i.e. type of actor in the column provides their view on the actor on the adjacent row. This is the same format used in Table 6.3, which describes the linkages.
Key to column headings: CA cl, CA club; CA cp, community promoter; BF, benchmark farm; AKF vf, AKF village facilitator; AKF dc, AKF distrist coordinator; AKF ad, AKF agriculture director; SDAE, member of agriculture directorate; SMG, seed multiplication group; SC, sales commission; RS, rural shops; FFS, farmer field school; VDO, village development organization; OF, other farmers non-FFS; WWF, World Wildlife Fund; Med, media.
a = Respondents suggested this partnership needs strengthening. Presence of other NGOs is limited in project area but other NGOs are present in other districts within the province.
b = Respondent suggested this needs strengthening. For example, plans for promoters to play more of a key role but respondent mentions this has just started so the promoters' role is minimal at this stage.
c = More use of ICT being used that could transmit information on prices and weather could be done through the use of mobile phone technology.
d = Mentioned benchmark farmers are too often very good at CA but are not so good in explaining technical aspects to farmers that are having difficulty with CA, etc.
e = AKF could also be invited to exchange ideas on CA with the agriculture directorate (SDAE). Agriculture extension staff could invite AKF facilitators to the demonstration plots/training they are involved in and agriculture extension staff could also attend AKF facilitator training on CA.

Table 6.3. Actor tie matrix and social network analysis indicators (1 = tie, 0 = no tie).[a]

Type of actor	CA cl	CA cp	BF	AKF vf	AKF dc	AKF ad	SDAE	SMG	SC	RS	FFS	VDO	OF	WWF	Med	TT	ODC	IDC
CA club	–	1	1	1	1	1	1	–	1	1	1	1	1	0	0	11	0.8	0.5
CA community promoter	–	0	0	0	1	1	1	1	1	1	1	1	1	0	0	9	0.6	1.0
Benchmark farmer	–	1	0	1	1	1	1	1	0	0	1	1	1	0	1	10	0.7	0.9
AKF village facilitator	1	1	1	0	1	1	1	1	1	1	1	1	1	–	–	12	0.9	0.9
AKF district coordinator	1	1	1	1	0	1	1	1	1	1	1	1	1	–	–	12	0.9	1.0
AKF agriculture director	1	1	1	1	1	0	1	1	–	1	1	1	1	1	0	11	0.8	0.9
Member of ag, dir. (SDAE)	–	1	1	1	1	1	0	1	1	0	1	1	1	0	0	10	0.7	0.9
Seed multiplication	1	1	1	1	1	1	1	0	1	1	1	1	1	0	0	12	0.9	0.8
Sales commission/association	–	1	1	1	1	1	1	0	0	1	1	0	–	0	0	9	0.6	0.8
Rural shops	–	1	–	1	1	1	1	1	1	0	–	1	0	0	0	7	0.5	0.8
FFS female	0	1	1	1	1	0	0	1	0	1	0	1	1	0	0	8	0.6	0.8
FFS male	1	1	1	1	1	1	1	1	1	1	0	0	1	0	0	12	0.9	0.9
VDO	1	1	1	1	1	1	1	1	1	1	1	1	0	0	0	11	0.8	0.8
Other farmers	1	1	1	1	1	0	1	1	1	1	–	1	0	0	0	10	0.7	0.1
World Wildlife Fund	–	1	1	1	1	1	1	–	1	–	1	–	1	–	–	10	0.7	0.1
Media	–	–	–	–	–	–	–	–	–	–	–		–	–	–	–	–	–
Total received ties	*7*	*14*	*12*	*13*	*14*	*12*	*13*	*11*	*11*	*11*	*11*	*12*	*11*	*1*	*1*	154	–	–
Av. score per actor (from actor matrix)	1.6	2.3	2.1	2.9	2.7	2.3	1.7	2.3	1.9	1.8	1.9	1.8	1.8	1.1	1.1	–	–	–

[a]Uses same formulas and definitions as Scott (2000) for network density and out-degree and in-degree centrality. Network density is calculated using the formula for a directed graph, i.e. $L/n\ (n-1)$ where L is the number of ties and n the number of nodes or actors. Out-degree/in-degree centrality calculated by number of ties initiated divided by $(n-1)$, i.e. n being number of nodes/actors. Network density overall is 69%. Network density for the FFS male group (70%) is slightly higher than for the FFS female group (68%). Though FFS groups were organized into FFS female groups and FFS mixed groups, interviews were conducted with males and females (irrespective of the particular grouping) to explore any differences in gender dimensions.

Key to column headings: CA cl, CA club; CA cp, community promoter, BF, benchmark farm; AKF vf, AKF village facilitator; AKF dc, AKF distrist coordinator; AKF ad, AKF agriculture director; SDAE, member of agriculture directorate; SMG, seed multiplication group; SC, sales commission; RS, rural shops; FFS, farmer field school; VDO, village development organization; OF, other farmers non-FFS; WWF, World Wildlife Fund; Med, media; TT, total ties (interactions mentioned by actor); ODC, out-degree centrality; IDC, in-degree centrality.

other nodes and is expressed as a proportion of all the possible ties within the network. The closer the figure is to 1, the greater the density of the network. In this case the value indicates that 69% of the possible ties in the network exist. A high network density figure is an indication that information is able to flow faster (Valente, 1995).

- Centrality analysis identifies those actors playing the most relevant roles within networks (Table 6.3). It is based on the extent to which the actors' network revolves around a single node. 'In-degree' relates to number of ties received by a node, which indicates its importance (prestige of an actor), and 'out-degree' relates to the number of ties initiated by the node, which is a signal of how influential an actor might be. For example, the government agriculture directorate (SDAE) has a lower out-degree centrality but a higher number of ties in-degree. This may be an indication of strong ranging formal ties/linkages with other actors but limited influence, as the weak score in the matrix suggests.
- AKF village facilitators, staff and community promoters scored very highly in the out-degree centrality, which signals their 'prestige' in the innovation system. FFS farmers also scored particularly high in this regard.
- The media has a very low number of linkages but has a higher score among CA users, signalling potential role of radio/TV (see Section 6.4).

6.3.2 Innovation system explored

Figure 6.2 shows the key interactions within the CA innovation system as identified by the different actors in the matrix. The arrows highlight different partnerships but also types of learning mechanisms. The thick bold arrows symbolize 'partnerships' that are formal, usually requiring a formal Memorandum of Understanding (MOU). These are likely to involve joint learning and innovation and can help to stimulate learning through 'interacting', 'imitating' and 'mastering' (Hall *et al.*, 2006b). The dash arrows signal partnerships that are termed contract based, i.e. usually requiring a formal contract. For example, AKF employees are village facilitators and agricultural extension staff are employees of the agricultural directorate. They, however, receive training through 'imitating', 'interacting' or 'searching' (Hall *et al.*, 2006b). Dot arrows highlight often 'self-constructed' networks that help to build social capital and which may be informal or formal in nature but are designed to improve information flows. Light black arrows signal linkages to supply of input or output markets. Some learning occurs here through interaction (Hall *et al.*, 2006b). Finally, the dash–dot arrows signal 'paternalistic' partnerships that are designed to spread knowledge, goods or services irrespective of preferences or agendas (Hall *et al.*, 2006b).

Role of NGOs – Aga Khan Foundation and others

The Aga Khan Foundation (AKF) plays a key role in technical assistance and training of the village facilitators and community promoters that are then

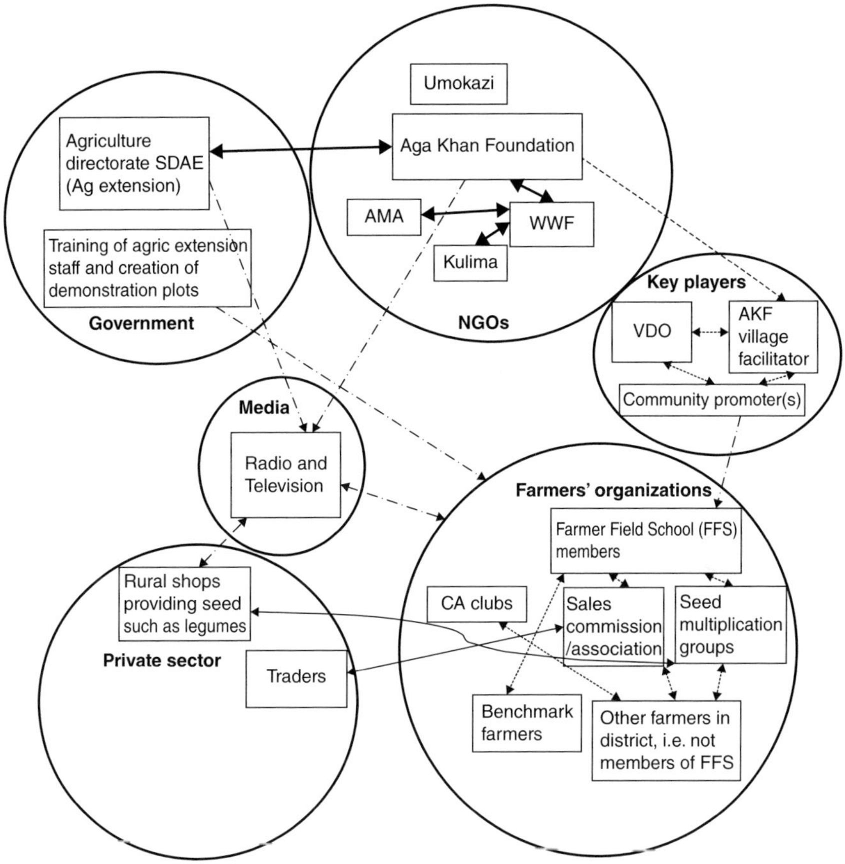

Fig. 6.2. Interaction of innovation actors in the Conservation Agriculture (CA) innovation system for Metuge District, Cabo Delgado. Arrows signal flow of information or goods and services related to CA. Thick bold arrow = partnerships may be formal which may require MOU. Dotted arrow = networks. Dash arrow = contract based. Light black arrow = goods and services that help to link to input and output markets. Dash-dot arrow = 'paternalistic' partnership, i.e. knowledge provided irrespective of agendas. AKF, Aga Khan Foundation; AMA, *Associação Meio Ambiente*; SDAE, *Serviços Distritais das Actividades Económicas*; VDO, village development organizations; WWF, World Wildlife Fund.

responsible for setting up farmer field schools (FFS). Prior to the role of community promoters being responsible for setting up FFS this role was done by AKF facilitators (Table 6.1). AKF has also helped to play a 'connecting' role by supporting the creation of farmer organizations and strengthening existing civil society-based organizations, e.g. village development organizations (VDOs). The VDO is now responsible for choosing the community promoter for CA from the village who is trained by the village-level facilitator.

Other NGOs, e.g. Umokazi and World Wildlife Fund (WWF), are not currently working in the district but Umokazi has done some previous work on 'good agricultural practices' in the district which involved, for example, training extension agents and farmer groups on eliminating burning of crop residue, application of mulch and planting in lines. WWF has a project on CA in the surrounding national park. There is currently no coordination with WWF on CA with AKF, though WWF have partnerships with Associação Meio Ambiente (AMA) and Kulima on CA, which are both national NGOs working in the surrounding national park. However, AKF and WWF have exchanged information and agreed on a common approach in terms of promoting the same key messages in the district/national park on the use of CA.

Local government organizations

The government organization SDAE has set up a number of farmer-led demonstration plots on CA. The matrix also reveals that SDAE's role is perceived to have less significance with respect to CA information flows in the district than other actors. There also seems to be very low interaction between AKF and SDAE, which could also be improved.

Private sector

There is little by way of input from suppliers of equipment and/or other agricultural inputs that may be of additional support to the CA innovation system; for example, locally made jab planters or implements that could be attached to animals to improve the scale of direct seeding. Rural shops, however, have been identified as providing seeds to farmers that offer options for the use of diversified crop rotations. Some of the actors in the matrix have noted that rural shops could play an increasingly important role by improving publicity and marketing efforts related to CA and the use of different seed types, e.g. legumes. Input (seed) supply companies also enter into contracts with rural shops. Thus different seed types can be found in a variety of rural shops. To improve information flows within the innovation system the government-led agricultural extension and NGOs such as AKF partner with the radio networks to provide information on seed availability in rural shops (location and type of seed available). This information is used by all producers (CA users and non-users). Rural shops also provide other equipment to farmers such as hoes, machetes, etc., which are particularly useful for the manual system of CA being used.

Media

District or national television programmes related to CA have been shown. Radio has also been used (as described earlier) though coverage in some districts is minimal. More use could be made of information and communications technologies (ICT) through mobile phone networks, which could transmit information on prices, climate information (e.g. forecasts/warnings) or meetings.

Benchmark farmers and other experienced farmers

A benchmark farmer acts as a conduit for other farmers ('lead' farmer) on best practices of CA in the district. These benchmark farmers, a few of whom then form the CA club, are involved in sharing their experiences with other farmers in villages where FFS groups have not been set up and represent the district at the regional level meetings on CA, e.g. the CA regional forum (Table 6.1). The exchange of innovative and pioneer farmers and their experiences through networks usually provides the fastest development of suitable technologies (Pretty, 2003). These groups can play a powerful role in encouraging others to join (Kassam *et al.*, 2014).

Though the benchmark farmers were seen in a positive light within the innovation system as a whole, they were described as being 'very good at CA but are not so good in explaining technical aspects to farmers that are having difficulty with CA' (Table 6.1).

Farmer field schools

Although farmers in FFS cover the main methods involved in CA, they might well choose some of these methods and not others, such as using the micro-pit as an entry point or trying without. Thus, the network partnerships between the CA clubs, benchmark farms and FFS encourage social learning between 'experienced' CA farmers and others. The FFS also acts as a platform for social learning by encouraging, for example, some farmers testing different components, e.g. different crops into the rotation, and/or finding different sources for mulch, solutions for pests/diseases and then sharing these findings with other farmers. Responses from FFS female farmers indicated that they felt they were not invited to the government agriculture extension demonstration plots whereas FFS male farmers were. This may be a reason why FFS female farmers scored the role of government as weak compared with males, who scored the government's role as higher.

The AKF staff highlighted (Table 6.1) that the members of the FFS were encouraged to spread information on CA to at least five other people and that farmers would often share information about CA in the local masjids (mosques) and among neighbours and family members. Local village-level and district-level staff (directly involved in formation of FFS and training facilitators) from AKF noticed that farmers played a key role in encouraging other farmers to use or try CA.

Associations and groups

Through FFS, AKF also shares market information that enables individual farmers to connect with producer associations that sell in bulk to traders. Among these groups sales commissions have been formed, which aim to improve market linkages with traders. For example, the sales commission then interacts with traders to negotiate the sale of certain crops, such as maize and sesame. Sales commissions and associations also play a reinforcing role as they have a vested interest in increasing quality and quantity of members, i.e. they also provide information to farmers on the use of treated seed, composting techniques, etc. The traders likewise interact with the sales commission on crops in demand, such as specific legumes. Kassam *et al.* (2014) noted that small groups can form into larger bodies such as associations and cooperatives that, if suitably organized, can have enormous clout in bargaining with traders and also pressurizing national and regional bodies for services such as extension and research in relation to CA. Other groups also play an important 'network' role. For example, seed multiplication groups play a strong role in providing information related to CA (Table 6.1).

Having explored the innovation system (the first aim of this chapter) the next section considers farmers' perceptions of the innovation itself and of aspects of the innovation system.

6.4 Farmers' Perceptions of CA and Different Social Referents

The first part of this section explores farmers' perceptions of CA and includes reasons for not using CA among non-users and an exploration of the specific beliefs surrounding CA (among users and non-users), drawing out differences between poorer and better-off farmers. The second part reports findings on social referents and motivation to comply with these referents regarding information on CA. This sheds further light on the innovation system and how poorer and better-off farmers view different actors.

6.4.1 Non-users of CA

Figure 6.3 shows the reasons farmers provided for not using CA. The majority of farmers (> 80% of those not using CA) cited lack of information as the primary reason. Only a handful of farmers considered lack of labour or concern over weeds as the reason for not using CA. Moreover, lack of equipment or inputs is unlikely to be an impediment in this setting. The fact that farmers using CA in this district are not using external inputs such as fertilizers and herbicides may provide more of an incentive to use CA (or at least experiment with or test CA on their land), given the low capital requirements needed to use CA.

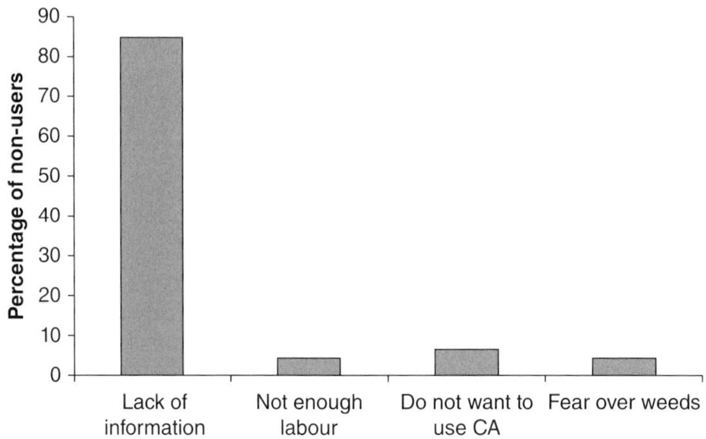

Fig. 6.3. Reasons for not using CA among non-users (n = 52).

On a Likert-type scale (1 = strongly agree, 5 = strongly disagree) users and non-users were asked about their perceptions of CA. Farmers were also asked about whom they would feel motivated to listen to regarding information on CA (1 = very motivated, 5 = not at all motivated) over the next 12 months and their perception towards using CA. The terms 'not sure' and 'cannot say' were used interchangeably to denote a neutral response.

Among non-users, poorer farmers had more favourable perceptions of CA than better-off farmers. This was particularly associated with the benefits associated with CA, i.e. benefits in the first year and a drought year, reduction in weeds, reduction in labour and increase in yields (Fig. 6.4). Farmers' perceptions were statistically different between the poorer and better-off farmers for all categories ($p<0.05$) except for the categories 'increase in pests' and 'cannot be used on all soil types'. Both groups of non-users of CA (poorer and better-off farmers) perceived CA to increase pests.

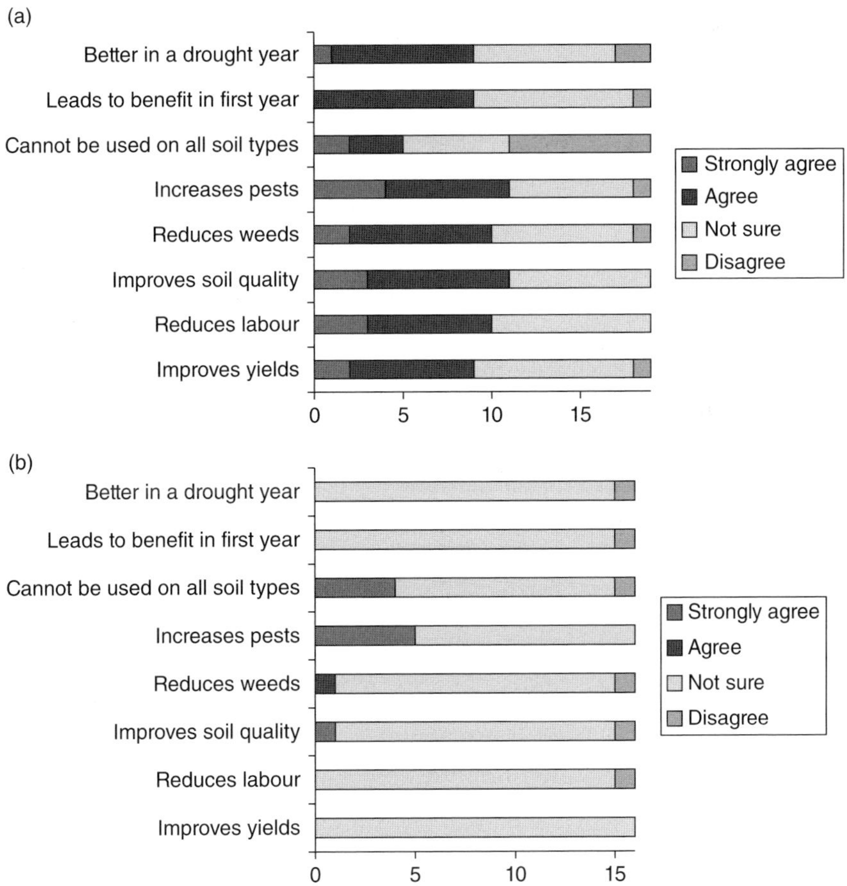

Fig. 6.4. (a) Frequency of responses related to perceptions of CA among the poorest wealth tercile of non-CA users (n = 19). (b) Frequency of responses related to perceptions of CA among the better-off wealth tercile of non-CA users (n = 16).

Interestingly, there were also differences among the wealth terciles with respect to the social referents with which farmers felt they were motivated to comply regarding information on CA (Fig. 6.5). Compared with better-off farmers, the poorer farmers signalled a higher motivation to comply with social referents including spouse, sibling, self-observation and self-initiative ($p<0.05$). Poorer farmers were also more likely to comply with radio and TV. Statistically significant differences were also found between the poorer and better-off farmers with respect to FFS, the AKF facilitator and the government ($p<0.05$). Strengths of motivation to comply with NGOs and other experienced farmers were not found to be statistically different. This suggests that farmers (irrespective of wealth category) equally value these two actors.

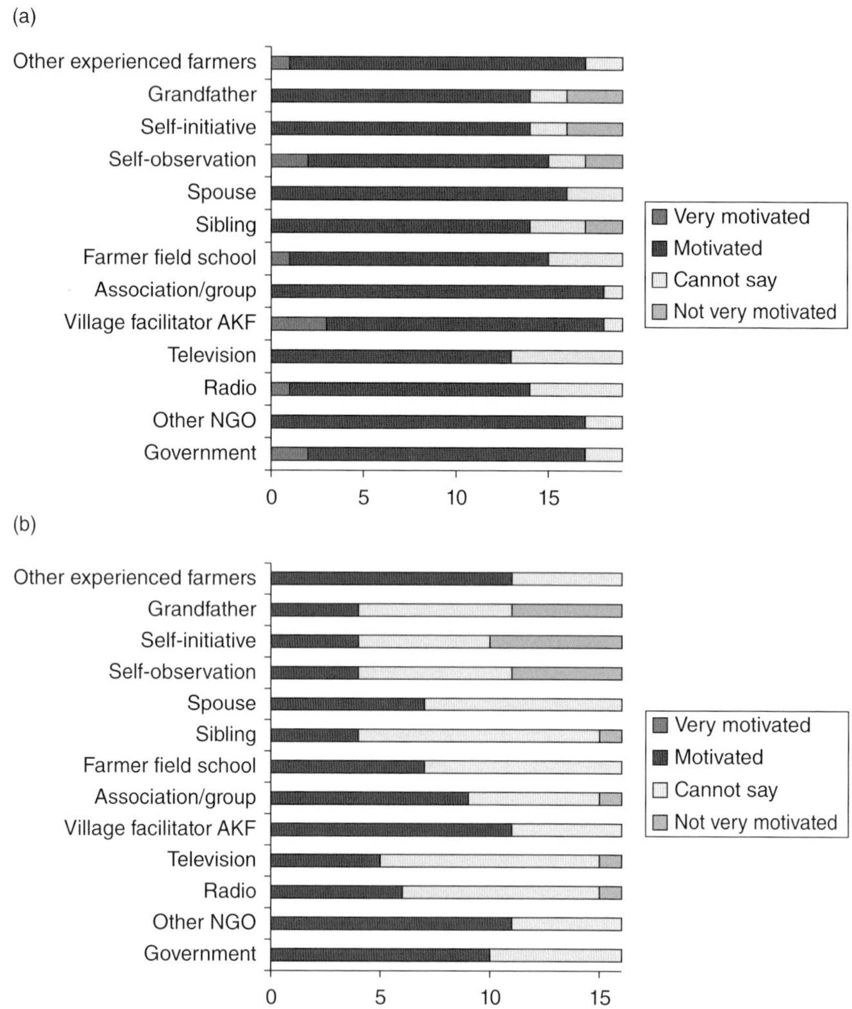

Fig. 6.5. (a) Frequency of responses related to motivation to comply with different social referents regarding using CA among the poorest wealth tercile of non-CA users (n = 19). (b) Frequency of responses related to motivation to comply with different social referents regarding using CA among the better-off wealth tercile of non-CA users (n = 16).

6.4.2 Users of CA

Overall both groups of farmers (poorer and better-off) who used CA had favourable perceptions (Fig. 6.6). These include the perception of increased yields, reduction in labour and reduction in weeds. Moreover, farmers also perceived benefits in the first year and during a drought year. The poorer, however, felt that CA did increase pests and this was significantly higher than the better-off farmers ($p<0.01$). Though perceptions were invariably positive for both groups, the comparisons also indicated that the poorer farmers had stronger overall perceptions of the benefits of CA ($p<0.01$). Both groups of users (poorer and better-off) perceived CA to be able to be used on all soil types.

Poorer farmers had a higher score for TV, radio and family referents such as elders, e.g. grandfather and siblings ($p<0.05$). Both groups were equally motivated by experienced farmers, associations/groups and other

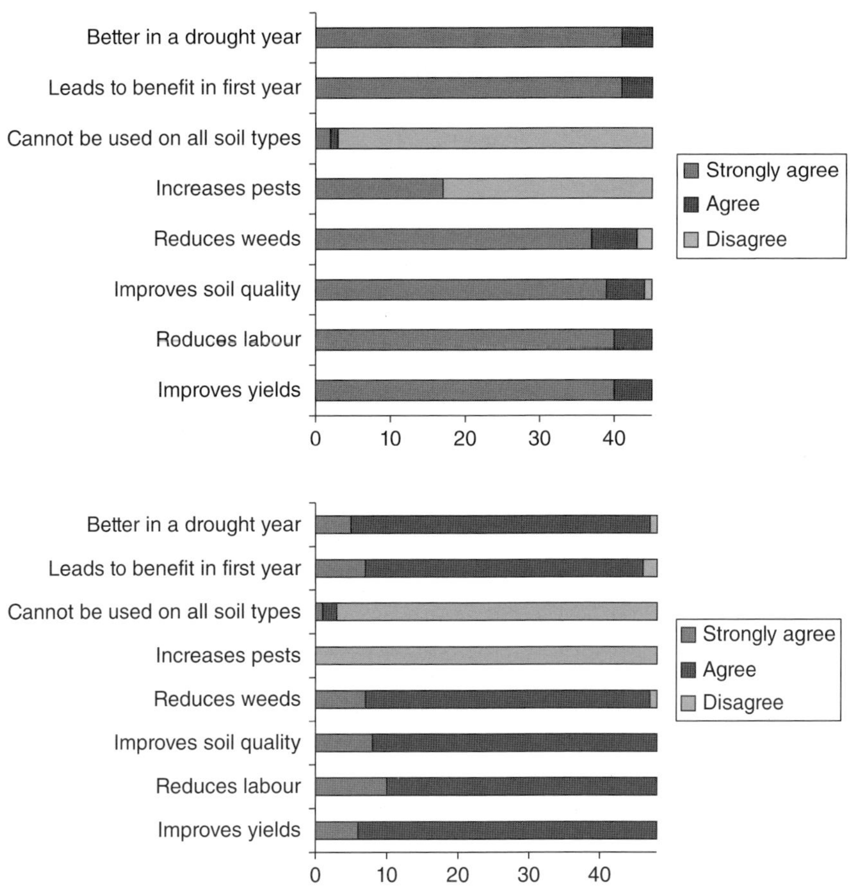

Fig. 6.6. (a) Frequency of responses related to perceptions of CA among the poorest wealth tercile of CA users ($n = 45$). (b) Frequency of responses related to perceptions of CA among the better-off wealth tercile of CA users ($n = 48$).

NGOs, i.e. no statistically significant difference was found (Fig. 6.7). Poorer farmers were also highly motivated by the AKF village facilitator, FFS and government compared with wealthier farmers ($p<0.01$). This is similar to the findings for non-users of CA. Furthermore, as with non-users, poorer farmers using CA also had a greater appreciation for self-observation and self-initiative ($p<0.05$). Self-observation and self-initiative were also found by Martínez-Garcia *et al.* (2013) to be strong social referents as farmers took up a technology after observations made or through taking the initiative through testing. Garforth *et al.* (2004) also found that contacts (local and personal) played an important role in engendering adoption of a technology.

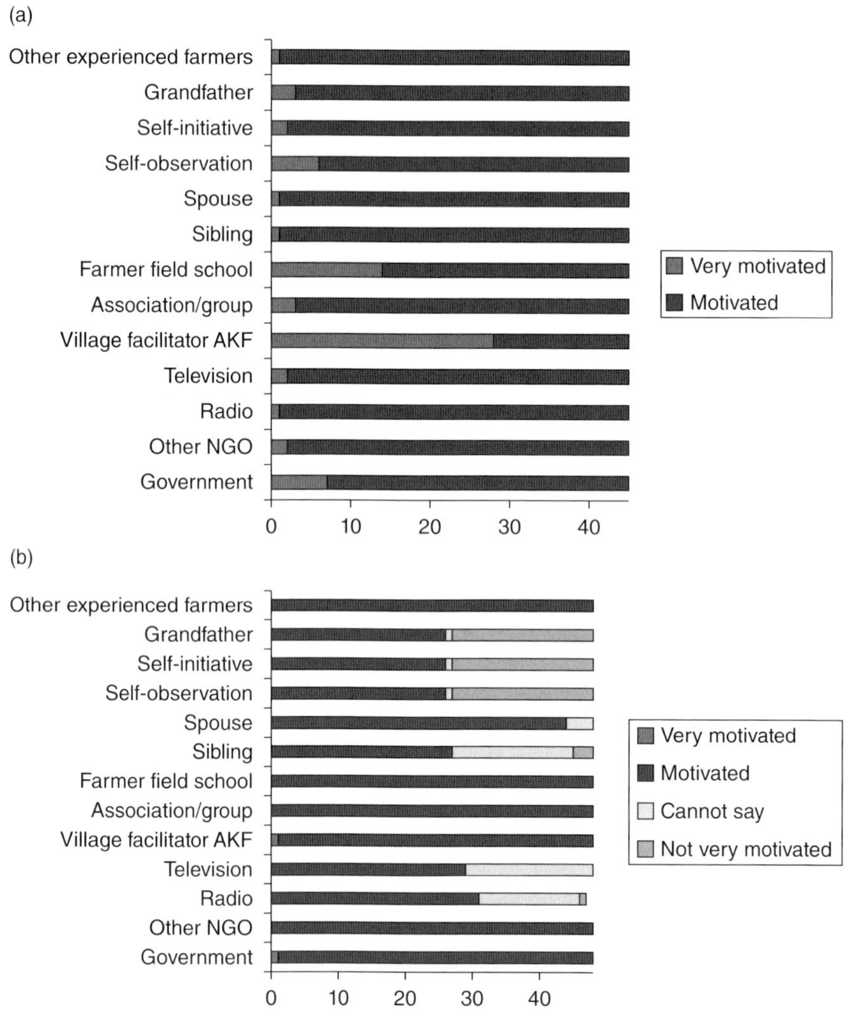

Fig. 6.7. (a) Frequency of responses related to motivation to comply with different social referents regarding using CA among the poor wealth tercile of CA users ($n = 45$). (b) Frequency of responses related to motivation to comply with different social referents regarding using CA among the better-off wealth tercile of CA users ($n = 48$).

Key findings among both users and non-users of CA were that poorer farmers have more favourable perceptions of CA. The poorer farmers not using CA also expressed a stronger intention to use CA than wealthier farmers ($p<0.05$) (Lalani *et al.*, 2016). It is interesting to note that the perceptions of farmers using CA were also matched, in part, by the farm budget data gathered in the household survey. For example, lower weeding time was found under CA. Evidence of short-term benefits under CA were apparent but these were largely dependent on crop mix and opportunity cost of labour assumed (Lalani, 2016).

6.5 Concluding Remarks

Two specific aims have been explored in this chapter relating to: (i) the construction and functioning of the innovation system, viewed through an AIS 'lens' by using timelines, stakeholder perceptions, typologies of partnerships and social network analysis; and (ii) perceptions of the users of the innovation itself and of whom they consider to be important actors in the innovation system (e.g. motivation to comply with information on CA). Furthermore, to ascertain whether the poor are beneficiaries of this innovation system, farmers' perceptions were disaggregated by wealth categories.

It is clear that the construction of the innovation system has been supported by a conducive policy environment including the MOU between AKF and the Ministry of Agriculture under the auspices of the strategic plan on agrarian reform. However, the modus operandi used to achieve a functioning innovation system has been the 'network' partnerships that have been developed through the support from AKF (e.g. FFS, community promoters, seed multiplication groups and village-level facilitators) and farmers themselves (e.g. benchmark farmers). These social learning mechanisms vis-à-vis other actors in the innovation system (e.g. traders, rural shops and radio) have further enabled farmers to 'innovate and change' (Hall *et al.*, 2006). The ability of NGOs to help to promote civil society-based organizations in the first instance is also in part due to a conducive policy/legal environment that allows formation of organizations such as village development organizations and associations and other market or civil society actors to operate. These have often been lacking in some countries in sub-Saharan Africa, which has stifled agricultural innovation. For example, the strong influence of public extension on smallholders in Ethiopia has limited the space for other potential actors, such as market or civil society actors, which has impeded innovation (Spielman *et al.*, 2011). Interestingly, where NGOs were operating (whilst the government had a central role) they had stronger and far-reaching ties within the public sector sphere (e.g. with public sector service providers) and wider afield such as with universities and research institutes, providing greater scope for new information to stimulate innovation (Spielman *et al.*, 2011). This supports the findings of the social network analysis presented in this chapter, i.e. the government agriculture directorate (SDAE) had strong formal ties or linkages (higher in-degree centrality) and lower out-degree centrality, suggesting limited influence in terms of knowledge flows within the innovation system, as the actor matrix also suggested. It has been well documented elsewhere that rapid uptake of CA will not take place without the appropriate enabling policy environment and institutional support to engage farmers at the community level (Kassam *et al.*, 2014, 2015).

There is the issue of sustainability when civil society actors may cease to operate in an area or if the overall effectiveness of community-based approaches in reaching the poor is reduced through, for example, 'elite capture' (as outlined earlier in this chapter). Notwithstanding this, the social learning mechanisms in this case study region (e.g. FFS, AKF village facilitator and community promoters), which have been seen to be playing a strong role in the innovation system among the actors themselves, have been central both to developing the capacity and confidence of local groups and individuals and to reducing the risk and uncertainty commonly associated with undertaking a 'new' farm management system among smallholder farmers.

This is supported by farmers' perceptions of the innovation (particularly favourable among the poorest) and motivation to comply with certain actors in the innovation regarding information on CA. Farmers expressed positive perceptions of key factors such as improved yields, reduction in labour and suppression of weeds. Farmers also feel that CA is able to contribute to benefits in the first year of implementation and in a drought year.

There is evidence to support the notion that this innovation system is reaching its target beneficiary, i.e. the poorest farmers. Interestingly, poorer farmers (among non-users and users of CA) had significantly stronger favourable perceptions of CA than wealthier farmers. There are also key differences in some of the social referents and key actors within the innovation system that farmers hold in high regard and are more likely to respond to in terms of receiving information on CA. For example, poorer farmers place a greater emphasis on social referents within the family, e.g. spouse, sibling, or grandfather, and others such as the village facilitator and FFS (p<0.05). They are also more likely to respond to media campaigns from either radio or TV (p<0.05). The village-level facilitator and FFS were also given particularly high scores in the actor matrix and in the social network analysis, which signals their pivotal role within this particular innovation system. Interestingly and importantly the results showed that poorer farmers value self-observation and self-initiative more than wealthier farmers (p<0.05). This indicates that, although the opinions of key referents are important to poorer farmers, their own experience also plays an important role in the adoption process. Ngwira *et al.* (2014) also found that CA adoption occurs within a strong social context whereby farmers learn by observing what other significant persons are doing. Thus, the formation of constructed networks such as seed multiplication groups and associations with sales commissions may further engender social capital coupled with CA adoption (as outlined earlier). However, a number of farmers did express an interest in using CA but a lack of desire to engage in group activities or networks, preferring one-to-one interactions instead. Similarly, other authors have found that group activities may be unattractive to some farmers and bilateral contacts may be more appealing (Ngwira *et al.*, 2014).

Overall, the farmers' perceptions also highlight that they value interactions with the government, other NGOs and the media with respect to using CA, which may signal potential entry points to strengthen the innovation system further given that these stakeholders have weaker roles at present in the innovation system. Alongside this, engagement of AKF (given its prominent role in the current innovation system) with research institutes and universities and in perhaps enabling local equipment to be manufactured (e.g. supporting local entrepreneurship) may also improve the potential reach of the innovation system. This will further improve the 'space' for farmers to innovate, which is crucial in developing locally relevant adaptations based on CA principles (Kassam *et al.*, 2014, 2015). For example, farmers have often found local solutions to issues surrounding mulch cover, weed management and equipment, etc.

Other issues that might have wider applicability to the discourse on CA in sub-Saharan Africa are as follows.

- Farmers' perceptions here indicate that CA alone, without external inputs such as fertilizers, herbicides, pesticides or compost, can improve yields relative to conventional tillage agriculture without the use of such inputs. Furthermore, the majority of CA users (72%) were using the local variety of maize, which illustrates that farmers are able to practise CA with the resources they have.
- This innovation system also shows subsidizing of inputs is not a necessary pre-condition to CA use and uptake.
- CA can be implemented in a variety of ways, be these manual systems as described here, or animal or mechanized systems. This case study of northern Mozambique highlights that manual forms of CA can be attractive for farmers, particularly those with very small plots of land (0.5 ha or less). Furthermore, where land-to-livestock ratios are low, such as in northern Mozambique, competition for mulch needed for livestock feed is not as pronounced as is perceived to be the case elsewhere.
- There is evidence to suggest that potential fears over weeds and labour in such a CA farming system are not key or absolute constraints to adoption, i.e. farmers have found CA to reduce labour requirements and weeding time. Nevertheless, weed control is a normal challenge for family farmers, irrespective of soil management. However, weed management under CA without herbicide has enormous implications for farmers in similar agroclimates in Southern Africa in particular. This is because, in manual CA systems, evidence from northern Mozambique shows that effective weed control is possible without the use of herbicides, since all three CA practices contribute to integrated weed control. However, for some farmers, while experience builds up during initial years of transformation to a CA system, managing some new cover crops or developing soil mulch cover in no-till systems can be a temporary challenge.

- Disaggregation of farmers by wealth shows that CA is being used by and benefiting the poorest farmers, rather than only benefiting wealthier farmers.
- Farmers' organizations are an important source of networking and learning for encouraging innovation. Social referents can also play a key role in this regard.
- Overall, agricultural innovation systems (like the one described here) that include dense networks of mutually supporting stakeholders and special extension processes, such as FFS, appear to be powerful enabling approaches for knowledge-intensive innovations such as CA.

References

Andersson, J. and Giller, K. (2012) On heretics and God's blanket salesmen: contested claims or Conservation Agriculture and the politics of its promotion in African smallholder farming. In: Sumberg, J. and Thompson, J. (eds) *Contested Agronomy: Agricultural Research in a Changing World*. Earthscan, London.

Biggs, S. and Matsaert, H. (2004) *Strengthening Poverty Reduction Programmes Using an Actor-Oriented Approach: Examples from Natural Resources Innovation Systems*. AgREN, Network Paper No. 134. ODI. Agricultural Research and Extension Network, London.

Bingen, J. and Gibbon, D. (2012) Early farming systems research and extension experience in Africa and possible relevance for FSR in Europe. In: Donhofer, I., Gibbon, D. and Dedieu, B. (eds) *Farming Systems Research in the 21st Century*. Springer, New York.

Borgatti, S.P., Mehra, A., Brass, D.J. and Labianca, G. (2009) Network analysis in the social sciences. *Science* 323 (5916), 892–895.

Chambers, R. and Jiggins, J. (1987) Agricultural research for resource-poor farmers part 11: a parsimonious paradigm. *Agricultural Administration and Extension* 27, 109–128.

Edirisinghe, J.C. (2015) Smallholder farmers' household wealth and livelihood choices in developing countries: a Sri Lankan case study. *Economic Analysis and Policy* 45, 33–38. Available at: http://dx.doi.org/10.1016/j.eap.2015.01.001 (accessed 20 September 2016).

Ekboir, J.M., Dutrénit, G., Martínez, V.G., Vargas, A.T. and Vera-Cruz, A.O. (2009) *Successful Organizational Learning in the Management of Agricultural Research and Innovation: The Mexican Produce Foundations*, IFPRI Report 162. European Initiative for Agricultural, Washington, DC. Available at: http://www.ifpri.org/sites/default/files/publications/rr162.pdf (accessed 20 September 2016).

FAO (2010) *The State of Food Insecurity in the World: Addressing Food Insecurity in Protracted Crisis*. FAO, Rome.

FAO (2015) *FAO CA website* [Online]. Available at: www.fao.org/ag/ca (accessed 20 September 2016).

Feder, G., Anderson, J.R., Birner, R. and Deininger, K. (2010) *Promises and Realities of Community-Based Agricultural Extension*. Discussion paper 959. IFPRI, Washington, DC.

Fox, I., Bardasi, E. and Van den Broeck, K (2005) *Poverty in Mozambique: Unravelling Changes and Determinants. Africa Region*. World Bank, Washington, DC.

Garforth, C.J., Rehman, T., McKerney, K., Tranter, R.B., Cooke, R.J., Yates, C.M., *et al.* (2004) Improving the design of knowledge transfer strategies by understanding farmer attitudes and behaviours. *Journal of Farm Management* 12(1), 17–32.

Grabowski, P.P. and Kerr, J.M. (2014) Resource constraints and partial adoption of conservation agriculture by hand-hoe farmers in Mozambique. *International Journal of Agricultural Sustainability* 12, 37–53.

Hall, A.J., Sivamohan, M.V.K., Clark, N., Taylor, S. and Bockett, G. (2001) Why research partnerships really matter: Innovation theory, institutional arrangements and implications for developing new technology for the poor. *World Development* 29(5), 783–797.

Hall, A., Janssen, W., Pehu, E. and Rajalahti, R. (2006a) *Enhancing Agricultural Innovation: How to Go Beyond Strengthening of Agricultural Research*. World Bank, Washington, DC.

Hall, A., Mytelka, L. and Oyeyinkka, A. (2006b) *Concepts and Guidelines for Diagnostic Assessments of Agricultural Innovation Capacity*. Working Paper 017. United Nations University – Maastricht Economic and social Research and training centre on Innovation and Technology, Maastricht, NL. Available at: http://www.merit.unu.edu/publications/wppdf/2006/wp2006-017.pdf (accessed 20 September 2016).

Hayami, Y. (2009) Social capital, human capital, and the community mechanism: toward a conceptual framework for economists. *Journal of Development Studies* 45, 96–123.

Hoffman, P.P.K. and Christinck, K. (2007) Farmers and researchers: how can collaborative advantages be created in participatory research and technology development? *Agriculture and Human Values* 24, 355–368.

Hounkonnou, D., Kossou, D., Kuyper, T.W., Leeuwis, C., Nederlof, E.S., Röling, N., *et al.* (2012) An innovation systems approach to institutional change: smallholder development in West Africa. *Agricultural Systems* 108, 74–83.

Johansen, C., Haque, M.E., Bell, R.W., Thierfelder, C. and Esdaile, R.J. (2012) Conservation agriculture for small holder rainfed farming: opportunities and constraints of new mechanized seeding systems. *Field Crops Research* 132, 18–32.

Kassam, A., Friedrich, T., Shaxson, F., Bartz, H., Mello, I., Kienzle, J. and Pretty, J. (2014) The spread of Conservation Agriculture: policy and institutional support for adoption and uptake. *Field Actions Science Reports*. Available at: http:// factsreports.revues.org/3720.

Kassam, A., Friedrich, T., Derpsch, R. and Kienzle, J. (2015) Overview of the worldwide spread of Conservation Agriculture. *Field Actions Science Reports*. Available at: http://factsreports.revues.org/3966 (accessed 20 September 2016).

Klerkx, L., Hall, A. and Leeuwis, C. (2009) Strengthening agricultural innovation capacity: Are innovation brokers the answer? *International Journal of Agricultural Resources Governance and Ecology* 8(5–6), 409–438.

Klerkx, L., van Mierlo, B. and Leeuwis, C. (2012) Evolution of systems approaches to agricultural innovation: Concepts, analysis and interventions. In: Darnhofer, I., Gibbon, D.P. and Dedieu, B. (eds) *Farming Systems Research into the 21st Century: The New Dynamic*. Springer, Dordrecht, The Netherlands.

Lalani, B. (2016) Economics and adoption of Conservation Agriculture in Cabo Delgado, Mozambique. Unpublished PhD thesis. University of Reading, UK.

Lalani, B., Dorward, P., Holloway, G. and Wauters, E. (2016) Smallholder farmers' motivations for using Conservation Agriculture and the roles of yield, labour and soil fertility in decision making. *Agricultural Systems*, 146, 80–90. Available at: http://dxdoi.org/10.1016/j.agsy.2016.04.002 (accessed 15 November 2016).

Leeuwis, C. (2011) Rethinking communication in innovation processes: creating space for change in complex systems. *The Journal of Agricultural Education and Extension* 17, 21–36.

Martínez-García, C.G., Dorward, P. and Rehman, T. (2013) Factors influencing adoption of improved grassland management by small-scale dairy farmers in central Mexico and the implications for future research on smallholder adoption in developing countries. *Livestock Science* 152 (2–3), 228–238. doi: 10.1016/j.livsci.2012.10.007.

Mendola, M. (2007) Agricultural technology adoption and poverty reduction: a propensity-score matching analysis for rural Bangladesh. *Food Policy* 32, 372–393.

Ngwira, A., Johnsen, F.H., Aune, J.B., Mekuria, M. and Thierfelder, C. (2014) Adoption and extent of conservation agriculture practices among smallholder farmers in Malawi. *Journal of Soil and Water Conservation* 69, 107–119.

Nkala, P. (2012) Assessing the impacts of conservation agriculture on farmer livelihoods in three selected communities in central Mozambique. PhD thesis, University of Natural Resources and Life sciences (BOKU), Vienna.
Nkala, P., Mango, N. and Zikhali, P. (2011) Conservation Agriculture and livelihoods of smallholder farmers in central Mozambique. *Journal of Sustainable Agriculture* 35, 757–779.
Norman, D.W. (2002) The farming systems approach: a historical perspective. Presentation at the *17th Symposium on the International Farming Systems Association in Lake Buena Vista*, Florida, USA, 17–20 November 2002. Available at: conference.ifas.ufl.edu/ifsa/papers/invite/norman.doc (accessed 15 November 2016).
OECD (1999) *Managing National Innovation Systems*. OECD, Paris.
Ortiz, O., Orrego, R., Pradel, W., Gildemacher, P., Castillo, R., Otiniano, R., *et al.* (2013) Insights into potato innovation systems in Bolivia, Ethiopia, Peru and Uganda. *Agricultural Systems* 114, 73–83.
Pimbert, M. (1994) The need for another research paradigm. *Seedling* 11, 20–25.
Pingali, P.L. (2012) Green revolution: Impacts, limits, and the path ahead. *Proceedings of the National Academy of Sciences USA* 109, 12302–12308, doi: 10.1073/pnas.0912953109.
Pretty, J (2003) Social capital and the collective management of resources. *Science* 302, 1912–1915.
Research for Development (EIARD) (2006) *Impact Assessment and Evaluation in Agricultural Research for Development*. The Task Force on Impact Assessment and Evaluation, London. Available at: http://www.eiard.org/docu/Eiard_Ia_paper.html.
Research into Use (RIU) (2010) Evaluation of innovation systems and agricultural research programmes, working paper, Department for International Development, UK. Available online at: https://assets.publishing.service.gov.uk/media/57a08b1bed915d622c000ae1/riu10innovsys-litreviewdrft.pdf (accessed 15 November 2016).
Röling, N. (1994) Platforms for decision making about ecosystems. In: Fresco, L. (ed.) *The Future of the Land.* Wiley, Chichester, UK.
Rubyogo, J.C., Sperling, L., Muthoni, R. and Buruchara, R. (2010) Bean seed delivery for small farmers in Sub-Saharan Africa: The power of partnerships. *Society & Natural Resources* 23, 285–302.
Sanginga, P., Chitsike, C., Njuki, J., Kaaria, S. and Kanzikwera, R. (2007) Enhanced learning from multi-stakeholder partnerships: lessons from the enabling rural innovation in Africa. *Natural Resources Forum* 31(4), 273–285.
Scott, J. (2000) *Social Network Analysis. A Handbook*, 2nd edn. Sage Publications, London.
Spielman, D.J. (2005) *Innovation Systems Perspectives on Developing-Country Agriculture: A Critical Review*. International Food Policy Research Institute, Washington, DC.
Spielman, D., Davis, K., Negash, M. and Ayele, G. (2011) Rural innovation systems and networks: findings from a study of Ethiopian smallholders. *Agriculture and Human Values* 28, 195–212.
Sumberg, J. (2005) Systems of innovation theory and the changing architecture of agricultural research in Africa. *Food Policy* 30(1), 21–41.
Temel, T., Janssen, W. and Karimov, F. (2002) *The Agricultural Innovation System of Azerbaijan: An Assessment of Institutional Linkages*. ISNAR Country Report 64. International Service for National Agricultural Research, The Hague, The Netherlands.
Valente, T.W. (1995) *Network Models of the Diffusion of Innovations*. Humpton Press, Cresskill, New Jersey.
Wall, P.C., Thierfelder, C., Ngwira, A., Govaerts, B., Nyagumbo, I. and Baudron, F. (2013) Conservation agriculture in Eastern and Southern Africa. In: Jat, R.A., Sahrawat, K.L. and Kassam, A.H. (eds) *Conservation Agriculture: Global Prospects and Challenges*. CAB International, Wallingford, UK.
Wiggins, S. and Cromwell, E. (1995) NGO's and seed provision to smallholders in developing countries. *World Development* 23, 413–422.
World Bank (2011) *Agricultural Innovation Systems: An Investment Sourcebook*. The World Bank, Washington, DC.

7 Conservation Agriculture in North Africa: Experiences, Achievements and Challenges

Oussama El Gharras,[1]* Mohamed El Mourid[2] and Hakim Boulal[3]

[1]*Institut National de la Recherche Agronomique, Settat, Morocco;* [2]*International Centre for Agricultural Research in the Dry Areas, Rabat, Morocco;* [3]*International Plant Nutrition Institute North Africa Program, Settat, Morocco*

7.1 Introduction

Traditionally, North Africa represents six countries: Algeria, Egypt, Libya, Mauritania, Morocco and Tunisia. However, in this chapter, North Africa refers only to Algeria, Morocco and Tunisia and will only consider rainfed regions. Agriculture plays an important role in this region, where the climate is mostly Mediterranean. The northern areas are characterized by wet winters and hot dry summers. The southern parts are mostly desert with < 150 mm annual rainfall where agriculture is based on rangeland pasture, cereals cultivation on flood plains and oasis farming systems. The middle or intermediate areas are typically semi-arid with < 450 mm annual rainfall and hot dry summers. Moderate temperatures from November to March and periods of sunshine interspersed between relatively short rainy periods are favourable for photosynthesis and efficient moisture use. However, since the 1980s the region has been experiencing inter- and intra-annual variability in precipitations, with a trend of decreasing precipitation. A very limited number of rainy days (fewer than 120 days per annum in general) and frequent droughts during the growing season combined with high temperatures are common constraints to plant growth (Agoumi, 2003). More than 85% of agricultural land is rainfed and is subject to these constraints. The Moroccan and Tunisian economies depend essentially on agricultural production. The Algerian economy is more reliant on oil production and export but the agricultural sector still plays an important role.

*oelgharras@yahoo.com

In rainfed regions of North Africa, agricultural production is primarily based on crop–livestock integration dominated by cereals and sheep. The main growing season extends from November to May–June in general. The rainy season coincides with the growing season of crops and temperatures sustain continuous vegetative development from planting to maturity. Cereals, mainly wheat and barley, are the dominant annual crops. Lately, monocropping and non-diversified cropping systems, especially under lower rainfall (< 350 mm), have become popular due to government policies that promote wheat production. The unavailability and cost of seeds, inadequate mechanization and weed-disease management for food legumes and oilseed crops are some of the main constraints for the practice of appropriate rotations. In fact, since manpower is becoming increasingly scarce and expensive in the rural areas, the acreage of food legumes tends to decrease. Government policies do not support legume crops either for seeds production or for marketing of the grain production. Similarly, annual oilseeds are neglected and have almost disappeared from the cropping systems in the whole region. Feed depends on imports from the international market and very few forage crops are grown for a sustainable integrated crop–livestock farming system.

North Africa's agricultural system is characterized by smallholder farmers and regional surveys showed that around 80% of farms are < 10 ha. Farms are getting smaller due to heritage fragmentation of the land. The crop–livestock production system helps farmers to survive during dry seasons and years of scarcity. Indeed, animals are considered to be the bank for farmers and their savings for investments generated during good crop years. Crop rotations have shortened and involve only the major mechanized crops within the region (wheat, barley and tilled fallow).

Considerable fluctuations in production are occurring, depending on weather conditions and farming practices. Even though North African countries have invested considerable efforts through large agricultural development programmes, they are still deficient in meeting their agricultural needs from domestic production. Furthermore, inadequate management of natural resources, largely promoted during the colonial period, has led to land degradation and depletion of soil water-storage capability. The introduction of heavy ploughs and tractors all over the arable lands has led to soil mining and degradation mainly in arid and semi-arid regions of North Africa. Certain summer tillage practices, in order to incorporate residues from the previous crop into the soil and to conserve water deep down in the soils, have no advantage under low rainfall and shallow soils. Undoubtedly, deep tillage, after harvest when soils are dried out on the surface and no rain events are probable, will only lead to depletion of soil organic matter (SOM) and soil water-storage capacity.

While rainfall amount and distribution are the main factors of wheat productivity in North Africa's rainfed wheat-based systems, yields could be improved through proper soil and crop management. Conservation Agriculture (CA), based on no-till (NT) production system with crop rotations and residue retention on soil surface, has been introduced in North Africa as a new technology to improve soil and water conservation and hence stabilize yields (Boulal *et al.*, 2014). The objective of this chapter is to review the experiences, achievements and challenges of CA in North Africa.

7.2 Experiences

In collaboration with international organizations and institutes, preliminary work on CA in North Africa began in semi-arid areas of Morocco and Tunisia in the early 1980s. Research trials comparing conventional tillage (CT) with NT and minimum tillage in terms of yield, water conservation and water use efficiency were conducted in agricultural research stations (Bouzza, 1990; Kacemi, 1992).The work focused on cereals (wheat and barley) grown in winter where grain yield, water use and water use efficiency have been compared. Trial results by Bouzza (1990) showed that NT could be adapted to cereal-based systems under less favourable rainfed conditions. The data on water availability to the crop in continuous cropping systems compared with wheat after fallow and the ability of the soils to store water during fallow and stored water effectiveness on yield were studied. NT management yielded more grain and straw than minimum till and CT in continuous wheat and in wheat after fallow. In areas averaging 240–300 mm annual rainfall, wheat after fallow used water as efficiently as continuous wheat. NT management during the summer dry period after harvest was a requirement for successful fallow. Further research in the 1990s confirmed the advantages of the CA NT system in semi-arid areas (Mrabet, 1997).

Due to the lack of direct drill machinery adapted to local conditions (cost and effectiveness), a programme to develop new NT seed drill prototypes was launched by *Institut National de la Recherche Agronomique* (INRA) Morocco in 1989. The research programme started with comparison of commercial soil-engaging components, single-disc, double-disc and hoe (furrow) openers (Bahri, 1992). Information was collected on the soil–seed environment characterized by soil disturbance, bulk density, soil moisture, seeding depth and plant stand. The hoe opener outperformed the other openers for all parameters except uniformity of sowing depth. Plant emergence, being the best parameter from the farmers' point of view, was statistically greater for the hoe than for the two disc openers. Bahri's study demonstrated mainly that hoe openers are more suitable for hard and dry soil conditions at sowing time, because of their better penetration. The furrow opener also created the greatest amount of soil disturbance compared with other types of openers. This may also be an advantage for creating better soil–seed contact and allowing water harvesting in the seeds zone. The furrow opener performed better on stony fields compared with disc openers that roll over the stones and leave seeds on the topsoil. Based on these findings, a programme to design and develop a low-cost direct seeding drill fitting farmers' needs was launched in 1995. The requirements of INRA soil scientists for the drill were:

- a machine capable of being operated by a 60 PTO HP tractor;
- operate in dry as well as humid soil conditions at seeding time;
- ensure minimum soil disturbance to minimize soil moisture losses; and
- leave most of the residues on the soil surface to control runoff and erosion and to conserve soil moisture.

The first farm experiments with the newly developed prototype were conducted during the 1996/7 cropping season in Morocco's Chaouia region. The objective of the on-farm research programme was to compare NT with CT under the farming system conditions of North Africa where residues are grazed. In fact, after harvest, all fields become open to common-land grazing. Therefore, wheat production was compared under NT management with farmers' CT practices. Experiments were conducted on a 12 ha plot where 4 ha was under NT and 8 ha under CT management. The soil is shallow, calcimagnesic, stony (about 15% soil cover) and on a sloping land (about 3%) that is vulnerable to erosion. All the treatments over the growing season were adapted to best fit the crop needs in order to achieve the potential yield levels. No pre-emergence herbicide (paraquat or glyphosate) was used except when emerging volunteers or weeds jeopardized the sown crop. Tillage operations were practised based on the farmers' experience and soil conditions: one or two tillage operations using a disc harrow at 100–150 mm depth if the soil moisture is adequate; or 200–250 mm depth followed by a second pass for seed bed preparation (100–150 mm depth). A conventional seed drill was used at a seeding rate of 150–160 kg of certified wheat seed/ha. For NT we used the same variety at a seed rate of 120–140 kg/ha. Crop rotations were similar under CT and NT treatments (wheat, barley and fallow or wheat, chickpea). The crop rotation was not fixed and the approach was on an opportunistic decision basis, made by the farmer depending on the climatic and market trends. Fertilization and disease treatments were the same over all CT and NT plots. Results are reported below in Section 7.3 (Achievements).

The first ten NT prototype seed drills were produced in 1999 in collaboration with a national private manufacturer in Morocco. Using these industrial prototype direct seed drills, national research and development institutions in collaboration with local extension services started to test the NT system in 2000 for a 3-year period in selected farms within three different agroclimatic regions: (i) Kelâa Esraghna Benguerir (arid regions with < 250 mm annual rainfall); (ii) Chaouia-Settat (semi-arid; 250–350 mm annual rainfall); and (iii) Khouribga (continental semi-arid with 250–400 mm annual rainfall combined with low temperatures around 0°C) within the framework of the Programme of Technology Development for Arid Regions (*Programme de Développement des Technologies en Aridoculture*, 2002).

Another programme was launched between INRA-Settat and the Regional Council (elected representatives of the populations) of Chaouia-Ouardigha for a better targeted project that was designed for implementation over a 6-year period (2000–2006). Unfortunately, the project was interrupted after 2 years of implementation, due to the ending of the term for regional council members. In 2006, the Arab Authority for Agricultural Investment and Development (AAAID), in collaboration with INRA–Settat, launched a large-scale research and development (R&D) programme starting with 900 ha in three communities in the Chaouia region (Jemaa Eriah, Khemissat Chaouia and Sidi Boumehdi). This experience marked a shift from a limited research managed programme covering 20–50 ha of on-farm research experimentation to an R&D programme over 1200 ha within 2 years of implementation. The programme aimed to develop a suitable approach for NT dissemination and adoption in Morocco as a first test bed, to be subsequently out-scaled to North Africa in general, as the same programme was conducted in Tunisia as well on a reduced scale of 150 ha for the first year and 250 ha for the second year. Unfortunately again, the economic crisis of 2008 prevented the continuation of the programme in both countries.

The small number of available NT seed drill prototypes manufactured in Morocco prevents their widespread use and cannot assure their cost effectiveness and the sustainability development of NT in North Africa. Further efforts are still needed involving researchers as well as local manufacturers to develop suitable low-cost drills adapted to North African farming conditions (El Gharras and Idrissi, 2006). In 2009, a private Moroccan company importing agricultural equipment imported a 'Kuhn' NT seed drill and conducted demonstration experiments with some potential buyers over 2 years. So far only seven such drills have been purchased by three institutions in Morocco (four for les Domaines Agricoles, two for INRA and one for West Maroc Farms), mainly because the price is too expensive for individual Moroccan farmers.

In 2011, within the Green Moroccan Strategy '*Plan Maroc Vert*' (PMV), a project supported by the Global Environment Fund (GEF) launched an NT programme targeting 1000 ha in Rabat, 400 ha in Ben Slimane and 2000 ha in Settat regions. INRA is in charge for the supervision and evaluation of the project. This is considered to be the first national development project for the use of NT (Balaghi *et al.*, 2010). Meanwhile, in 2011 the International Centre for Agricultural Research in the Dry Areas (ICARDA) in collaboration with the Australian Centre for International Agricultural Research (ACIAR) developed a project for North Africa, 'Adapting Conservation Agriculture for Rapid Adoption by Smallholder Farmers in North Africa' (CANA), aiming to reach more than 5000 farmers and stakeholders in the region by 2015. The CANA research project is based on the innovation platform concept where all the stakeholders are invited to participate and share their experiences and contribute to the development and promotion of NT and CA practices. The target platform in Morocco is Oued Zem, in Khouribga region, with new communities that are experimenting with NT and CA for the first time. This platform represents the semi-arid regions and the results are to be out-scaled to similar environments within all North Africa regions through a similarity study coordinated by ICARDA. The project field work was launched in 2012/13 with 11 farmers on an area of 40 ha; by the third year (2014/15), the area had expanded to more than 400 ha.

All the efforts deployed by INRA Morocco, in collaboration with the Moroccan Ministry of Agriculture departments and international organizations, only reached an NT area of about 9000 ha. This is mainly due to the long time devoted to research experimentation under controlled environments, with limited efforts invested in development for adaptation and promotion of CA adoption.

In Tunisia, an initial programme on CA started in the late 1990s with the support of French cooperation agencies, including technical support from the French Agricultural Research Centre for International Development (CIRAD) and financial support from the French Agency for Development (*Agence Française de Développement*, AFD). The programme was based on the assessment of direct sowing, especially for wheat, in sub-humid and semi-arid regions of Tunisia. In the early 2000s, the involvement of Tunisian research/development institutions under the Integrated Rural Development Project (*Projet de Développement Agricole et Rural Intégré*, PDARI), with financial support from AFD, promoted the NT system using an imported direct seed drill developed by the Brazilian company Semeato. Optimistic results, in terms of yield and economic costs through the initial projects, encouraged an increase of areas with the involvement of national farmer associations and the creation of a National Field Crops Institute (INGC). Hence, in Tunisia, during the project lifetime, NT was more actively promoted among large-scale farmers with an estimated area of 12,000 ha by the end of the PDARI project. However, there has not been any increase in NT area in the country nor in the number of NT seed drills sold or manufactured except machines purchased within other ongoing projects. The main reasons for the stagnation are the high cost of the commercialized machinery that is prohibitive to smallholder farms and the absence of any support or strategy from the agricultural ministry departments.

The AAAID project was also implemented in Tunisia during the period 2006–2008 in collaboration with INGC and Kef Agricultural Institute (ESAK). Also, the ICARDA/ACIAR (CANA) project was conducted in the Fernana zone of Jendouba Governorate, Tunisia, representing the sub-humid platform for the North Africa region. Another 4-year (2013–2016) CA-based project entitled 'Integrated Crop–Livestock Conservation Agriculture for Sustainable Intensification of Cereal-based Systems in North Africa and Central Asia' (CLCA) is being implemented with support from ICARDA and IFAD, focusing on crop–livestock integration. This project includes Tunisia and Algeria in North Africa and Tajikistan in Central Asia.

In Algeria, the assessment of NT systems was initiated in the mid-2000s, also using imported NT drills. The *Institut Technique des Grandes Cultures* (ITGC) is the implementing institution of all NT projects within this country, where a double-disc multi-purpose drill from a multinational manufacturer and a Brazilian double-disc no-till drill were imported. On-farm trials, with private and public land managers, are conducted for the promotion of CA technology. Over a 6-year period, attempts to introduce CA technology among farmers resulted in a total area of 5500 ha under NT by the year 2011 (Zaghouane and Makhlouf, 2011). In addition to these efforts, the CANA project implemented in Algeria covers the region of Sétif, representing the highlands platform of the North Africa region. The same region has been the target of the CLCA project within Algeria.

7.3 Achievements

Early work on conservation tillage showed that the introduction of NT systems reduces evaporation and increases soil moisture availability for crops (Bouzza, 1990; Kacemi, 1992; Mrabet, 2000). In Bouzza's experiment, different tillage equipment and timing periods were compared with NT in continuous wheat and wheat–fallow rotations. Part of the residues (2 t) was kept on the soil surface in order to reduce soil moisture evaporation for the NT treatments. Results are reported in Table 7.1.

Experimental trials showed that, by keeping residues on the soil surface and eliminating tillage (Mrabet, 2011), soil can keep its moisture within the seed zone up to 5 weeks above the wilting point, whereas it remains only 15 days in tilled plots. Long-term trials conducted in Morocco (Mrabet, 2011), in Tunisia (Angar *et al.*, 2010) and more recently in Algeria (Chennafi *et al.*, 2011; Benkherbache *et al.*, 2012; Benniou, 2012) showed that grain yields of wheat can be maintained or improved with NT systems compared with CT. However, crop rotations play an important role in NT systems. Best performance of NT versus CT were observed in semi-arid areas (Table 7.2) mainly when wheat is planted after clean fallow (Mrabet, 2002) resulting from the increase of water conservation (Bouzza, 1990).

On-farm trials conducted in Morocco showed that wheat production under NT outperformed the CT practices every year, no matter what the rainfall amount and distribution were over the cropping season (El Brahli *et al.*, 2009). During the cropping season 1999/2000, the total annual rainfall over the experimental site was about 200 mm. All crops failed and the region was considered an area stricken by drought. The only harvested field was the 4 ha of NT wheat sown on a clean fallow that yielded more than 1 t of grain/ha and about 50 bales/ha with 14 kg of straw per bale. In 2003/4, the total amount of rain was more than 400 mm, but the CT wheat plot was

Table 7.1. Storage efficiency and amount of stored water for different types of fallow in semi-arid area of Morocco (Bouzza, 1990).

Type of fallow	Storage efficiency[a] (%)	Amount of stored water in 1.2 m profile (mm)
Chemical	28	84
Clean	18	54
Stubble mulch	21	63
Weedy	10	30

[a]Calculated as the ratio of stored water and the rainfall received during fallow period.

Table 7.2. Effect of cereal rotation and tillage on wheat yield (t/ha) in experimental stations Sidi El Aïdi and Jemaa Shim (Bouzza, 1990; Mrabet, 2000).

	Sidi El Aydi[a]		Jemaa Shaim[b]	
Type of tillage	Wheat–wheat	Wheat–fallow	Wheat–wheat	Wheat–fallow
No-till	1.9	3.5	1.7	3.0
Minimum till	1.6	3.4	1.5	3.0
Conventional tillage	1.4	2.4	1.6	2.4

[a]Average grain yield 1983–1992 in clay soil Vertisol, average rainfall 370 mm.
[b]Average grain yield 1983–1998 in clay soil Vertisol, average rainfall 270 mm.

significantly affected by 50 days of drought combined with high temperatures (> 20°C during the day) for the period of January and February (Fig. 7.1).

Furthermore, long-term research has shown that several years of CA leads to improved soil quality. The soil organic carbon (SOC) content is improved in the topsoil horizon (Ibno-Namr and Mrabet, 2004; Bouzrara *et al.*, 2010; Jemai *et al.*, 2012; Mekhlouf *et al.*, 2013). Data from an 11-year study under NT versus CT of a Calcixeroll soil in Morocco showed that soil under NT contained 3.4 t more of SOC/ha than CT in the top 20 cm layer (Bessam and Mrabet, 2003). Also, the lack of soil disturbance increases the availability of nutrients near the surface (Ibno-Namr and Mrabet, 2004; Ben Moussa-Mechraoui *et al.*, 2010) and significantly improves water infiltration rate compared with CT (Mekhlouf *et al.*, 2013).

From on-farm trials conducted in Morocco (El Gharras *et al.*, 2009), after a rainfall event of 25 mm, the tilled soil was completely washed and runoff sealed the soil surface. In the NT soil, the surface was covered by small pieces of straw even after baling and all-summer grazing on residues of the preceding wheat crop. Evidently, there are also small stones on the soil surface that contribute, along with surface mulch from crop residues, to decreasing the impact of raindrops on the ground surface, reducing soil erosion and water evaporation as well (Fig. 7.2).

Economic studies confirm that NT decreases the cost of land preparation, saving time and energy by reducing machine passes for soil preparation (Ben-Salem *et al.*, 2006; Boughlala *et al.*, 2013; Fenni, 2013). This is a good entry point to convince farmers to undertake a high-risk economic activity and a reason to minimize the investment at the beginning of the cropping season. Unfortunately, this saving is not always acquired in the first years of NT implementation. During the transition years, when significant rains occur early (in September

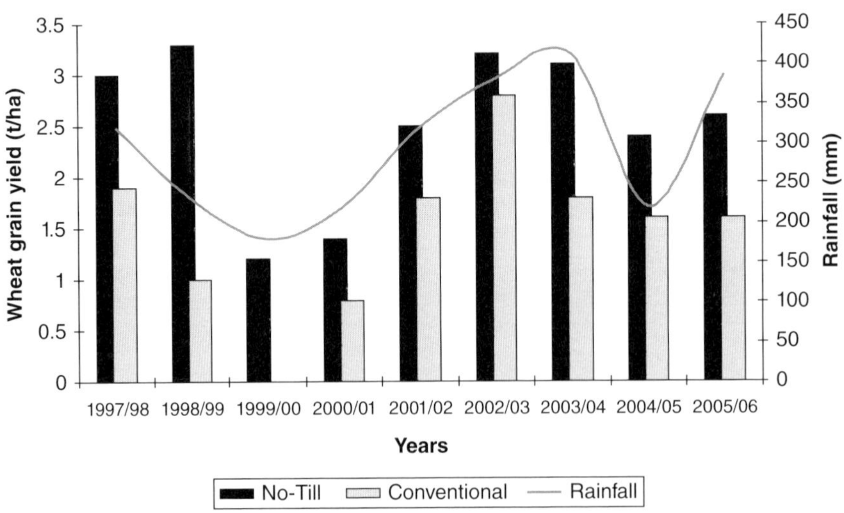

Fig. 7.1. Grain yield at farmer's field under conventional and no-tillage (El Gharras *et al.*, 2009).

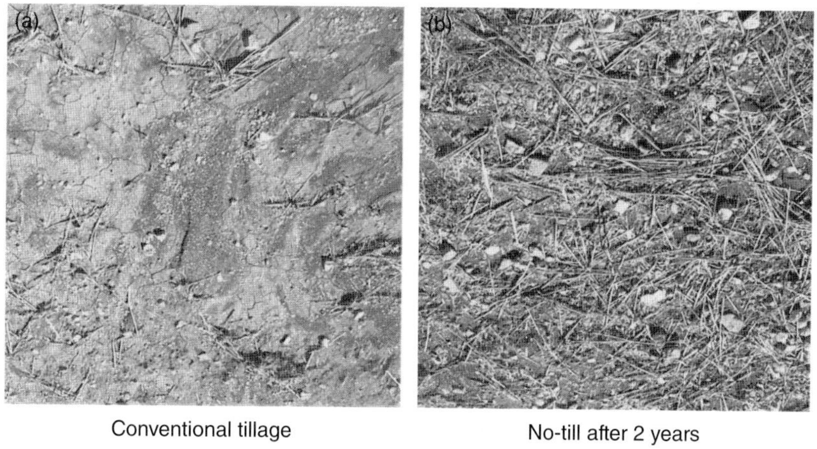

Fig. 7.2. Soil surface at farmer's field after 25 mm rainfall (El Gharras *et al*., 2009). (a) Conventional tillage; (b) no-till after 2 years.

and/or October, for example), application of pre-sowing non-selective herbicides to control weed infestations is required. In this case farmers spend as much as for the tillage operation for mechanical weed control. Nonetheless, in semi-arid Morocco, pre-sowing herbicide was applied only six times over a period of 17 years. Moreover, economic savings are also achieved in seed savings and efficient use of fertilizers that are placed with the seeds in the soils.

The first and only development programme in Morocco aiming to improve CA development and to enhance adoption was implemented in collaboration with the World Bank. This project, entitled '*Projet d'Integration des Changements Climatiques dans le Plan Maroc Vert*' (PICC-PMV) (Balaghi *et al*., 2010), was implemented in two essential regions known for cereal-based production systems. The project was launched in 2011 in Rabat and Ben Slimane regions, with the objective of reaching 1500 ha, and in Settat region targeting 2000 ha in 2012. The pilot phase of this project finished in the cropping season 2014/15 and all the partners agree that this is still at an experimental stage and they have to move to a wider-scale development programme.

7.4 Challenges

Adoption of CA systems faces a number of cultural, economic and technical constraints in North Africa. The small farm size, weak local farmers' organizations and institutions, the local knowledge and heritage traditions and the non-availability of appropriate technologies are the major constraints to widespread application of this promising system for the region. In fact, affordable NT seed drills are one of the main constraints raised during discussions with farmers. Since the small farmers and mechanization service providers that are the basis of the farming system and agricultural production cannot manage to pay for reliable international drills, NT adoption will be constrained (Boulal *et al.*, 2014). The CANA project is working within the three North African countries – Algeria, Morocco and Tunisia – on development and manufacturing of low-cost seed drills for smallholder farmers and some promising results are expected within the coming years. These seed drills will not solve the problems of adoption but will expose any other hidden problems. Lack of trained farmers and mechanization service providers for use, maintenance and repair of agricultural equipment are also real limiting factors. The improper use of herbicides or seed drills may have an undesirable outcome on the adoption of CA technology.

Certified seeds for food legumes, oil crops and forage crops required to improve crop rotations and feed resources are neither available nor affordable to farmers. All other inputs (pesticides and fertilizers), with the appropriate mechanization services, are a challenge for small individual farmers. Because of inefficient organization and management they have to pay more and wait until last to be served. As this is considered a real constraint today, it offers an opportunity to develop small businesses and job opportunities within rural communities. Particular focus should be on farmers who are the target technology beneficiaries and on resource staff for innovative suggestions to adapt farming systems based on CA to their local agroclimatic and socio-economic conditions and constraints.

The small increase in water conservation and water use efficiency obtained by NT system has a tremendous effect on yield improvement and agricultural production stability. Increased weed infestation during the early years of NT and development of root diseases when continuous cereal crops are grown may constitute serious challenges to NT adoption. Lack of tillage increases the rate of weed and disease infestation and hence pre-emergence herbicides are used in most cases if adequate crop rotations are not implemented on a long-term basis. Moreover, post-emergence herbicides are not effective if times and rates of their application are not adequate. The cost of herbicides is a challenge for CA in general and in particular for smallholder farming systems in North Africa. Use of forage crops for early soil coverage and highly competitive species and varieties against weeds is another strategy to reduce the use of herbicides. This is especially effective when the forage crops are harvested green and for hay-making before they set seed. This practice also allows reduction in the pressure on cereal-straw residue through grazing and improves the sustainability of the crop–livestock integration system. The challenge is to make appropriate seeds available to farmers through on-farm seed multiplication and marketing programmes.

CA adoption could be also hindered by lack of appropriate and timely financial resources. Smallholder farmers cannot afford to take chances and put their savings, which are normally invested in animal production, into risky and uncertain crop production activities that are subject to climate change and unfavourable consequences. Therefore, these smallholder farmers need support from government institutions during the initial phase of NT system implementation until they improve their incomes and gain confidence and expertise to develop and manage good quality CA cropping systems.

7.5 Conclusion

North African farming systems were developed over thousands of years. Crop–livestock integration is one of the main pillars of the resilience of these farming systems. The developed knowledge over history proved that, despite many drought cycles, farmers were able to make it through and cover their expenses or minimize their losses. Technologies introduced during the colonial period are more adapted to humid agroclimatic conditions, when there is no erosion risk. However, successive tillage operations and use of heavy machinery generate soil degradation with low fertility over time. The conveyed messages about tillage and tilled fallow advantages are some of the main constraints for any change. On the other hand, there is no specific research or development CA programme designed for these conditions. That is why, despite promising research results, adoption is still very low. After all, the successful experiences for CA promotion in Algeria, Morocco and Tunisia have shown that it is possible to adopt CA. The major constraint to rapid CA adoption and its wide-scale dissemination among farmers in North Africa remains the unavailability of affordable NT drills in the region and the absence of a clear strategy to implement a CA development programme for wide adoption by all types of farmers.

A participatory management approach that will involve all stakeholders in an on-farm research and demonstrations programme in order to fit the CA system to local needs is the alternative solution for the expected wide-scale dissemination and adoption of CA among North African farmers.

References

Agoumi, A. (2003) *Vulnerability of North African Countries to Climatic Changes: Adaptation and Implementation Strategies for Climate Change*. International Institute for Sustainable Development (IISD), Winnipeg, Canada, 11pp.

Angar, H., Ben Hadj Salah, H. and Ben-Hammouda, M. (2010) Semis direct et semis conventionnel en Tunisie: les résultats agronomiques de 10 ans de comparaison. *Options Mediterranéennes*: Serie A 96, 53–59.

Bahri, A. (1992) Evaluation of opener and presswheel combinations on a No-Till grain drill when seeding wheat. MSc thesis. University of Nebraska, Lincoln, Nebraska.

Balaghi, R., Jlibene, M. and Benaouda, H. (2010) *Projet d'Intégration des Changements Climatiques dans le Plan Maroc Vert: Rapport de faisabilité du PICCPMV*. Available at: http://www.inra.org.ma/docs/environ/piccpmvfaisab.pdf (accessed 15 November 2016).

Benkherbache, N., Benniou, R., Merat, M., Mekhalfia, T. and Makhlouf, M. (2012) From minimum tillage to no tillage, reaction of Waha, Variety of Durum Wheat in Algerian semi-arid region. *The Online Journal of Science and Technology* 2(3), 17–22.

Ben Moussa-Mechraoui, S., Errouissi, F., Ben-Hammouda, M. and Nouira, S. (2010) Comparative effects of conventional and no-tillage management on some soil properties under Mediterranean semi-arid conditions in north-western Tunisia. *Soil & Tillage Research* 106, 247–253.

Benniou, R. (2012) Agriculture conservation role of moisture and soil organic matter semi-arid. *Journal of Materials and Environmental Science* 3(1), 91–98.

Ben-Salem, H., Zaibet, L. and Ben-Hammouda, M. (2006) Perspectives de l'adoption du semis direct en Tunisie, une approche économique. *Options Mediterranéennes*, Serie A 69, 69–75.

Bessam, F. and Mrabet, R. (2003) Long-term changes in soil organic matter under conventional and no-tillage systems in semiarid Morocco. *Land Use and Management* 19, 139–143.

Boughlala, M., El Gharas, O. and Dahan, R. (2013) Economic comparison between conventional and no-tillage farming systems in Morocco. In: *57th AARES Annual Conference*, Sydney, New South Wales, 5th-8th February, 2013.

Boulal, H., El Mourid, M., Ketata, H. and Nefzaoui, A. (2014) Conservation Agriculture in North Africa: past and the future. In: Jat, R.A., Sahrawat, K.L. and Kassam, A.H. (Eds) *Conservation Agriculture: Global prospects and challenges*. CAB International, Boston, Massachusetts, pp. 293–310.

Bouzrara, S., Ould Ferroukh, M.E.H. and Bouguendouz, A. (2010) Influence du semis direct et des techniques culturales simplifiées sur les propriétés d'un sol de la ferme pilote Sersour (Sétif). *Options Méditerranéennes*, Serie A 96, 123–129.

Bouzza, A. (1990) Water conservation in wheat rotations under several management and tillage systems in semi-arid areas. PhD dissertation, University of Nebraska, Lincoln, Nebraska.

Chennafi, H., Hannachi, A., Touahria, O., Fellahi, Z.E.A, Makhlouf, M. and Bouzerzour, H. (2011) Tillage and residue management effect on durum wheat [*Triticum turgidum* (L.) Thell. ssp. *turgidum conv. durum* (Desf.) MacKey] growth and yield under semi-arid climate. *Advances in Environmental Biology* 5(10), 3231–3240.

El Brahli, A., El Gharras, O. and El Hantaoui, N. (2009) Le système semis direct: nouveau mode de production et modèle d'agrégation pour une agricultura pluviale. *Transfert de Technologie* 182, 1–4.

El Gharras, O. and Idrissi, M. (2006) Contraintes technologiques au développement du semis direct au Maroc. *Options Mediterranéennes*: Serie A 69, 120–124.

El Gharras, O., El Brahli, A. and El Mourid, M. (2009) No-Till system applied to Northern Africa rain fed agriculture: Case of Morocco. In: *Proceedings of the 4th WCCA, New Delhi, India*.

Fenni, M. (2013) Socio-economic and environmental benefits of direct seeding of wheat in Sétif high plains (north east of Algeria). In: *Proceedings WEI International Academic Conference*, Istanbul, Turkey. West East Institute (WEI), Westchester, PA, USA, p.47.

Ibno-Namr, K. and Mrabet, R. (2004) Influence of agricultural management on chemical quality of a clay soil of semi-arid Morocco. *Journal of African Earth Sciences* 39, 485–489.

Jemai, I., Ben Aissa, N., Ben Guirat, S., Ben-Hammouda, M. and Gallali, T. (2012) On-farm assessment of tillage impact on the vertical distribution of soil organic carbon and structural soil properties in a semiarid region in Tunisia. *Journal of Environmental Management* 113, 488–494.

Kacemi, M. (1992) Water conservation, crop rotations and tillage systems in semiarid Morocco. PhD dissertation, Colorado State University, Fort Collins, Colorado.

Mekhlouf, A., Rouag, N., Boukhadra, R., Chenni, S., Fenni, M. and Makhlouf, M. (2013) Influence of cropping systems on soil properties in semi-arid conditions of Setif, Algeria. *Journal of Agricultural Science and Technology* A3, 653–658.

Mrabet, R. (1997) Crop residue management and tillage systems for water conservation in a semi-arid area of Morocco. PhD dissertation, Colorado State University, Fort Collins, Colorado.

Mrabet, R. (2000) Differential response of wheat to tillage management systems under continuous cropping in a semi-arid area of Morocco. *Field Crops Research* 66, 165–174.

Mrabet, R. (2002) Conservation agriculture: for boosting semiarid soil's productivity and reversing production decline in Morocco. In: *Proceedings of International Workshop on Conservation Agriculture for Sustainable Wheat Production in Rotation with Cotton in Limited Water Resource Areas*, Tashkent, Uzbekistan. International Center for Agricultural Research in the Dry Areas (ICARDA), Aleppo, Syria, pp. 56–61.

Mrabet, R. (2011) Effects of residue management and cropping Systems on wheat yield stability in a semiarid Mediterranean clay soil. *American Journal of Plant Sciences* 2, 202–216.

Zaghouane, O. and Makhlouf, M. (2011) Promoting no-till practices to stabilize durum wheat yields and sustain agricultural production in semi-arid regions of Algeria. In: 5th *World Congress on Conservation Agriculture (WCCA), Australia*, full text of Poster, 4 pp. WCCA/FSD Local Organising Committee, Brisbane, Australia.

8 How Can We Cover Millions of Hectares with Conservation Agriculture in Africa?

Roland Bunch*

Independent consultant, Juneau, USA

8.1 Introduction

Let it be admitted from the start that Conservation Agriculture (CA) in Africa has spread faster in the past 5 years than in previous years but it is still not spreading as fast as it should. Also, the great majority of African smallholder farmers who have adopted CA use it on less than half a hectare. However, neither of these phenomena will turn CA into just a temporary fad that will die a sad and lonely death. In fact, there is evidence that CA, with a few simple improvements, could someday become the dominant way of producing food in much of the tropical world.

To achieve such a goal, we need to look at CA with a cold eye to shortcomings in some of the present practices and then find practical solutions that will: (i) significantly increase basic grain productivity; (ii) require less labour than do other farming systems; (iii) use only local resources that are plentiful; and (iv) increase net benefits for the farm family.

Of course, a number of techniques could help to move CA ahead. Planting techniques adapted to CA (even very simple ones such as dibble sticks) and more effective market chains that extend into the villages for such things as the pigeon peas, which now have a huge market in India, would certainly help.

A surprising number of the changes needed to increase productivity and reduce the need for labour and scarce resources can be achieved with the incorporation of green manure/cover crops (GMCC) into the CA system. GMCC are defined as 'any species of plant, usually leguminous, whether it is a tree, a bush, a climbing vine, a crawler, a grain legume or a water-borne plant, that farmers grow to maintain or improve their soil fertility or

*rdbunchw@gmail.com

control weeds, even when they have many other reasons for growing these plants'. In Africa, the CA concept does not include GMCC, so they are often neglected.

However, in southern Brazil, over 2 million farmers are now using CA on over 30 million hectares (Calegari, 2012). By far the most important difference between Brazilian CA and CA as practised in Africa is that virtually no one in Brazil would think of using CA without GMCC, whereas in Africa the use of GMCC in CA is relatively rare. The remainder of this chapter will explain why the incorporation of GMCC into CA in Africa and elsewhere could fairly easily achieve the same dramatic success that the Brazilians have achieved.

8.2 Material and Methods

The material in this case comes from almost exactly 30 years of personal experience in working with both CA and GMCC across nearly 50 nations in the global south. The methods have been diverse. Together with long-term colleague Gabino López, the author has searched out no-tillage systems, mulch systems and GMCC systems across the world. Many of these systems are now practised by over 10,000 farmers and a few are practised by well over 100,000 farmers (Bunch, 2012). The latter tend to be traditional zero-till and GMCC systems that farmers have used for centuries. Thousands of smallholder farmers have been interviewed who have originated or adopted these systems. Repeated visits have also been made to the programmes of EPAGRI (*Empresa de Pesquisa Agropecuária e Extenção Rural de Santa Catarina*) in the State of Santa Catarina, Brazil, to learn from that extremely valuable experience.

In Honduras, work was carried out for two decades with over 4000 farmers, introducing different possible no-tillage, mulch and GMCC systems, identifying farmer experimenters and keeping track of which systems were adopted or dis-adopted by farmers. Although most of this effort consisted of learning from other people's experiences and experiments, occasionally certain top-quality farmer experimenters were encouraged to carry out specific experiments that were needed to answer critical questions.

In the early 2000s, two technical conferences were organized, attended by selected smallholder experimenters from across Honduras, most of them trained by one or another of six non-governmental organizations (NGOs). At these conferences more than 75 star farmer experimenters were brought together and each presented their most important discoveries, much as professionals do at international conferences.

8.3 Results and Discussion

8.3.1 How well is CA fulfilling the criteria of a scalable technology?

To begin, it is necessary to look at how CA rates according to the four criteria mentioned above: does it (i) significantly increase basic grain productivity,

(ii) require less labour than do other farming systems, (iii) use only those local resources that are plentiful and (iv) increase net benefits for the farm family?

Yields

In terms of yields, CA has a patchy record. In some cases, maize yields have increased from about 1 t/ha to over 3 t/ha. In other cases, yields are hardly over 1.5 t/ha.

Labour demands

This is probably where CA, as practised in some cases, starts to look a little sickly. Hauling mulch material on to fields requires a lot of work. There are estimates of anywhere from 5 to 12 days to gather the mulch for a quarter of a hectare. That means it would require approximately a whole month for one person to collect enough mulch material for just 1 ha – and that is only if the highly optimistic assumption is made that they do not have to go a lot farther to gather the mulch for the last 0.75 ha. Further, if the whole community starts gathering the mulch, it could take more than twice that long, since a person might often have to walk more than 1 km to obtain a decent load of grass. Furthermore, the easiest mulch material to gather is all grass. This means that the farmers' young crops may have a very difficult first month or so, because when the rains come and crops most need nitrogen, much of the soil's nitrogen will be tied up while most of that grass is decomposing.

Gathering all this grass means that there is not much left for the village's grazing animals. The mulching therefore may be reducing the overall incomes of animal owners, because grass normally provides more income when fed to cattle than when it is used to cover the soil.

Another constraint often mentioned in connection with CA is that, without tillage, weeds can become more numerous initially. This increase in weeds either costs additional labour (usually provided by already overworked women) or additional costs in herbicides, once again reducing the profitability of CA.

Local resources

Many misunderstood CA applications can have some serious consequences in its use of local resources. Grass for mulching is available for the small plots of a few farmers, but if everyone in a village decides to use CA on all of their land, there would usually be a very serious shortage of mulch material. Also, a lot of the higher yields achieved with CA depend on the use of animal manure or compost. These resources, too, become very limiting when farmers want to expand their CA to more than a fraction of a hectare. Further, the cost of making compost, from bringing the material together to making the compost pile, turning it over, transporting the material out to the field and spreading it across the land, is prohibitive for use on basic grains (except for rice). It takes a minimum of about 20 t/ha per annum of biomass (fresh weight) to maintain yields over time (Bunch, 2012). I have never seen a smallholder farmer apply 20 t of compost to a hectare of cropland, nor have many other people.

Profit

Profits from CA can be very attractive for about 0.25 ha, but if the use of CA is expanded to even just 1 ha, the cost of mulching and enriching the soil without GMCC can become prohibitive. It is therefore not at all surprising that farmers nearly always have less than 0.5 ha of CA.

These problems are not part of the nature of CA, therefore one should not just be content that a farmer is only planting 0.25 ha of CA. At least in doing so, farmers are probably maximizing the output of their animal manure, compost and labour, even if only on a small part of their land. But neither we nor they need settle for such meagre returns from CA.

However, there is incontrovertible evidence to the contrary. In Brazil, which is by far the best and most extensive example of CA anywhere in the developing world, over 2 million farmers are using CA. Another 1 million farmers use CA in Paraguay. Many of the Brazilian farmers use CA on anywhere from 5 ha to 20 ha. Some wealthy well educated farmers in Brazil use CA on thousands of hectares. Thus it is not necessary to be resigned to the idea that CA can only be practised profitably on a small scale, or that its dissemination must be slow and difficult.

8.3.2 What can be done?

There is one very simple and obvious difference between CA as practised in southern Brazil and that practised in most of Africa. That difference is the inclusion in CA of GMCC. Most Brazilian farmers would never think of using CA without using GMCC along with it. In fact, for decades Brazilian farmers who were adopting CA always planted GMCC for 2 years before they would even start using no-tillage. Now they are more likely to start using GMCC and no-tillage at the same time, but GMCC are always included as part of CA (Eilitta, 2004).

It has been proved that, in Africa, such a process is not necessary. Nor would it be particularly desirable in areas that are more drought-prone than southern Brazil. In drought-prone areas, mulch is of tremendous importance. Operating for 2–3 years without the mulch would not be advisable.

But the incorporation of GMCC into CA is of the essence. There are a good dozen very important synergies between GMCC and CA. In fact, GMCC can make considerable strides toward solving every one of the four major problems identified above. GMCC can produce prodigious amounts of *in situ* mulching material. They can greatly improve soil fertility and soil quality, so that yields in CA will rise even higher than the best yields achieved so far; and after farmers have a handful of seed, they can produce all the seed they need. No other local resources are essential except the land itself. Furthermore, many GMCC will cover and smother virtually every weed in a field, thereby reducing the weed problem. Many GMCC systems will actually reduce the weeding necessary in a field to only one weeding per crop of maize or cassava, instead of two or three.

Lastly, because of the reduced labour and higher yields achieved with GMCC, the net profit, or benefits, from CA will increase dramatically. GMCC can also make farming systems completely sustainable over decades, provide high-protein food for the family, improve soil quality, reduce weed populations and completely rid people's land of particularly noxious parasites and weeds like striga (*Striga hermonthica*) and spear grass (*Imperata cylindrica*).

All these advantages are the good news but no cure-all like this comes free. There is a need to do a lot of learning to find the best GMCC systems for the people with whom we are working. Incorporating the right GMCC into CA is not just a matter of planting a lot of mucuna (*Mucuna* spp.) or lablab beans (*Dolichos lablab*) everywhere. The best GMCC systems have to be appropriate to the climate and the local farming system, if these species can improve or decrease populations of pests, diseases (nematodes), etc., the needs of the farmers, the topography of their land, their food preferences and their major crops, among other things. There is a lot of homework to be done.

8.3.3 Using GMCC

What can be expected from GMCC for CA in terms of the four criteria above?

YIELDS Different yield increases will be achieved by different systems among the 130 or so known GMCC systems. Nevertheless, it can reasonably be expected that over 5 years GMCC should raise yields of maize by at least 100% if they are presently under 1.5 t/ha, by 50% if they are between 1.5 t/ha and 2.5 t/ha and by 30% of they are higher than that.

These increases in yield will be produced by any of perhaps a dozen improvements brought about by GMCC in farmers' production systems. Probably the most important issue here, especially for people who are as committed to mulching as CA proponents are, is the role that a mulch can play in a GMCC system.

Many humid tropical forests inhabit some of the poorest soils on the planet, with pH values of 5.0 or less, virtually no available phosphorus and toxic levels of aluminium. Yet they go on, year after year for millennia, producing phenomenal amounts of biomass. A smallholder farmer cuts down a piece of these forests and within 3–5 years the soil has become so infertile that they have to let the forest grow back again to restore the soil. Why is the forest so able to do what any farmer wishes they could do, but cannot? The answer given, if anyone does give an answer, is that the farmer has used up the few nutrients left in the soil. But the nutrients used by a smallholder farmer in 5 years are insignificant. Furthermore, if those lost nutrients were so important, how does the forest go about restoring the fertility of the soil once again without them?

The trees of a humid tropical forest obtain their nutrients mostly from what foresters call the litter layer. Agronomists call the same thing a mulch. If one digs up the top 20 cm of a forest floor, one will find a mat of tree roots several centimetres thick. These roots are not feeding from the soil; they are feeding from the mulch. Why? The soil, with high acidity, aluminium toxicity and virtually no available phosphorus, is basically a hostile environment for plant roots. Feeder roots always go to where the environment is more favourable and the nutrient supply more abundant and well balanced. In this case, that environment is the mulch.

Given two caveats, the same will happen in CA. Crops will grow extremely well, but only if the mulch is both biodiverse and moist – conditions that are both found in the litter layer of a humid tropical forest. The moisture content of the mulch is an issue that will be taken up below, when dispersed shade is discussed. The biodiversity of the mulch will be achieved when GMCC is used along with the farmer's crops.

Unfortunately, if the mulch is entirely or largely composed of grasses, it will lack nitrogen (the C/N ratio will be too high) and the crop's feeder roots will not feed there very well. That is, by not including significant amounts of leguminous material in mulches, farmers are denied by far the best and most efficient manner of feeding their crops. Once there is a build-up of a healthy amount of leguminous biomass in the mulches, 'dinner is served'. Crops will be able to take advantage of the best feeding environment this side of a scientific laboratory.

However, nitrogen is not the only issue. Acid soils tie up phosphorus in minutes and do not leave more than about 0.5% of the soil phosphorus in forms that are available to plants. This means that the vast majority of the generally low amounts of phosphorus in African soils is unavailable to crops. Thus, crops that feed from the soil will be starved of phosphorus, even when there is quite a bit there. In a mulch, however, virtually all the phosphorus that is not available immediately will be available sometime within the next few months, when the organic matter that contains it decomposes (i.e. mineralizes), and also when the soil management could increase soil and plant bioactivity (actions of microorganisms), in order to access the soil nutrients.

Furthermore, GMCC have proved that they can, like the tropical forest, produce enough biomass to maintain soil fertility for decades. The standard of 20 t/ha (green weight), mentioned in Section 8.3.1 above, is a fairly easy target. Lablab beans (*Dolichos lablab* or *Lablab purpureum*), runner beans (*Phaseolus coccineus*), mucuna (*Mucuna* spp.), ratooned pigeon peas (*Cajanus cajan*) and many other GMCC species can all produce more than twice that much biomass in one season. So farmers will produce a lot more, and more sustainably, if they feed their crops through a mulch that includes legumes. In that way, millions of smallholder farmers can do exactly the same as a tropical forest does: produce huge amounts of edible biomass for decades, if not centuries, without in any way damaging the environment. Some authors, such as Bationo (2012), maintain that tropical soils will inevitably deteriorate during 10–20 years of continuous cropping. Nevertheless, the experience of many farmers, perhaps best documented by Triomphe (1996), contradicts that statement by increasing and then maintaining crop yields, without any use of chemical fertilizer, for the same or longer periods of time. The crucial difference is probably that only one of the trials reported by Bationo (2012) produced 20 t/ha or more of leguminous biomass per annum, and that system was initiated on virgin farmland. In fact, if the original sources in Bationo (2012) are consulted, this same system – an alley-cropping system using leucaena and mother of cacao – did improve and then maintain yields when applied to a degraded soil. Thus, if Bationo (2012) had included this latter experiment in his book, it would have disproved the book's main conclusion about the long-term maintenance of soil fertility.

LABOUR DEMANDS Even when CA reduces the labour input involved in soil preparation, CA as it is practised in some parts of Africa can have huge labour demands that come from hauling grass for mulching and hauling biomass to fertilize the soil, whether it is animal manure, compost, kitchen scraps or compound sweepings. GMCC will produce high-nitrogen biomass that, weight for weight, fertilizes the soil roughly as well as animal manure and can provide over 40 t/ha (green weight) of mulch material, with absolutely no transportation costs whatsoever, because it is produced *in situ*. This is the general condition under which the soil cover in a mature CA system should be produced.

The labour required by GMCC is rarely more than that required to plant the plants, harvest the seeds and cut them down. Planting is a very simple operation that can usually be done together with the planting of the maize or whatever species the GMCC is intercropped with (i.e. often by throwing the GMCC seeds into the same hole as the maize). Cutting down the GMCC, though a major task, requires much less labour than cutting down a forest fallow, or cutting down and hauling maize stalks around to pile them up and burn them. GMCC also require a good deal less labour than that required to haul mulch material and organic fertilizers out to the field. Thus, the labour requirements of using GMCC are approximately 20–40% less than those for the practices presently being used.

LOCAL RESOURCES The materials required for most GMCC systems are nothing more than a handful of GMCC seeds for the first planting (which in some cases may still be difficult for programme personnel to access). After that, the farmers produce their own seed, year after year. If farmers cannot easily produce their own seed from a particular species of legume, they simply switch to another species for which they can. There is no material involved in growing GMCC that is in short supply, that becomes scarcer if everyone in the village uses CA, or that becomes more labour-intensive if everyone decides to grow 1 ha of CA. The cost of using GMCC remains almost exactly the same per hectare planted, whether the farmer does CA on 0.25 ha, or on 25 ha, unless they can mechanize, in which case the cost/ha of CA will be reduced as the size of the plot expands, rather than being increased.

PROFIT The net profit of CA using GMCC will vary a good deal but will almost always be better than the net profit of doing CA without them. This happens because, as mentioned above, yields increase and labour costs per hectare decrease.

Additional benefits

In addition to those already mentioned, GMCC provide a huge number of additional benefits for CA, including: increased soil organic matter and soil nutrients; nitrogen fixation; weed control; and other benefits.

INCREASED SOIL ORGANIC MATTER AND SOIL NUTRIENTS There is occurring, all around us, a crisis of soil depletion. This is because of a series of unprecedented factors that are working together in a sort of 'perfect storm'. Firstly, and most important, fallowing periods have now dropped in much of Africa and Latin America from 15 years to 8, to 4 and now down to 2 years and, for many farmers, to zero. Since fallowing has been the primary way in which farmers kept their land fertile for millennia, this is a major tragedy. At the same time, animal manure is scarcer because large amounts of common pasturelands have been turned into fields. Chemical fertilizers have increased in price over the last couple of decades and global warming, among other things, means that even the weeds produce less biomass. In all, soil fertility, especially soil organic matter levels, is taking a beating that is unprecedented in the history of smallholder agriculture.

GMCC, by comprising what amounts to an improved fallow, can repair the damage done by the loss of fallowing. In traditional fallowing, farmers had to leave about ¾ of their farms idle. Now, with GMCC, they can bring the fallow process right into their fields, cropping their land at the same time that they feed it with fertile leaves, creating, in a sense, a 'simultaneous fallow'.

NITROGEN FIXATION The most common species of GMCC that are used – cowpeas (*Vigna unguiculata*), green beans (*V. radiata*), pigeon peas, lablab beans, mucuna, jack beans (*Canavalia ensiformis*), tephrosia (*Tephrosia vogelii* or *T. candida*) and runner beans – fix anywhere from 80 kg to 250 kg N/ha/season (Monegat, 1991). That means they can all lose even half of their nitrogen to volatilization (which will inevitably occur when they are left on the soil as part of a mulch) and still have the 40 kg N/ha/season needed to feed most smallholders' crops.

WEED CONTROL Another factor that occasionally causes problems in CA is that, without tillage, weeds can become a problem. Of course, CA mulches reduce weeding labour significantly, but the 'green mulches' and dry mulches of GMCC can usually help to reduce weed problems even more. GMCC species like mucuna, lablab, jack beans and runner beans are excellent at controlling weeds and in many cases can rid fields entirely of very noxious weeds. Striga and nutgrass are two noxious weeds that can be eliminated entirely with an adequate management of GMCC.

OTHER BENEFITS In addition to everything above, GMCC can provide high-protein food, wasteland restoration, a light shade for other crops ('dispersed' shade), soil moisture conservation, high-quality fodder for grazing animals, a reduction in pests and plant diseases (including nematodes and corn borer worms), medicinal herbs and firewood.

What are the challenges?

OPPORTUNITY COST Non-food-producing GMCC cannot be grown on land that has an opportunity cost. Farmers will never give a higher priority (nor should they) to GMCC than they do to food or cash crops. Thus, the GMCC must be grown on land, or at times, or in ways, such that the GMCC do not interfere with the other uses of the land. The GMCC have to fit into the farming system, rather than the farming system having to accommodate the GMCC. This sounds very difficult to achieve but GMCC can be intercropped with other crops, or grown during the dry season, or when there is too much rain for other crops, or when there are frosts. They can also be grown under trees or on wastelands that are being recuperated. Maize fields, so very common in Africa, are an ideal place for intercropping GMCC, unless the fields are already being intercropped with something else.

SLOW RESULTS Normally, the results of GMCC on increasing yields are not seen until the following cropping season. This means that in many cases no increase in yields is observed for 15 months or longer. Farmers can lose patience with technologies that take so long. However, some of the benefits of CA were not seen for some time, either, so those farmers using CA are already accustomed to waiting to see benefits. Also, there are ways of demonstrating to farmers what GMCC can do. Even given all this, farmers prefer to see concrete and significant results sooner.

DRY SEASON PROBLEMS The most common problem with GMCC is that of growing GMCC like cowpeas, green beans, groundnuts, or mucuna, and then letting the residues lie on the ground through a 6-month dry season before the rains return. During 6 months of hot weather, most of the nitrogen is volatilized and much of the biomass burned off, so that when the crops are planted again, almost nothing is left to fertilize the crop. Either shade must be provided (i.e. a cooler environment), or GMCC be used that will survive the dry season and still be green when it is time to cut them down before the crops are planted – like lablab beans, jack beans, tephrosia, late-planted mucuna or ratooned pigeon peas. With some of these plants, like lablab beans, this means that they will also provide a very good 'green mulch' throughout most of the dry season.

DIFFICULT GROWING CONDITIONS Smallholder farmers often have to work under very difficult conditions, including in drought-prone areas, or on extremely acid and depleted soils, etc. These conditions affect GMCC species just as much as they do subsistence or cash crops. Each of these problems must be solved in a different way.

SYNCHRONIZATION Sometimes the nutrients provided by the GMCC are not available when farmers' crops most need them, so the crops cannot take full advantage of them. Sometimes this problem can be solved by changing the GMCC system; at other times it must be solved by supplementing the nutrients available to the plants, using very small amounts of foliar sprays made from plant extracts, animal manure or small amounts of chemical fertilizer.

8.3.4 The most promising GMCC systems for farmers using CA

Choosing the right system for each situation is probably the most difficult factor in incorporating GMCC into CA. The number of different possible GMCC systems for smallholder farmers is over 90. By far the best way of choosing the best system(s) for a given area is to study this great variety of systems before choosing any single one. Bunch (2012) described most of these systems and under what conditions each one can be used. Nevertheless, there are a few systems that will be of some use fairly widely.

The easiest case is that of areas above 1500 m in elevation. In this case, runner beans (*Phaseolus coccineus*) can often be intercropped with maize. The runner bean produces a great deal of biomass, covers the soil well, can maintain the soil for 20 years of growing maize every year and produces a bean the taste of which is preferred in most parts of the world over common beans – a fact that is usually reflected in a higher price for the farmer. The only problem with runner beans is that most varieties are climbers and the bean produces so much biomass that it can cause the maize to lodge. If possible, it would be advantageous to procure seeds of bushy-type runner beans found in Kenya, around the town of Thika, or in Zimbabwe. The white-seeded varieties also have a very good international market as green pods (Joseph Mureithi, Kenyan Agricultural Research Institute (KARI), personal communication, 2014). Otherwise, this bean should be planted at a rate of only one seed for every 20 m^2 of land.

For lower altitudes, the best possibilities will depend on a whole series of factors. For areas where grazing animals are not common or they are not allowed to graze freely during the dry season, probably the legume with the greatest potential is the lablab bean. It produces a good deal of biomass and produces an edible bean eaten in parts of Peru, Malawi, Mozambique, Uganda, Kenya and Thailand. The leaves and immature beans (used like common peas) are also quite tasty. In Kenya the lablab bean is prized in much of the country and is sold in even the most upscale supermarkets. It is also an excellent, palatable fodder, with the whole plant having a protein content of 23%. Lablab beans can perfectly well be intercropped with maize and will raise maize yields quite quickly. The main drawback of lablab beans is that they require a fairly fertile soil. In a poor soil, they will not grow well until the second or third year.

Where cattle roam free and CA plots are not protected, legumes that are resistant to cattle will have to be used. The best candidates for this situation include tephrosia (*Tephrosia vogelii* or *T. candida*) and jack beans (*Canavalia ensiformis*). Both of these legumes can be intercropped with maize and allowed to grow throughout the dry season.

In particularly difficult situations, such as drought-prone areas or where the soil is highly depleted, or even on wastelands, jack bean is by far the best species. It produces a large amount of biomass (though it does not control weeds as well as mucuna, lablab or runner beans) and usually grows clear through the dry season. It fixes around 240 kg N/ha/season in many situations and is highly resistant to drought, even when only a few weeks old. It is also highly resistant to degraded soils, which makes it ideal for recuperating wastelands. It can be associated with maize, sorghum, millet, or even cassava, as long as care is taken to use only the bushy type. Jack bean has no other uses (though the long pod can be used as firewood), but after 2–3 years it can restore even the worst land to the point that other, more useful GMCC can be used.

In all lowland areas (below 1000 m in elevation), programmes should also disseminate the use of fertilizer trees in what is called CA with trees (CAWT), which, in fact, is another way of incorporating GMCC into CA. This practice is highly recommended in the lowland tropics, where the hot sun dries out the soil because it increases evaporation and transpiration rates. It also burns off organic matter, volatilizes nitrogen and causes crops to stop growing in the middle of the day, which just by itself can decrease crop production by 30%.

Probably the best technology to use in this case is to plant mother of cacao trees (*Gliricidia sepium*) in rows about 10 m apart, with the trees spaced each 5 m apart within the row. Mother of cacao is highly drought-resistant and within 2 years under favourable conditions will produce a large tree 6 m tall. In poor soils with no irrigation and around 500 mm annual rainfall, it will still produce a tree 6 m tall in about 5 years. The leaves are very good for fodder and for fertilizing the soil. The branches provide good firewood, the flowers are edible by humans (they are widely eaten in El Salvador and southern Honduras) and the bark can be used to kill rats and mice. It is best to plant the tree using stakes (also called 'cuttings'), to avoid the labour of making a nursery. Furthermore, trees planted by stakes will grow out of reach of grazing animals by the second dry season. The largest problem with mother of cacao is that it has to be treated for termites when it is planted by stakes; it also has to be protected from animals during the first dry season, or maybe two dry seasons, if conditions are particularly difficult and seedlings are used.

8.4 Conclusions

A lot will have to be learned for GMCC to be used extensively all over Africa. Knowledge on the part of extension agents is usually the limiting factor in their spread. However, once programmes have identified the best systems and have obtained initial seed supplies, GMCC are relatively simple for farmers to adopt and manage and will usually become a major factor in motivating farmers to plant more and more of their land to CA. Even more important, the tremendous advantages of CA with GMCC should cause CA to spread spontaneously from farmer to farmer, just as it has done in the past in countries from Brazil to Cameroon and from Kenya to Vietnam.

References

Bationo, A. (2012) *Lessons Learned from Long-term Soil Fertility Management Experiments in Africa*. Springer, Dordrecht, The Netherlands.

Bunch, R. (2012) *Restoring the Soil. A Guide for Using Green Manure/Cover Crops to Improve the Food Security of Smallholder Farmers*. Canadian Foodgrains Bank, Winnipeg, Manitoba, Canada, pp. 57–75.

Calegari, A. (2012) *Conservation Agriculture in Latin America and the Caribbean*. Available at: https://dl.sciencesocieties.org/publications/meetings/download/pdf/2012am/70994. (Accessed 22 September 2016.)

Eilitta, M. (2004) *Green Manure/Cover Crop Systems of Smallholder Farmers. Experiences from Tropical and Subtropical Regions*. Kluwer Academic Publishers, Dordrecht, The Netherlands, pp. 1–36.

Monegat, C. (1991) *Plantas de Cobertura del Suelo, Caracteristicas y Manejo en Pequenas Propiedades*. Claudino Monegat, Chapeco, Brazil, pp. 140–141.

Triomphe, B. (1996) Seasonal nitrogen dynamics and long-term changes in soil properties under the mucuna/maize cropping system on the hillsides of Northern Honduras. Dissertation for PhD, Cornell University, Ithaca, New York, pp. 110–122.

9 Climate-smart Push–Pull: A Conservation Agriculture Technology for Food Security and Environmental Sustainability in Africa

ZEYAUR R. KHAN,[1]* CHARLES A.O. MIDEGA,[1] JIMMY O. PITTCHAR,[1] ALICE MURAGE[2] AND JOHN PICKETT[3]

[1]*International Centre of Insect Physiology and Ecology (ICIPE), Nairobi, Kenya;* [2]*Kenya Agricultural and Livestock Research Organization, Naivasha;* [3]*Rothamsted Research, Harpenden, UK*

9.1 Introduction

Developing adaptable and productive agricultural systems that are resilient in the face of the risks and shocks associated with long-term climate variability is essential to maintaining food production into the future (Pretty *et al.*, 2011; Khan *et al.*, 2014), but resilience is not enough. Climate-smart agricultural systems also need to protect and enhance natural resources and ecosystem services in ways that mitigate future climate change effects (Tittonel and Giller, 2012). The International Centre of Insect Physiology and Ecology (ICIPE, or *icipe*) and partners developed a conservation agricultural technology, push–pull (www.push-pull.net), which responds to these needs in mixed farming systems practised in Africa (Midega *et al.*, 2015a). The push–pull technology fits Conservation Agriculture (CA) principles of minimum soil disturbance in its minimum soil tillage agronomic management, continuous soil cover with a perennial cover crop and plant residue, as well as a diversified cereal–legume–fodder intercropping strategy.

9.1.1 Africa's challenge of low productivity and poverty

Africa faces particular problems in feeding its population and remains the only continent in which per capita food production has been on the decline over

*zkhan@icipe.org

the past decades (World Bank, 2008). In Africa, agriculture still accounts for over 65% of full-time employment, 25–30% of gross domestic product (GDP) and over half of total export earnings (IFPRI, 2004; World Bank, 2008). It underpins the livelihoods of over two-thirds of Africa's poor, as smallholder agriculture remains the main source of household nutrition and incomes. Although some improvements have been made in African agriculture, high population growth and pressure on land and resources have reduced the per capita availability of domestically grown food (Pretty *et al.*, 2011). As a result, hunger and poverty remain widespread because of low agricultural productivity, with estimates indicating that about 265 million people in sub-Saharan Africa (SSA) face hunger (FAO, 2004).

Land degradation, pests and weeds hamper efficient production of cereals, particularly maize and sorghum, the main staple and cash crops for millions of smallholder farmers in SSA. Low and declining yields are affecting food security, nutrition and incomes, trapping farmers in poverty and poor health. The resource-constrained smallholder farmers living in the arid and semi-arid regions who practise mixed crop–livestock systems are particularly badly affected (FAO, 2004; Khan *et al.*, 2014). Indeed, projections indicate that, unless drastic steps are taken, SSA will have more than 500 million food-insecure people by 2020 (USDA, 2010).

In addition to widespread poverty (with more than 60% of the SSA population living on less than $40 per month), population pressure on land is high. Landholdings commonly amount to just 1 ha or less. Soils are severely degraded and have low organic matter as a result of continuous monocropping (Sanchez, 2002; Oswald, 2005; Rodenburg *et al.*, 2005). Many fields are heavily infested with parasitic striga weeds, while insect pests – principally stem borers – devastate cereal crops, commonly causing over half the harvest to be lost (Kfir *et al.*, 2002). In addition, farmers are now facing unpredictable rainfall and rising temperatures. Food insecurity is common, with a critical shortage of cereals in almost 70% of rural households. These constraints are expected to increase during the next few decades as agriculture intensifies to meet the extra food demand from a growing population and as a result of climate change (Fischer *et al.*, 2005).

9.1.2 The compounding effects of climate change

Climate change is anticipated to have far-reaching effects on the sustainable development of SSA and global efforts to achieve the Millennium Development Goals (MDGs) and post-MDG targets (IPCC, 2007; Khan *et al.*, 2014). Studies have predicted the effects of climate change on Africa's major cereal-growing regions by calculating the percentage overlap between historical growing-season temperatures (for 1960/2002) and the values projected for 2025, 2050 and 2075 (IPCC, 2007). The results indicate that temperatures will overlap, on average, by 58% with the historical observations by 2025, 14% by 2050 and 3% by 2075. This suggests that, within two decades, growing-season average temperatures will be warmer than those of 1960/2002 for 4 years in 10 for the majority of Africa's maize growing areas by 2025, growing to nearly 9 years in 10 by 2050 and nearly 10 years out of 10 in 2075 (Burke *et al.*, 2009).

Furthermore, climate predictions indicate that rainfall will become progressively more unpredictable, with a decline in yearly totals and an increasing incidence of floods and droughts. Together with the rising temperatures, this will lead to worsening land degradation and insect pest and weed pressure, with crop failure occurring more often, exacerbating food insecurity. To adapt to these adverse conditions, resource-constrained smallholder farmers will need to change their systems to incorporate cereal crops with greater drought resistance (e.g. sorghum and millet) and replace cattle with small ruminants for dairy production (Pretty *et al.*, 2011; Khan *et al.*, 2014).

9.1.3 Holistic agricultural intensification

From the foregoing, there is an urgent need for a significant, and sustainable, increase in grain yields and animal production. The need for sustainability requires ecologically sound ways of managing weeds and pests, with a strong focus on maintaining soil, crop and water resources. Therefore the solution lies in sustainable agricultural intensification that maximizes soil quality and crop productivity, adopting a systems approach (social, economic and environmental) to agricultural development, and making specific recommendations based on integrated analyses of specific agroecologies, locations and cultural preferences (IAASTD, 2009; Pinstrup-Andersen, 2010). Sustainable agricultural intensification is defined as producing more output from the same area of land while reducing the negative environmental impacts and at the same time increasing contributions to natural capital and the flow of environmental services (Royal Society, 2009; Conway and Waage, 2010; Godfray *et al.*, 2010).

Sustainable agriculture in this context will require a more holistic approach, reflecting the multifunctionality of agriculture, promoting value chain approaches, improving market access and developing the rural non-farm sector. Integration of a variety of packages of resource-conserving technologies and practices – conservation agriculture, integrated pest management (IPM), integrated soil fertility management (ISFM), agroforestry, aquaculture, water harvesting and livestock integration – is likely to increase productivity significantly and sustainably (Pretty *et al.*, 2006).

9.2 Push–Pull: A Broad-based Conservation Agriculture (CA) Solution

The push–pull technological innovation, developed by the International Centre of Insect Physiology and Ecology (ICIPE) (http://www.icipe.org) with Rothamsted Research (www.rothamsted.bbsrc.ac.uk) in the UK and partners in East Africa, addresses smallholder agricultural constraints, food insecurity, environmental degradation and loss of biodiversity. ICIPE and partners recognized that addressing the interrelated problems of soil degradation, high temperatures and water stress, and reduced yields due to biotic constraints, notably insect pests and weeds, through improved management strategies would significantly increase farm productivity, resulting in better nutrition and alleviating poverty for millions of farmers in SSA.

Push–pull is an innovation that holistically combines multifunctional resource-conserving IPM and ISFM approaches while making efficient use of natural resources to increase farm productivity by addressing most aspects of smallholders' constraints (Cook *et al.*, 2007; Hassanali *et al.*, 2008). Developed for systems based on minimum or no tillage, it provides continuous soil cover with a perennial cover crop and plant residue, as well as a diversified cereal–legume–fodder intercropping practice. The perennial intercrop provides live mulching, thus improving above-ground and below-ground arthropod abundance, agrobiodiveristy and the food web of natural enemies of stem borers (Khan, 2002; Khan *et al.*, 2006c; Midega *et al.*, 2008 2015a). The technology effectively controls the major insect pests of cereals in SSA, i.e. lepidopteran stem borers, and the devastating parasitic striga weeds, both of which can cause total yield loss to cereals. Furthermore, it improves soil health and conserves soil moisture. The technology involves the use of intercrops and trap crops in a mixed cropping system (Khan *et al.*, 2006a). These companion plants release behaviour-modifying stimuli (plant chemicals) to manipulate the distribution and abundance of stem borers and beneficial insects for management of the pests (Fig. 9.1). The system relies on an in-depth understanding of chemical ecology, agrobiodiversity, plant–plant and insect–plant interactions (http://www.push-pull.net/publications.shtml) and is well suited to African socio-economic conditions.

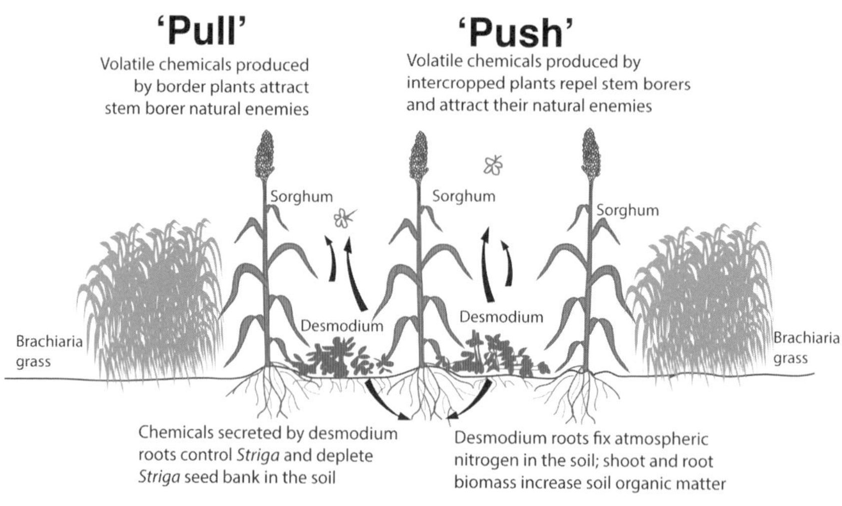

Fig. 9.1. How the push–pull technology works. Aerial responses are mediated by volatile organic compounds. In the rhizosphere, chemicals secreted by desmodium roots inhibit attachment of germinated striga to maize roots and cause rapid depletion of striga seeds in the soil.

The main cereal crop is planted with a repellent intercrop such as desmodium (push) and an attractive trap plant such as Napier grass (pull) planted as a border crop around this intercrop. Gravid stem borer females are repelled from the main crop and are simultaneously attracted to the trap crop (Cook *et al.*, 2007). The companion plants are valuable themselves as high-quality animal fodder, thereby facilitating livestock production.

The Napier grass trap crop produces significantly higher levels of volatile cues (stimuli), used by gravid stem borer females to locate host plants, than maize or sorghum (Birkett *et al.*, 2006). There is also an increase of approximately 100-fold in the total amounts of these compounds produced in the first hour of nightfall (scotophase) by Napier grass (Chamberlain *et al.*, 2006), the period at which stem borer moths seek host plants for oviposition (Päts, 1991), causing the differential oviposition preference. However, about 80% of the stem borer larvae do not survive (Khan *et al.*, 2006b) as Napier grass tissues produce sticky sap in response to feeding by the larvae, which traps them, causing their mortality. The intercrop legumes in the *Desmodium* genus (silverleaf, *D. uncinatum* and greenleaf, *D. intortum*), on the other hand, produce repellent volatile chemicals that push away the stem borer moths. These include (*E*)-ocimene and (*E*)-4,8-dimethyl-1,3,7-nonatriene, semiochemicals that are produced during damage to plants by herbivorous insects and are responsible for the repellence of desmodium to stem borers (Khan *et al.*, 2000). Desmodium also effectively controls striga weed, resulting in significant yield increases in maize from 1 t/ha to 3.5 t/ha per cropping season (Khan *et al.*, 2006c, 2008a).

In the elucidation of the mechanisms of striga suppression by *D. uncinatum*, in addition to benefits derived from increased availability of nitrogen and soil shading, an allelopathic effect of the root exudates of the legume, produced independently of the presence of striga, was found to be responsible for the dramatic reduction of striga weeds in an intercrop of desmodium with maize. Blends of secondary metabolites found in the root exudates of *D. uncinatum* have striga seed germination stimulatory and post-germination inhibitory isoflavones that directly interfere with striga parasitism (Tsanuo *et al.*, 2003). This combination thus provides a novel means of *in situ* reduction of the striga seed bank in the soil through efficient suicidal germination even in the proximity of graminaceous host plants. Other *Desmodium* spp. have also been evaluated and have similar effects on stem borers and striga (Khan *et al.*, 2006c) and are currently being used as intercrops in maize, sorghum and millets.

Extensive research and development (R&D) efforts revealed that not only were stem borers and striga effectively controlled by the technology under farmers' conditions, but farmers also reported additional benefits such as increased soil fertility, up to threefold increases in grain yields, and improved availability of animal fodder resulting in increased milk production (Khan *et al.*, 2008a,b; Midega *et al.*, 2014). All these gave significantly higher economic returns to the farmer, with cost–benefit analyses showing significantly higher returns to both land and labour than conventional farmers' practices.

The CA framework provides a set of principles in terms of yield improvements while conserving and enhancing the resource base (Lin *et al.*, 2011), including the principles of continuous minimum mechanical soil disturbance, permanent soil cover with plant residue and cover crops, and diversifying crop rotations, sequences and combinations, adapted to local socio-economic and environmental conditions, which contribute to maintaining biodiversity above and in the soil and help to avoid build-up of pest populations within the spectrum of soil inhabitants (FAO, 2008). In application of these CA principles, push–pull has been proven to have a positive contribution to cereal and livestock productivity through natural control of biotic constraints to cereal production, and soil health improvement. The technology has been shown to provide multiple benefits and other ecosystem services such as reduced soil erosion and improved water resources, lower production costs, and climate change adaptability and mitigation.

The technology thus has significant complementary and synergistic effects to CA-based agricultural systems. The push–pull technology fits in the CA production framework as it enables sustainable cereal crop and livestock production intensification through ecosystem or agroecological approaches based on natural biological processes (Khan *et al.*, 2014). This includes minimum tillage, as both desmodium intercrop and border crop are permanent, thus offering reduced tillage, the natural mulching effect of desmodium and improvement of soil biota and agrobiodiversity (Midega, 2015b). The technology controls insect pests using natural behavioural manipulation, not inorganic pesticides, and manages parasitic striga weeds through allelopathy.

Desmodium also fixes atmospheric nitrogen (110 kg N/ha) (Whitney, 1966), adds organic matter to the soil, conserves soil moisture and enhances soil biodiversity, thereby improving soil health and fertility. Additionally, it provides ground cover and, together with the surrounding Napier grass, protects the soil against erosion. It therefore improves agroecosystem sustainability and resilience, with great potential to mitigate the effects of climate change. The reliance on locally adapted companion plants and reduction of dependence on inorganic fertilizer reduces its carbon footprint, while enhancing both environmental and economic sustainability of the system.

Both desmodium and Napier grass provide valuable year-round quality animal forage whilst the sale of desmodium seeds generates additional income for the farmers. So far, on-farm uptake by over 130,000 farmers in eastern Africa has confirmed that the technology is very effective and has significant impacts on food security, human and animal health, soil fertility, the conservation of agrobiodiversity, provision of agroecosystem services, empowerment of women and income generation for resource-poor farmers (Fig. 9.2).

A number of studies have demonstrated that push–pull is more profitable than farmers' own practices and some of the practices designed to improve soil fertility. Significantly higher benefit–cost ratio was realized with push–pull compared with maize monocrop and/or use of pesticides, posting a positive return on investment of over 2.2 compared with 0.8 obtained with the maize monocrop, and slightly less than 1.8 for pesticide use (Khan *et al.*, 2001). Additionally, push–pull with no additional fertilizer had the best gross returns, while less profit was recorded with the use of fertilizer, implying it was economically propitious for poor smallholders who

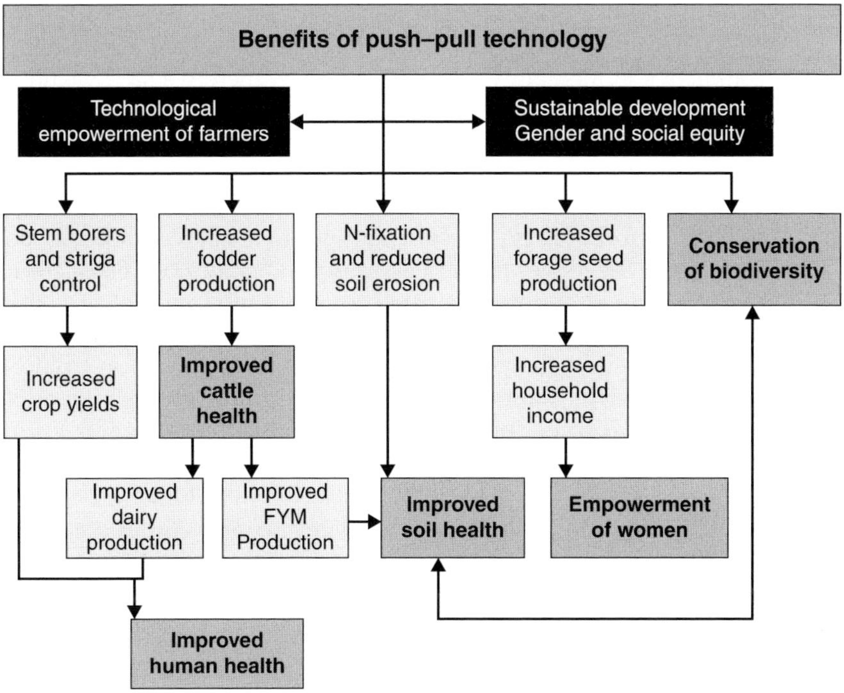

Fig. 9.2. Schematic presentation of the benefits of the push–pull technology. FYM, farmyard manure.

could not afford external inputs to invest in push–pull. In a more detailed economic analysis utilizing data of over seven cropping seasons, returns to investment for the basic factors of production under push–pull were significantly higher compared with those from maize–bean intercropping and maize monocrop systems (Khan *et al.*, 2008c). Positive total revenues ranged from \$351/ha in low potential areas to \$957/ha in the high potential areas, with general increases in subsequent years. The returns to labour that were recovered within the first year of establishment of push–pull ranged from \$0.5/man-day in the low potential areas to \$5.2/man-day in the higher potential areas, whereas in the maize monocrop this was negligible or even negative. Furthermore, the net present value (NPV) from push–pull was positive and consistent over several years. A more recent study (De Groote *et al.*, 2010), which used discounted partial budget and marginal analysis, corroborated these findings and concluded that push–pull earned the highest revenue compared with other soil fertility management technologies, including green manure rotation.

9.3 Climate-smart Technological Innovation

The new 'climate-smart' adaptation of the technology has unrivalled potential to equip farmers with the resilience and adaptability they need to deal with the additional problems associated with climate change (Khan *et al.*, 2014). With rising uncertainties in the region's rainfed agriculture due to the continent's vulnerability to climate change, there was a demand and urgent need to adapt the conventional push–pull to withstand increasingly hotter and drier conditions. The trap and intercrops used in conventional push–pull were rainfall and temperature limited, as the initial system was developed under average rainfall (800–1200 mm) and moderate temperatures (15–30C°). In order to ensure that push–pull continues to affect food security positively in Africa over the longer term, new drought-tolerant trap (*Brachiaria* cv. Mulato) and intercrop (drought-tolerant species of desmodium, e.g. *D. intortum*) plants have been selected from research undertaken with funding from the European Union. The new companion plants also have the appropriate chemistry in terms of stem borer attractancy for the trap component, and stem borer repellence and striga suppression, and ability to improve soil fertility and soil moisture retention, for the intercrop component (Khan *et al.*, 2014). Both trap and repellent plants provide high-quality livestock fodder over long periods of drought. In addition, they provide other ecosystem services such as biodiversity improvement and conservation, and organic matter improvement.

9.3.1 Identification of drought- and temperature-tolerant trap and repellent plants

ICIPE, Rothamsted Research and national partners in Ethiopia, Kenya and Tanzania identified and selected new drought- and temperature-tolerant trap and intercrop plants suitable for drier agroecologies. From a total of 400 grasses previously identified under different agroecologies in those countries, an initial 21 candidate drought-tolerant trap plants were selected. *Brachiaria* cv. Mulato was chosen as the trap plant for the climate-adapted push–pull, given also its ability to attract stem borers, farmers' preference for it as livestock fodder, and the commercial availability of its seed that would allow faster dissemination and uptake. Additionally, it allowed minimal survival of stem borer larvae, a suitable characteristic of a trap plant that would support populations of natural enemies within season and when the cereal crop is not in season.

Desmodium spp. were identified that are drought tolerant and emit volatiles that repel stem borers, fix nitrogen to improve soil fertility, produce high biomass but have low growth habit to cover the soil and improve soil health. Forty-three accessions of 17 *Desmodium* spp. initially collected from dry and hot areas in Africa and other arid environments were obtained from germplasm repositories[1] and from field survey samplings. Greenleaf desmodium (*D. intortum*) was observed to be more drought tolerant compared with silverleaf desmodium (*D. uncinatum*) and was chosen as the intercrop species for immediate integration into a climate-adapted push–pull. Greenleaf desmodium was chosen given its known ability to control striga and stem borers (Khan *et al.*, 2006c) coupled with commercial availability of its seed that would enable its wider testing by farmers within the project target areas. The work to isolate and purify all the active compounds in the desmodium root exudates and fully elucidate their effects on striga suppression is ongoing at ICIPE and Rothamsted Research. Similarly, the full mechanism of stem borer control by the new companion plants is currently being elucidated with the aim of providing both sustainability and quality assurance as more companion plants are selected for new agroecologies.

9.3.2 Field implementation of climate-smart push–pull technology

Currently over 53,000 smallholder farmers in drier parts of Kenya, Tanzania, Uganda and Ethiopia have taken up the climate-smart push–pull and have reported effective control of stem borers and striga weed resulting in significant increases in grain yields of both maize and sorghum (Midega *et al.*, 2015b). In both maize and sorghum trials the yield increase was more than threefold compared with the corresponding control plots (Fig. 9.3).

The new companion plants also ensure availability of high-quality fodder at the farms. This improves the productivity of livestock, while surplus fodder is preserved as hay. Validation of gross return from the climate-smart push–pull showed $1075.8/ha and $1289/ha gross benefits and a marginal rate of return (MRR) of 109% and 143% for sorghum and maize, respectively, implying that the net increase in benefits of climate-smart push–pull outweighed the net increase in costs compared with farmers' practices (Murage *et al.*, 2015). Additionally, over 80% of the interviewed farmers (350) reported effective control of striga and stem borers, improvement in soil fertility and improved grain yields. Other benefits mentioned included increases in fodder and milk production.

Climate-smart push–pull thus opens up significant opportunities for smallholder growth in drier areas and represents a platform technology around which new income generation and human nutritional components, such as livestock keeping, can be added. It therefore affords the smallholder farmers in these areas an opportunity to enter into cash economy. The technology can

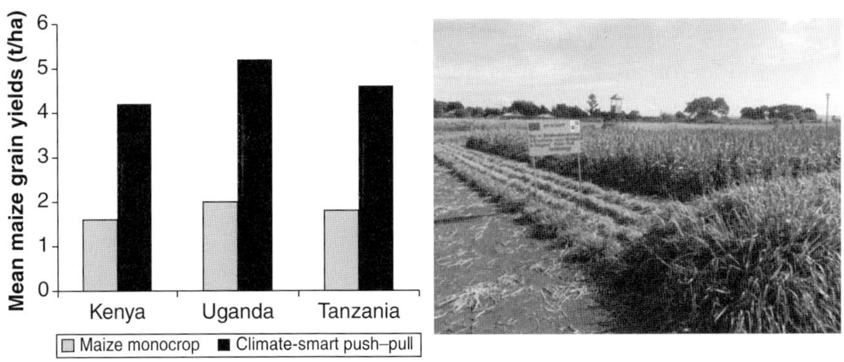

Fig. 9.3. On-farm results of maize yields from climate-smart push–pull plots planted with the drought-tolerant companion plants *Brachiaria* cv. Mulato as the trap plant and greenleaf desmodium as the intercrop plant.

also be tailored to specific farmer needs. Smallholder farmers in SSA typically practise multiple cropping, where cereals are intercropped with food legumes. The technology has been adapted to these farming systems by incorporating edible beans, planted either in the same hole with maize or between maize plants within a row. This has increased the technology's appeal to the farmers, as it guarantees an additional protein source in the diet (Khan *et al.*, 2009), resulting in higher technology adoption rates in the region.

9.4 Maintaining Push–Pull's Multiple Benefits

The new trap plants and intercrops have met the expectations of farmers and scientists, giving effective control of stem borer and striga weed. As a result, yields of maize and sorghum have increased significantly, with up to a fivefold increase over control plots (Midega *et al.*, 2015b) (Fig. 9.3). The new companion plants have also provided ample good-quality livestock fodder, producing enough to allow farmers to make hay for the dry season, and the better-fed dairy animals have produced more milk. In all, the system gives significantly higher economic returns to the farmer.

On-farm uptake of conventional and climate-adapted push–pull by over 130,000 farmers in eastern Africa has confirmed that push–pull has significant impacts on food security, human and animal health, soil fertility, income generation, empowerment of women, conservation of agrobiodiversity and provision of agroecosystem services. Climate-smart push–pull will spread these benefits wider, conferring the benefits to additional crops and agroecologies.

9.4.1 Food security, diet and health

Dealing with striga and stem borer means a healthier, more productive cereal crop. Farmers who have adopted climate-smart push–pull reported steady increases in the amount of grain they harvested. On-farm research has confirmed yield increases, with the climate-smart push–pull routinely producing more than double the amount harvested from an equivalent area of monocrop maize across fields in eastern Africa (Midega *et al.*, 2015b). The practising farmers report that their health and that of their families has improved since adopting the technology, because it has resulted in an improved diet, particularly through drinking more milk. Dietary diversity has also increased. Due to increased income from the sale of push–pull products and by-products, many farmers report that they are in a better position to purchase foods that they are not able to produce for themselves.

9.4.2 Protecting the soil

The technology contributes to the CA principle of conserving and enhancing the resource base, with continuous minimum mechanical soil disturbance, continuous soil cover, addition of organic matter, prevention of loss of top soil through soil erosion, improved water conservation and other ecosystem services. Practising farmers describe improvements in soil fertility after adopting the technology. Protected from erosion by the trailing low-growing stems and leaves of the desmodium intercrop, this soil is one of the keys to the increased grain yields that farmers experience. The majority of adopters incorporate the technology into an integrated crop–livestock production system. This close association with livestock means that farmyard manure can be added to the soil, increasing the fertility benefits already gained from the fixation of nitrogen by the desmodium intercrop. Most of the farmers thus notice an improvement in their soil within a very short time of adopting the technology.

9.4.3 Providing nutritious fodder

Livestock fulfil many purposes in the livelihood systems of farming households. They provide milk, meat, manure and draught power. Their outputs become goods for sale and exchange and the animals themselves represent a form of savings. Keeping animals well fed and healthy is often pivotal in maintaining soil fertility, paying school fees and eating a nutritious, balanced diet.

These farmers use their fodder crops to feed goats, sheep, cattle, pigs, poultry and even rabbits. Many farmers report positive changes in the health and productivity of their animals, particularly due to the nutritional qualities of desmodium. Because it is rich in protein, desmodium fodder often has a positive effect on milk yields, frequently doubling or even tripling them.

9.4.4 Generating income

There are a number of ways that the technology generates cash income, including the sale of cereals, milk and fodder. Increased income is often spent on school fees or home improvements, or invested in diversifying livelihood enterprises. There are also many examples of income being invested in the community and upholding social safety nets to protect the vulnerable.

9.4.5 Gender equity

Once established, the technology reduces the drudgery of digging and weeding, a task performed most often by women, freeing up their time and labour for more productive tasks like selling milk or starting a poultry enterprise (Fig. 9.4). Diversified farm income means that there is more money available to buy medicines, household goods and other essentials. Stall-feeding dairy cattle also frees the women and children from the task of herding cattle to graze.

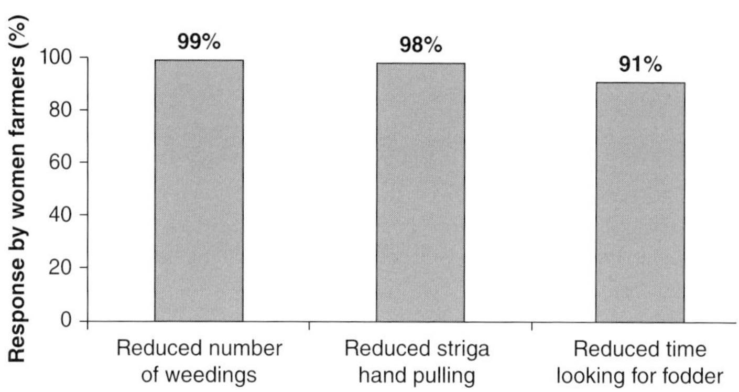

Fig. 9.4. Perception of 600 female farmers in East Africa on impact of climate-smart push–pull technology on labour dynamics.

9.4.6 More stable and resilient agroecosystems

As far as possible, climate-resilient agroecosystems maintain the functions and services provided by natural systems. This means integrating instead of segregating, closing water and nutrient cycles, increasing biological and genetic diversity and regenerating instead of degrading bio-resources. Push–pull technology contributes to stable and climate-resilient agroecosystems by providing farmers with a tool for on-farm diversification that is in line with these underlying principles.

Producing a good grain harvest is increasingly becoming a question of gambling with the weather as climatic patterns become more unpredictable. Farmers have for many years habitually diversified the crops they plant as an insurance strategy against climate uncertainty. Push–pull reinforces this strategy, because it can equally usefully be applied to maize, sorghum, millet and rice. Furthermore, having both conventional and climate-smart variants widens the range of planting material that farmers can use to tailor their cereal cropping practices to local climatic conditions.

9.5 Partnerships in Implementation

The push–pull development programme has brought together a wide range of contributors from disciplines as diverse as chemical ecology, organic chemistry, socio-economics and weed science to develop a systemic approach to solving the problems of stem borer and striga that works for farmers.

While push–pull is in many ways an elegantly simple technology, it is based on a set of complex ecological and chemical relationships between plants, insects and soil. Scientists at ICIPE have found that it works best in practice when farmers understand clearly why it works and what each component in the system contributes to its overall success. This means that disseminating push–pull has been a knowledge-intensive process, with a strong emphasis placed on building farmers' capacities.

As well as being fully trained in all aspects of push–pull, farmers have been involved at every stage of the research and dissemination over the past 15 years. The result is a cohort of farmers who understand basic agroecology, supported by a vibrant network of farmer-leaders and farmer-teachers who have become experienced peer educators. Most of them have visited the push–pull garden at ICIPE's Mbita Point research station and many have taken part in technology development trials.

The district-based ICIPE field workers usually work with farmer groups set up for mutual support and self-help and these have become a well established medium for delivering agricultural extension and other development interventions. In some places, push–pull triggers the formation of new farmers' groups, while in others existing groups incorporate push–pull into their own portfolio of activities.

One hallmark of the successful spread of push–pull has been ICIPE's capacity to identify and work in harmony with the groups and organizations it meets in the field. Exploiting synergies with other active research and development organizations has created new channels for sharing knowledge. Particularly important is ICIPE's partnership with Heifer Project International (HPI) and Send a Cow, non-governmental organizations (NGOs) whose livestock-focused work has proved to be a good fit with push–pull. A strong informal partnership was cemented when the NGOs formally became implementing partners for climate-smart push–pull in East Africa.

9.6 Developing Sustainable Adaptable Practices for the Future

One of push–pull's strengths is the way the programme has been managed as a learning process. Because farmer participation is built into the processes of research and dissemination, contextual changes encountered in the field can be communicated, discussed and responded to. Achieving this level of reflexivity to contextual change is a vital aspect of the climate-smart qualities of push–pull and is thanks in no small part to the flexible approaches of the many donor organizations that have funded its development and spread. Both the technology and its model for dissemination represent a substantial resource for the future.

The development of climate-smart push–pull has made it possible for the technology to be implemented in new areas with less rainfall and to increase the potential number of farmers who might find it a useful and profitable addition to their livelihood strategies. Work has begun to extend climate-smart push–pull in Tanzania, Uganda and Ethiopia and trials are being carried out in Nigeria, Zambia and Malawi. Push–pull technology is expanding to a broader range of agroecosystems and farming practices in Africa.

There is still great need for adaptive agricultural practices that can cope with increasingly variable climatic conditions and still produce food for people and livestock. Similarly there is need for development pathways that respect ecological limits and restore ecosystem health. Experiences with push–pull offer important lessons about developing and implementing the kind of climate-smart technologies that are needed to meet these challenging goals.

Note

[1] Desert Legumes Program (DELEP) at the University of Arizona, USA; ILRI forage plants gene bank in Ethiopia; and the US Department of Agriculture (USDA) Agricultural Research Service (ARS) Plant Genetic Resources Conservation Unit (PGRCU).

References

Birkett, M.A., Chamberlain, K., Khan, Z.R., Pickett, J.A., Toshova, T., Wadhams, L.J. and Woodcock, C.M. (2006) Electrophysiological responses of the lepidopterous stemborers *Chilo partellus* and *Busseola fusca* to volatiles from wild and cultivated host plants. *Journal of Chemical Ecology* 32, 2475–2487.

Burke, M.B., Lobell, D.B. and Guarino, L. (2009) Shifts in African crop climates by 2050, and the implications for crop improvement and genetic resources conservation. *Global Environmental Change* 19, 317–325.

Chamberlain, K., Khan, Z.R., Pickett, J.A., Toshova, T. and Wadhams, L.J. (2006) Diel periodicity in the production of green leaf volatiles by wild and cultivated host plants of stemborer moths, *Chilo partellus* and *Busseola fusca*. *Journal of Chemical Ecology* 32, 565–577.

Conway, G.R. and Waage, J. (2010) *Science and Innovation for Development*. UK Collaborative on Development Sciences (UKCDS), London.

Cook, S.M., Khan, Z.R. and Pickett, J.A. (2007) The use of 'push–pull' strategies in integrated pest management. *Annual Review of Entomology* 52, 375–400.

De Groote, H., Vanlauwe, B., Rutto, E., Odhiambo, G.D., Kanampiu, F. and Khan, Z.R. (2010) Economic analysis of different options in integrated pest and soil fertility management in maize systems of western Kenya. *Agricultural Economics* 41, 471–482.

Fischer, G., Shah, M., Tubiello, F. and van Velthuizen, H. (2005) Socioeconomic and climate change impacts on agriculture: an integrated assessment, 1990–2080. *Philosophical Transactions of the Royal Society of London. Series B, Biological Sciences* 360, 2067–2083.

Food and Agriculture Organization (2004) *Status of Food and Agriculture*. FAO, Rome, Italy.

Food and Agriculture Organization (2008) *Investing in Sustainable Intensification – The Role of Conservation Agriculture*. FAO, Rome.

Godfray, C., Beddington, J.R., Crute, I.R., Haddad, L., Lawrence, D., Muir, J.F., *et al.* (2010) Food security: the challenge of feeding 9 billion people. *Science* 327, 812–818.

Hassanali, A., Herren, H., Khan, Z.R., Pickett, J.A. and Woodcock, C.M. (2008) Integrated pest management: the push–pull approach for controlling insect pests and weeds of cereals, and its potential for other agricultural systems including animal husbandry. *Philosophical Transactions of the Royal Society of London. Series B, Biological Sciences* 363, 611–621.

IAASTD (2009) *Agriculture at a Crossroads, International Assessment of Agricultural Knowledge, Science and Technology for Development*. Island Press, Washington, DC.

IFPRI (2004) *Ending Hunger in Africa: Prospects for the Small Farmer,* Policy Brief. International Food Policy Research Institute, Washington, DC.

IPCC (2007) *Climate change 2007: Impacts, Adaptation and Vulnerability*. Contribution of Working Group II to the Fourth Assessment Report of the Intergovernmental Panel on Climate Change.

Kfir, R., Overholt, W.A., Khan, Z.R. and Polaszek, A. (2002) Biology and management of economically important lepidopteran cereal stemborers in Africa. *Annual Review of Entomology* 47, 701–731.

Khan, Z.R. (2002) Cover crops. In: Pimental, D. (ed.) *Encyclopedia of Pest Management*. Marcel Dekker Inc., New York.

Khan, Z.R., Pickett, J.A., Van den Berg, J., Wadhams, L.J. and Woodcock, C.M. (2000) Exploiting chemical ecology and species diversity: stemborer and striga control for maize and sorghum in Africa. *Pest Management Science* 56, 957–962.

Khan, Z.R., Pickett, J.A., Wadhams, L.J. and Muyekho, F. (2001) Habitat management strategies for the control of cereal stemborers and striga in maize in Kenya. *Insect Science and Its Application* 21, 375–380.

Khan, Z.R., Hassanali, A. and Pickett, J.A. (2006a) Managing polycropping to enhance soil system productivity: a case study from Africa. In: Uphoff, N., Ball, A.S., Palm, C., Fernandes, E., Pretty, J., Herren, H., *et al.* (eds) *Biological Approaches to Sustainable Soil Systems*. CRC Press, Taylor and Francis, Boca Raton, Florida, pp. 575–586.

Khan, Z.R., Midega, C.A.O., Hutter, N.J., Wilkins, R.M. and Wadhams, L.J. (2006b) Assessment of the potential of Napier grass (*Pennisetum purpureum*) varieties as trap plants for management of *Chilo partellus*. *Entomologia Experimentalis et Applicata* 119, 15–22.

Khan, Z.R., Pickett, J.A., Wadhams, L.J., Hassanali, A. and Midega, C.A.O. (2006c) Combined control of striga and stemborers by maize–*Desmodium* spp. intercrops. *Crop Protection* 25, 989–995.

Khan, Z.R., Midega, C.A.O., Amudavi, D.M., Hassanali, A. and Pickett, J.A. (2008a) On-farm evaluation of the 'push–pull' technology for the control of stemborers and striga weed on maize in western Kenya. *Field Crops Research* 106, 224–233.

Khan, Z.R., Amudavi, D.M., Midega, C.A.O., Wanyama, J.M. and Pickett, J.A. (2008b) Farmers' perceptions of a 'push–pull' technology for control of cereal stemborers and striga weed in western Kenya. *Crop Protection* 27, 976–987.

Khan, Z.R., Midega, C.A.O., Njuguna, E.M., Amudavi, D.M., Wanyama, J.M. and Pickett, J.A. (2008c) Economic performance of 'push–pull' technology for stemborer and striga weed control in smallholder farming systems. *Crop Protection* 27, 1084–1097.

Khan, Z.R., Midega, C.A.O., Wanyama, J.M., Amudavi, D.M., Hassanali, A., Pittchar, J. and Pickett, J.A. (2009) Integration of edible beans (*Phaseolus vulgaris* L.) into the push–pull technology developed for stemborer and striga control in maize-based cropping systems. *Crop Protection* 28, 997–1006.

Khan, Z.R., Midega, C.A.O., Pittchar, J.O., Murage, A.W., Birkett, M.A., Bruce, T.J.A. and Pickett, J.A. (2014) Achieving food security for one million sub-Saharan African poor through push–pull innovation by 2020. *Philosophical Transactions of the Royal Society of London. Series B, Biological Sciences* 369, 20120284.

Lin, S.J., Friedrich, T., Hansmann, B. and Kassam, A. (2011) *A Framework for International Standard for Conservation Agriculture: Towards Stewardship for Sustainable Production*. Australian Centre of International Agricultural Research, Canberra, Australia.

Midega, C.A.O., Khan, Z.R., van den Berg, J., Ogol, C., Dippenaar-Schoeman, A.S., Pickett, J.A. and Wadhams, L.J. (2008) Response of ground-dwelling arthropods to a 'push–pull' habitat management system: spiders as an indicator group. *Journal of Applied Entomology* 132, 248–254.

Midega, C.A.O., Salifu, D., Bruce, T.J.A., Pittchar, J., Pickett, J.A. and Khan, Z.R. (2014) Cumulative effects and economic benefits of intercropping maize with food legumes on *Striga hermonthica* infestation. *Field Crops Research* 155, 144–152.

Midega, C.A.O., Bruce, T.J., Pickett, J.A. and Khan, Z.R. (2015a) Ecological management of cereal stemborers in African smallholder agriculture through behavioural manipulation. *Ecological Entomology* 40, Suppl. S.1, 70–81.

Midega, C.A.O., Bruce, T.J, Pickett, J.A., Pittchar, J.O., Murage, A. and Khan, Z.R. (2015b) Climate-adapted companion cropping increases agricultural productivity in East Africa. *Field Crops Research* 180 (2015) 118–125.

Murage, A.W., Midega, C.A.O., Pittchar, J.O., Pickett, J.A. and Khan, Z.R. (2015) Determinants of adoption of climate-smart push–pull technology for enhanced food security through integrated pest management in eastern Africa. *Food Security* 7, 709–724.

Oswald, A. (2005) Striga control technologies and their dissemination. *Crop Protection* 24, 333–342.
Päts, P. (1991) Activity of *Chilo partellus* (Lepidoptera: Pyralidae): eclosion, mating and oviposition time. *Bulletin of Entomological Research* 81, 93–96.
Pinstrup-Andersen, P. (ed.) (2010) *The African Food System and Its Interaction with Human Health and Nutrition*. Cornell University Press, New York.
Pretty, J., Noble, A.D., Bossio, D., Dickson, J., Hine, R.E., Penning de Vries, F.W.T. and Morrison, J.I.L. (2006) Resource-conserving agriculture increases yields in developing countries. *Environmental Science and Technology* 40, 1114–19.
Pretty, J., Toulmin, C. and Williams, S. (2011) Sustainable intensification in African agriculture. *International Journal of Agricultural Sustainability* 9, 5–24.
Rodenburg, J., Bastiaans, L., Weltzien, E. and Hess, D.E. (2005) How can selection for striga resistance and tolerance in sorghum be improved? *Field Crops Research* 93, 34–50.
Royal Society (2009) *Reaping the Benefits: Science and the Sustainable Intensification of Global Agriculture*. Royal Society, London.
Sanchez, P. (2002) Soil fertility and hunger in Africa. *Science* 295, 2019–2020.
Tittonel, P. and Giller, K. (2012) When yield gaps are poverty traps: the paradigm of ecological intensification in African smallholder agriculture. *Field Crops Research* 143, 76–90.
Tsanuo, M.K., Hassanali, A., Hooper, A.M., Khan, Z.R., Kaberia, F., Pickett, J.A. and Wadhams, L. (2003) Isoflavanones from the allelopathic aqueous root exudates of *Desmodium uncinatum*. *Phytochemistry* 64, 265–273.
USDA (2010) *Food Security Assessment 2010–2020 Report*. Washington DC, USA: USDA Economic Research Service.
Whitney, A.S. (1966) Nitrogen fixation by three tropical forage legumes and the utilization of legume-fixed nitrogen by their associated grasses. *Herbage Abstracts* 38, 143.
World Bank (2008) *World Development Report 2008: Agriculture for Development*. The World Bank, Washington, DC.

10 How to Make Conservation Agriculture EverGreen

Dennis P. Garrity*

World Agroforestry Centre, Nairobi, Kenya

10.1 Introduction

After decades of research, and the sustained efforts of many pioneering farmers, the concept of Conservation Agriculture (CA) has been steadily expanding (Kassam *et al.*, 2015). Globally, more than 155 million hectares of annual cropland are now managed under zero-tillage CA systems. Meanwhile, worldwide concerns about the potentially devastating effects of climate change on food production continue to intensify. CA has been highlighted as an important component of climate-smart agriculture (CSA) (FAO, 2013; Lipper *et al.*, 2014).

Investments in CA in the developing world are increasing. However, the uptake of CA in Africa, and in the rainfed upland areas of Asia, has been quite modest so far. Evidence from research, and from widespread indigenous practice, indicates that successful CA systems for tropical smallholders benefit substantially from the integration of trees into these systems (Garrity *et al.*, 2010). EverGreen Agriculture (EVA) is defined as a particular type of agroforestry practice in which woody perennials are integrated directly into crop fields. When EVA is combined with CA practices it is known as Conservation Agriculture with Trees (CAWT). Such an EverGreen Conservation Agriculture (ECA) addresses a number of the critical constraints to sustained smallholder CA uptake, by increasing and sustaining crop productivity in these systems and by buffering them in the face of drought, higher temperatures and long-term climate change. The success of ECA is now beginning to be observed at scale in several countries in Africa (Fig. 10.1).

There are three long-established principles in CA as described on part of the comprehensive FAO website dedicated to CA (www.fao.org/ag/ca):

*d.garrity@cgiar.org

Fig. 10.1. EverGreen Conservation Agriculture with *Faidherbia albida* trees is now practised by tens of thousands of farmers in Southern Africa.

(i) no or minimum mechanical soil disturbance; (ii) maintenance of soil mulch cover with crop residues, stubbles and cover crops; and (iii) cropping system diversification with annuals and perennials (including legumes) in rotations/sequences and associations. There are short-term advantages to crop yields where CA is currently practised. In particular, CA enables the farmer to plant earlier, because primary tillage is not necessary to prepare a seedbed. This enables timely use of seasonal rainfall. There is also enhanced rainwater conservation in the soil to better tide crops over during drought periods (Rockstrom *et al.*, 2009).

No-till in combination with the other two principles significantly increases rainfed crop productivity in dry climates, suggesting that it may become an important climate-change adaptation strategy for ever drier regions of the world. However, expansion of CA should be done with attention to the details in these areas, as implementation of the other two principles can be challenging in resource-poor and vulnerable smallholder farming systems (Kassam *et al.*, 2015).

There are a number of unique constraints to smallholder adoption of CA that are retarding its rapid uptake. Most important among these in initial years are: (i) competing use of crop residues for fodder where livestock production is common; (ii) inadequate biomass accumulation of cover crops in the off-season; (iii) different approaches to weeding when herbicides are not used; (iv) variable yield results across soil types; and (v) the perceived need for greater application of organic and inorganic nutrients (Giller *et al.*, 2009).

This chapter will first discuss the role of trees in CA systems and highlight examples of promising ECA models. It will review the experiences of developing and applying ECA models that are tuned to the specific contexts of farming in Zambia, Malawi and the Sahelian countries. It will then examine the implications of ECA systems for developing a more climate-smart agriculture and opportunities for more effectively scaling them up by linking their extension to the ongoing fertilizer subsidy programmes in many countries. Finally, the chapter will highlight the road ahead for expanding ECA through international scientific and policy collaboration, spearheaded by the Partnership to Create an EverGreen Agriculture.

10.2 EverGreen Conservation Agriculture Systems in Africa

Most African smallholders are engaged in both crop and livestock production but their fodder resources are usually very inadequate, particularly in the late dry season. Conventional tillage farmers typically use most of their available crop residues for animal fodder or fuel and usually they cannot afford to retain them as a soil cover (Giller *et al.*, 2009). Thus, in order for farmers to adopt CA, it is imperative that they have alternative ways to increase plant biomass in their farming system. That is critical in order for them to have enough dry-season fodder to enable some crop residues to be retained on the soil surface. In addition, some three out of four African smallholders are not applying any inorganic fertilizers, often because of non-availability, cash constraints, high climatic risk and poor fertilizer response on soils that are low in organic matter or severely degraded. Low yields and declining soil fertility are inevitable in this situation and options for overcoming these constraints are limited. Thus, greater use of the biological process of nitrogen fixation and more efficient nutrient cycling are critical but suitably practical options are needed. The question is: how can biomass production be increased to enhance soil surface cover and to generate more organic nutrients to complement whatever amounts of inorganic fertilizers a smallholder farmer can afford to apply?

The CA and agroforestry research and development communities have now mutually recognized the value of integrating fertilizer trees and fertilizer shrubs into CA systems in order to further enhance both fodder production and soil fertility (FAO, 2010, 2011). Practical systems for intercropping fertilizer trees in maize farming have been developed and are being extended to hundreds of thousands of farmers in Malawi and Zambia (Garrity *et al.*, 2010). The portfolio of options includes: (i) intercropping maize with fast-growing nitrogen-fixing trees, including *Gliricidia sepium* and *Tephrosia candida*, or *Cajanas cajan* (pigeon peas); using trees such as *Sesbania sesban* as an improved fallow; or integrating full-canopy fertilizer trees such as *Faidherbia albida* into the CA system (Akinnifesi *et al.*, 2010).

The integration of *F. albida* into CA systems has proven to be a particularly effective practice (see conservationagriculture.org; evergreenagriculture.net). *Faidherbia* is an indigenous African acacia that is widespread on millions of farmers' fields throughout the eastern, western and southern sub-regions of the continent. It is highly compatible with food crops because it is physiologically dormant during the rainy season. It sheds its nitrogen-rich foliage at the beginning of the wet season and only re-foliates at the beginning of the dry season. Thus, it exhibits minimal competition with food crops grown in association with it, while enhancing their yields and improving soil health (Barnes and Fagg, 2003; Garrity *et al.*, 2010). Several tonnes of additional biomass can be generated annually per hectare by the intercropped trees. This high-quality biomass accelerates soil fertility replenishment. It also provides additional high-protein fodder for livestock from both the leaves and the pods. Numerous publications have recorded increases in maize grain yield when it is grown in association with *Faidherbia,* ranging from 6% to more than 200% (Barnes and Fagg, 2003), depending on the age and density of the trees, the agronomic practices used and the weather conditions.

The effects of *Faidherbia* tend to be most remarkable on soils of low inherent fertility. In semi-arid cropping systems based on millet and sorghum, double-storey production systems with medium to high densities of fertilizer trees, including *Faidherbia* and a number of other indigenous species, are now observed across more than 5 million hectares in Niger and on hundreds of thousands of hectares in Mali, Burkina Faso, Senegal and Ethiopia (Garrity *et al.*, 2010). Depending upon which woody perennial species are used, and how they are managed, their incorporation into CA helps to maintain vegetative soil cover. The trees increase nutrient supply through nitrogen fixation, improve nutrient cycling, suppress insect pests and weeds, enhance soil structure and water infiltration, increase carbon storage and soil organic matter and enhance above- and below-ground biodiversity (Garrity *et al.*, 2010).

ECA systems reinforce the principle of adequate residue retention on the soil surface. The integration of trees and shrubs throughout the crop fields supplies increased high-quality foliage from the tree biomass, enhancing the amount of organic sources of nutrients. The specific benefit of this kind of agroforestry is not just the fertilizer trees but also the fact that the soil and cropping system management is under CA.

This applies the principle of crop diversification to tree–crop diversification as described in the definition of CA principles. It incorporates the role of fertilizer/fodder trees to enhance soil fertility more effectively and to provide needed biological and income diversity in the farming system. Tree intercropping or tree–crop association is therefore a type of simultaneous rotation. Indeed, the synergies that will result need to be better exploited by creating more tightly linked crop–tree–livestock systems (Fig. 10.2).

10.3 Conservation Agriculture in Zambia Champions *Faidherbia*

In Zambia, maize production is the foundation of agriculture and the basis for the country's food supply. However, the average maize yield is only 1.1 t/ha. Since 1996, a coalition of stakeholders from the private sector, government and donor communities has promoted a package of agronomic practices based on the principles of no-till conservation farming (Haggblade and Tembo, 2003). The effort is spearheaded by the Zambian Conservation Farming Unit (CFU) (see conservationagriculture.org). During the past decade, CA has been introduced over large areas of the country.

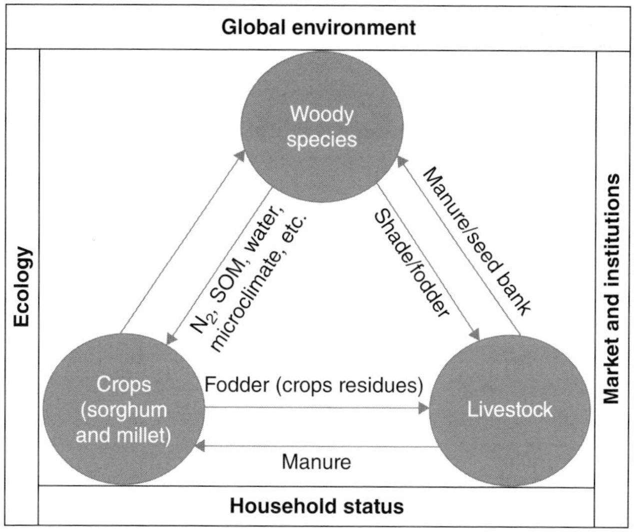

Fig. 10.2. EverGreen Conservation Agriculture strengthens the integration of trees into crop and livestock production systems by exploiting numerous interactions among the components. It combines the best of conservation farming and agroforestry to reverse stagnation in agricultural production and enhance resilience to climate change.

As the Zambian CFU worked to make conservation farming as feasible as possible for smallholder farmers, they encountered a problem that defied conventional solutions. More than two-thirds of the country's smallholder farmers were unable to afford inorganic fertilizers and they had little or no access to livestock manure or other nutrient sources. This fundamentally limited smallholder maize yields and their soil fertility was declining each year. To address the problem, the CFU investigated the incorporation of *F. albida* trees into CA-based maize production systems. They found that maize yields were dramatically increased when the crop was associated with these trees (Fig. 10.3).

The Zambian CFU incorporated *Faidherbia* into its CA extension programme, recommending that *Faidherbia* seedlings be planted in a grid pattern at a density of 100 trees/ha. Fields with *Faidherbia*–maize systems managed with such a planting pattern (trees at a spacing of 10 m × 10 m) can accommodate full mechanization (see Fig. 10.1). The result is a maize farming system in a parkland of *Faidherbia* trees. The trees may live for 70–100 years, providing inter-generational benefits for a farm family, with a very modest initial investment. As the trees mature and develop a spreading canopy, they are gradually thinned down to about 25–30 trees/ha.[1] Currently, some 68,000 farmers are estimated to have *Faidherbia* trees on their farms (Nkatiko, 2014).

10.4 EverGreen Agriculture in Malawi

Farming households in Malawi highly value trees within their farming systems. The regeneration of trees in fields is encouraged by protecting naturally

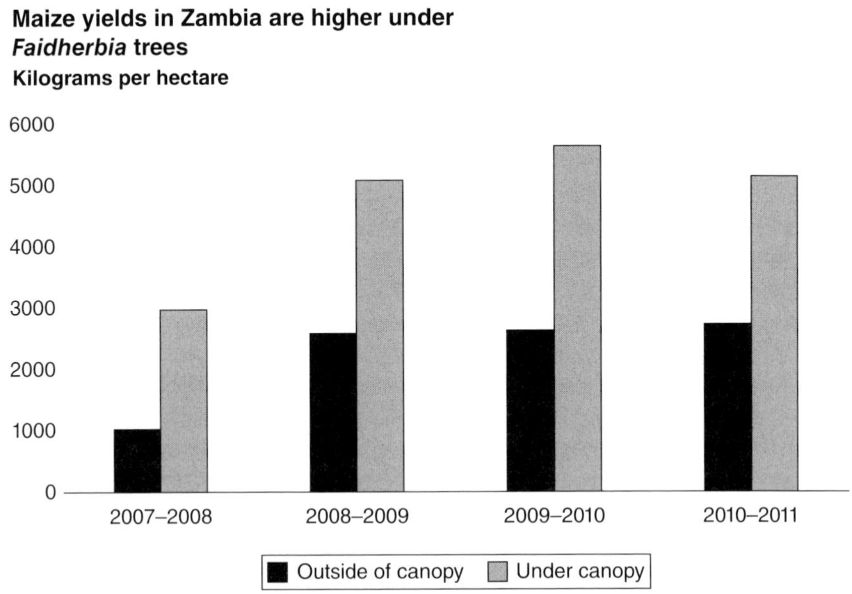

Fig. 10.3. Increases in yields of maize grown in association with *Faidherbia* trees. Average of 40 trials on farmers' fields in Zambia. (Reprinted from Winterbottom *et al.*, 2013.)

regenerating indigenous trees, by planting tree seedlings and by leaving favoured trees in the fields when woodlands are cleared prior to cultivation (Dewees, 1995). Recent surveys show that the default option for farmers is to protect a wide variety of trees occurring on their land. Recent observations and mapping studies across the country indicate that farmer-managed natural regeneration of trees on croplands is becoming very widespread in Malawi (Winterbottom, 2015).

There is a long history of research on suitable fertilizer tree practices in Malawi and in the neighbouring countries in Southern Africa (Sileshi *et al.*, 2008). Cultivation of crops under *F. albida* has been traditionally practised in Malawi for generations. Formal research on the tree began in Malawi in the 1980s when it was carried out as part of the activities of the Agroforestry Commodity Team under the government's Department of Agricultural Research and Technical Services (DARTS). Saka *et al.* (1994) reported 100–400% maize yield increases under *Faidherbia* trees in the Lakeshore plain of Malawi. Several agencies have been promoting its cultivation in Malawi for the last two decades (Akinnifesi *et al.*, 2008), particularly the non-governmental organization (NGO) Total LandCare. It is estimated that currently about 500,000 Malawian farmers (out of a total of about 2.7 million farmers) have *Faidherbia* trees on their farms (H. Phombeya, MAFE Land Resource Centre, Lilongwe, personal communication 2009 cited in Garrity *et al.*, 2010). The majority of these stands were developed through assisted natural regeneration of seedlings that emerged in farmers' fields.

Since the early 1990s, the World Agroforestry Centre and its partners in East and Southern Africa have been developing a range of agroforestry systems that improve soil quality and significantly boost crop yields, providing high returns to both land and labour. The most popular system in southern Malawi, where land holdings are very small (< 0.5 ha), is intercropping maize with nitrogen-fixing trees (*G. sepium*, *S. sesban* and *T. candida*) or pigeon peas. In these systems, farmers plant the trees in rows between their crops. *Gliricidia* is pruned back two or three times a year and the leaves and the biomass are returned to the soil. A long-term experiment spanning more than a decade, involving the continuous cultivation of maize with *Gliricidia* at Makoka Research Station, Malawi, yielded more than 5 t/ha in good years, and an average of 3.7 t/ha overall, in the absence of mineral fertilizers. This compared favourably with an average of 0.5–1.0 t/ha in control plots without *Gliricidia* or mineral fertilizer (Makumba *et al.*, 2006; Akinnifesi *et al.*, 2007). Also, *Tephrosia (T. vogelii* and *T. candida*) and pigeon peas are often relay-intercropped with maize (Snapp *et al.*, 1998; Akinnifesi *et al.*, 2008).

Rotational fallows that incorporate nitrogen-fixing trees are also suited to areas where land holdings are somewhat larger (> 1 ha). In this case, during the fallow period farmers grow short-lived shrubs such as *S. sesban* and *T. candida*, rather than long-lived intercropped trees like *Gliricidia*. Rotational fallows of *S. sesban* and *T. candida* have been widely tested in farmer participatory research in Malawi. Results from 152 farms show that agroforestry increased the yield of maize by 54–76% compared with unfertilized sole maize, which is the *de facto* farmer practice (Akinnifesi *et al.*, 2009). When supplemented with inorganic fertilizer, the yield increase over the control was 73–76% across tree species (Akinnifesi *et al.*, 2009).

In addition to increasing soil fertility and crop yields, these practices suppress weeds (Sileshi and Mafongoya, 2006), improve water filtration (Chirwa *et al.*, 2007) and dramatically increase the amount of soil carbon (Makumba *et al.*, 2007). There is evidence that production systems that incorporate *Gliricidia*, *Tephrosia*, *Faidherbia* and other leguminous cover crops assist rural populations in adapting their agriculture to the adverse effects of climate change. Research results and farmer interviews indicated that these systems increased the grain harvest during serious droughts (Akinnifesi *et al.*, 2010; Sileshi *et al.*, 2010). Farmers obtain at least a modest yield during seasons when those who are not using these practices may experience crop failure.

Malawi launched an Agroforestry Food Security Programme in 2007 based on these results. The programme is managed by the World Agroforestry Centre, the Ministry of Agriculture, the Malawian Smallholders Farmers' Association (NASFAM) and a number of NGOs. It provides tree seeds, nursery materials and training for a range of agroforestry species, including fertilizer trees. Over 200,000 farmers have received training and tree materials through the programme. A diverse range of fruit, timber, fuel wood and tree cash crops is also being integrated into maize farming systems to enhance enterprise diversity and income generation. This national effort relies on whole-village mobilization, particularly through farmers and women's groups, to accomplish the scale of action targeted. Women and the rural poor are the major beneficiaries. The poor are often observed to adopt agroforestry systems more rapidly than wealthier households (Ajayi *et al.*, 2005; Place *et al.*, 2005; Pye-Smith, 2008).

The programme is assisting the uptake of three types of nitrogen-fixing tree legumes: (i) short-term species such as *T. candida, S. sesban* and pigeon peas, which are planted and incorporated within 1 year; (ii) medium-term solutions such as *Gliricidia,* which can be continuously pruned for organic fertilizer for at least one to two decades; and (iii) long-term full canopy trees of *F. albida,* which provide benefits for many decades. These species are often combined in the same fields. The optimum combinations are being tailored to the range of variation in agroecological conditions and farm circumstances across the country. Research to date has indicated that such forms of EVA may generally increase yields from 1 t/ha to 2–3 t/ha, even if farmers cannot afford commercial nitrogen fertilizers. However, with an application of a quarter-dose of mineral fertilizer, maize yields may surpass 4 t/ha (Akinnifesi *et al.*, 2010; Sileshi *et al.*, 2010).

A current opportunity is to link Malawi's fertilizer subsidy programme directly to agroforestry investments on the farm. This linkage will provide for long-term sustainability in nutrient supply and the build-up of soil health as the basis for sustained yields and improved fertilizer response efficiency. This can be done in the short term by combining the provision of limited amounts of subsidized fertilizer with the provision of seed and technical advice to establish fertilizer tree systems. Farmers can thus be further encouraged to produce more of the nitrogen required by their crops on farm, increasing and sustaining their maize yields and improving their soils. This would foster a gradual shift of investments from fertilizer subsidies to sustainable on-farm fertility regeneration through integrated soil fertilizer management (ISFM). Discussions are underway with the government of Malawi to map out such a 'subsidy to sustainability' pathway. A pilot of this approach is currently being implemented with the participation of 10,000 farmers.

10.5 EverGreen Agriculture in the Sahel

There is increasing recognition of the potential of *Faidherbia* in many other parts of Africa, in conventional tillage and CA systems. In Niger, millet production in combination with *Faidherbia* is practised using non-inversion tillage methods. The majority of Nigerien farmers do not use a plough or hoe for land preparation on their typically sandy soils. Rather, they use a hand-drawn implement (a type of manual sweep) for land preparation that is passed just underneath the soil surface, loosening the soil and undercutting the weeds. Thus, agriculture in parts of Niger is now essentially an agroforestry system (Garrity *et al.*, 2010) with minimum-tillage crop production in a parkland of full-canopy fertilizer/fodder trees, dominated in most parts of the country by *F. albida.*

Recently, the World Agroforestry Centre studied the economics of farmer-managed natural regeneration practices in the Sahel across 1000 farms in four countries (Place and Binam, 2013). The estimated increase in crop yields due to the fertilizer trees was 15–30%. The estimated value of the additional basic fruit, pod, leaf and wood tree products harvested was valued at about $200 per year per household.

Fuel wood availability has now become a critical constraint in many farming systems. ECA farms have a ready supply of fuel wood for household use, with a surplus for sale that is obtained from the annual prunings of the trees grown on the farm. The creation of medium- to high-density agroforest parklands on the crop fields of Niger has also stimulated the widespread development of wood markets, where excess wood is being marketed by farmers as an additional source of cash income. Some of this wood is now being exported to Nigeria.

Faidherbia is a common tree on croplands in Ethiopia, particularly in the semi-arid parts of the country. The Ethiopian government has been promoting its wider use on farms for many years. In 2012, a National Faidherbia Programme was launched, whose goal is the establishment of 100 million additional *Faidherbia* trees in the coming years. This effort is being integrated into the country's strategies on climate-smart agriculture and to deploy sustainable land management practices to restore greater productivity on 15 million hectares of land during the next 15 years.

The indigenous shrub *Guiera senegalensis* coexists with crops to varying degrees in farmers' fields throughout the Sahel, creating islands of fertility in the sandy, often degraded soils. When it is coppiced in crop fields it dramatically increases crop productivity (in some cases by 250%) and provides ecological benefits to landscapes (Dossa *et al.*, 2012.) The results suggest that these shrubs should also be recognized as an important local biological resource with agronomic potential and that they should be promoted as a component of cropping systems in the Sahel, particularly in the more arid zones.

In Burkina Faso, the *zaï* technique of cultivation with planting pits may be considered a variation of ECA. This practice has been steadily expanding for decades. The planting pits have intensified both cereal and tree production. Biomass production in these systems is dramatically increased, for soil amelioration as well as for livestock fodder (Reij *et al.*, 2009).

The experiences of Zambia, Malawi, Niger, Ethiopia and Burkina Faso indicate that the principles of ECA are applicable to a wide range of food crop systems in Africa, if accompanied by adequate testing and farmer engagement. This has stimulated expanded investments in upscaling the integration of trees into cropping systems in these and other countries, both in conventional annual tillage systems and in conjunction with CA annual cropping practices.

10.6 EverGreen Conservation Agriculture is Climate-smart

Incorporating trees into annual crop farming may confer important sustainability benefits through ecological intensification. It may also increase the resilience of the farm enterprise to climate change through greater adaptation to drought at the crop level and at the household level.

At the crop level there are three key processes in play for drought resilience. The first is that the presence of the trees enhances the effects of rainwater capture and storage. This improved rainfall infiltration and soil moisture storage are particularly valuable on farmlands where rainfall runoff and soil erosion are a problem. According to farmers in Niger, the presence of the trees not only provides more soil moisture to their sorghum and millet crops, but also elevates the water-table levels in the village wells.

Secondly, tree cover on crop fields reduces wind speeds at the canopy level, providing a windbreak effect that reduces the deleterious effects of desiccating winds. This is an acute problem in the Sahel. Nigerien farmers report that high winds and sand-blasting often destroy crop seedlings as they emerge from the soil after germination. This makes it necessary to plant the crop repeatedly to achieve a successful stand establishment. With a moderate density of fertilizer trees they have found that they no longer have to plant more than once a season.

Thirdly, daytime and night temperatures are increasing as a result of climate change. Higher temperatures increase crop heat stress, particularly at midday, and they are particularly devastating during the crop flowering stage. Higher temperatures also reduce the length of the grain-filling period, which is now being observed to lower crop yield potential directly in Europe as well as Africa. The dispersed light shade that is provided by the trees in an ECA system reduces crop canopy temperatures significantly during the midday period, thus providing a helpful canopy-temperature buffering effect (CIMMYT, personal communication, 2013). Global temperatures will continue to rise, as predicted by global climate models. This will make the utility of this microclimate buffering effect increasingly important in the future. Thus, the value of trees in CA systems to buffer crop temperatures is an area that is expected to attract much greater attention.

ECA systems also increase drought resilience at the household level (de Leeuw *et al.*, 2014). Trees on croplands serve as an additional household asset that can be harvested for cash during periods when severe drought or other emergencies are experienced. This was observed to be an important means by which families coped with household food deficits during the 2009/10 drought in the Sahel (Reij and Winterbottom, 2015).

The climate change mitigation potential of ECA systems is also significant. They accumulate much more carbon than is possible with CA based on annual crops alone. Conventional CA systems tend to sequester an annual maximum of 0.2–0.4 t C/ha. ECA systems accumulate carbon both above and below ground in the range of 2–4 t C/ha, roughly an order of magnitude higher than with CA alone. This is particularly true for systems incorporating fertilizer trees such as *Faidherbia* or *Gliricidia* (Makumba *et al.*, 2007; Kaonga and Bayliss-Smith, 2008).

Consequently, there is considerable interest in the development of reward systems for farmers that will channel carbon offset payments from developed countries to stimulate more carbon sequestration in African food crop systems, while simultaneously enhancing the livelihoods of smallholders and the environment. These investments will encourage development pathways resulting in higher carbon stocks at a whole-landscape scale.

Making CA evergreen could therefore be one of the most significant ways to help climate-proof agriculture in the future and restore degraded farmlands to higher sustained levels of productivity, while also helping agriculture to reduce the level of its carbon dioxide emissions and thus become part of the solution to climate change. Indeed, in many situations where smallholders find it quite challenging to adopt zero tillage and other components of conservation farming, they may nevertheless begin the process by incorporating fertilizer trees into their croplands as a first step. Then, as their biomass production increases and as their soil fertility improves, they may follow on with the implementation of the more complex CA practices.

10.7 From Fertilizer Subsidies to Sustainability

The incorporation of fertilizer trees into CA systems offers a major opportunity for countries to increase food production by enhancing the biological fixation of nitrogen in farmers' fields. This in-field biological fertilizer production can reduce the costs of fertilizer purchases at the farm level and help to offset fertilizer importation and subsidies at the national level. For example, the government of Malawi launched an input-subsidy programme in 2004 that generated large maize surpluses and helped to improve rural welfare. This success caused a surge of interest among African governments in deploying fertilizer subsidies as a means of enhancing food security. However, in Malawi itself, the recurrent costs of the programme later contributed to the country's recent near-bankruptcy, which has brought massive economic difficulties. The fertilizer subsidy programme is now being scaled back, while an alternative strategy for the long term is taking root.

The Malawi Agroforestry Food Security Programme has been assisting farmers to deploy bio-fertilizer trees on about 200,000 farms across the country. These practices have doubled farm yields without inorganic fertilizer inputs, though modest additional fertilizer applications may further increase yields (see Section 10.4 above). A pilot programme is currently being implemented to link the fertilizer subsidies with these evergreen agriculture investments to provide long-term sustainability in nutrient supply and to build up soil health. This 'subsidy to sustainability' pathway for integrated soil-fertility management has provided a medium-term solution to the fertilizer subsidy conundrum (Garrity *et al.*, 2010).

A recent evaluation of the performance of fertilizer subsidies among the 12 countries that are currently implementing them has emphasized their generally poor return on investment and the major burdens that they place on incurring national trade deficits and budget deficits (IFPRI, 2013). By shifting attention to upscaling fertilizer tree technologies in CA systems, governments can reap substantial benefits and create a sustainable crop nutrient supply situation. This has prompted the 19 member states of the Common Market for Eastern and Southern Africa (COMESA) to launch an initiative with the EverGreen Agriculture Partnership and the World Agroforestry Centre to build the capacity of the countries to link the upscaling of fertilizer tree technologies to their fertilizer subsidy programmes.

10.8 Looking Forward

CA systems involving trees referred to as ECA systems in this chapter should attract much more research and extension attention than has been the case so far. Their success will depend on more knowledge and practical solutions in a number of areas, including: (i) the identification of a wider range of tree species for varied agroecologies; (ii) higher-quality tree germplasm; (iii) better tree seed dissemination systems; and (iv) further improvements in tree propagation and establishment methods. The optimum tree densities for different ECA systems have yet to be fully established; and the best practices in exploiting the soil fertility synergies between organic and inorganic nutrient sources also need to be elucidated.

Likewise, more determined efforts are needed to develop replicable models of community mobilization to create grassroots momentum for upscaling, and to overcome the challenges of insecure land and tree tenure in some areas and the difficulties of establishing trees on croplands where free-grazing has been a traditional practice. Effective models for addressing these challenges have been implemented by communities in Niger and other countries; and the lessons they have learned and the modalities that they have developed need greater attention and adaptation to other contexts (Winterbottom *et al.*, 2013).

CIMMYT and ICRAF are now actively collaborating in a number of projects to document the effects of trees incorporated into maize and wheat cropping systems and to determine the best management practices for a range of cereal-based farming systems in East and Southern Africa. Pioneer Hi-Bred Seeds Corporation has recognized the future importance of ECA systems and has been evaluating its maize hybrids under ECA systems in order to recommend the best varieties for these management systems. The company has also entered into a collaboration with ICRAF to promote ECA as a key direction in creating a more sustainable and climate-smart agriculture.

Targeting and scaling-up methodologies for ECA deserve particular attention. These need to be supported by work to reverse the policy frameworks in some countries that currently discourage farmers from cultivating trees. Farmer organizations have always been instrumental in the development and spread of CA. They will play an increasingly important role in expanding the practice of ECA. For example, the Malawi National Smallholder Farmers Association (NASFAM) has played a key role in the scaling-up of these practices. There is a growing interest in community-based grassroots mobilization through Landcare in Africa and Asia (http://www.silc.com.au/landcare-international/). Landcare can provide a particularly suitable approach for the engagement of farming communities in the refinement and spread of ECA.

Farmer-managed natural regeneration (FMNR) based on CA systems that include trees is increasingly considered to be a 'foundational practice' for use on all farms in the dryland zones of Africa, since it complements all other improved farming practices and cropping and livestock systems. The value of the FMNR practice is reinforced by the fact that it requires no cash investment by farmers and thus has enormous advantages compared with the costs and establishment risks of planting trees in semi-arid and sub-humid environments.

The 2nd African Drylands Week, convened by the Africa Union (AU) in Ndjamena, Chad, in August 2014, reviewed the evidence of the massive scale of transformations in land regeneration that are occurring in the Sahel, and in parts of East and Southern Africa, by applying the principles of FMNR on croplands and through other forms of EVA, such as the integration of trees in annual CA systems and in conventional tillage annual cropping systems. The conference noted that the upscaling of these practices can contribute substantively towards meeting the commitment of the African Heads of State Summit in Malabo to ensure that 30% more households in Africa are resilient to risks related to climate change (Malabo Decision, June 2014) and to achieving Vision 25×25 within the context of the African CSA Alliance and the Comprehensive Africa Agriculture Development Programme (CAADP) Climate Change Programme, which is aiming to achieve 25 million farmers practising climate-smart agriculture by 2025.

The conference therefore acted to: 'Recommend and propose that the drylands development community, through the African Union, and all collaborating and supporting organizations, commit seriously to achieving the goal of enabling every farm family and every village across the drylands of Africa to be practising FMNR and assisted natural regeneration by the year 2025.'[2] The AU is also currently developing a strategy to end hunger in Africa by 2025 that builds on the Malabo Declaration. In this context, a target is being developed for the scaling-up of fertilizer tree technologies (FTTs) to contribute to a doubling of food production by 2025. This target will include a range of FTTs in addition to FMNR that encompasses CA with trees, and other nitrogen-fixing tree systems, such as those deployed in the Malawi National Agroforestry Food Security Programme, the Zambian Conservation Farming Programme and the Ethiopian National Faidherbia Programme.

Finally, it is noted that the issues of land and landscape restoration are rapidly rising in the consciousness of policymakers at the global level. The UN Convention on Combating Desertification is championing the goal of attaining a land degradation-neutral world; the Bonn Challenge and 2014 New York Summit on Climate Change have exhorted the global community to restore 350 million hectares of degraded land by the year 2030. This has encouraged many countries to declare formal commitments to this goal. The restoration of degraded agricultural land is a major part of this effort and the incorporation of trees into CA cropping systems is now becoming recognized as an important vehicle to achieving this degree of restoration in the future.

10.9 The EverGreen Agriculture Partnership

Currently, there are efforts under way to upscale evergreen agriculture, including ECA, in 17 countries in Africa and several countries in Asia. An accelerated effort is needed to expand the reach of these systems to transform the farms of tens of millions of the poorest small-scale farmers. Therefore, a global partnership has been launched to support governments, farmers' organizations, the NGO community and civil society to achieve a massive scaling-up movement. It is known as the Partnership to Create an EverGreen Agriculture[3] (ICRAF, 2012). The Partnership is supporting the information needs, capacity building and knowledge generation required to assist in this effort. Many international and regional organizations have endorsed this work and are supporting it and embedding it into their programming. The Partnership is collaborating with the African Conservation Tillage (ACT) Network to expand the engagement of the CA community in scaling-up ECA. Many NGOs are now engaged in implementing this work on the ground. The momentum that has been generated is encouraging. We are therefore beginning to glimpse a future of more environmentally sound and productive farming where much of our annual food crop production occurs in CA incorporating trees and livestock.

Notes

[1] Extension directions for creating a small farm-scale *Faidherbia* tree nursery, and for out-planting and managing the trees on cropland, are contained in an illustrated leaflet http://conservationagriculture.org/conservation-farming-information (Sowing & Transplanting Faidherbia, 2012).
[2] See: http://www.au.int/en/newsevents/13983/second-africa-dry-land-week-n%E2%80%99djamena-chad
[3] See: evergreenagriculture.net

References

Ajayi, O.C., Place, F., Kwesiga, F., Mafongoya, P. and Franzel, S. (2005) *Impact of Fertilizer Tree Fallows in Eastern Zambia*. World Agroforestry Centre, Nairobi.

Akinnifesi, F.K., Makumba, W., Sileshi, G., Ajayi, O.C. and Mweta, D. (2007) Synergistic effect of inorganic N & P fertilizers & organic inputs from *Gliricidia sepium* on productivity of intercropped maize in Southern Malawi. *Plant and Soil* 294, 203–217.

Akinnifesi, F.K., Chirwa, P.W., Ajayi, O.C., Sileshi, G., Matakala, P., Kwesiga, F.R., *et al.* (2008) Contributions of agroforestry research to livelihood of smallholder farmers in Southern Africa: 1. Taking stock of the adaptation, adoption and impact of fertilizer tree options. *Agricultural Journal* 3, 58–75.

Akinnifesi, F.K., Sileshi, G., Franzel, S., Ajayi, O.C., Harawa, R., Makumba, W., *et al.* (2009) On-farm assessment of legume fallows & other fertility management options used by smallholder farmers in southern Malawi. *Agricultural Journal* 4, 260–271.

Akinnifesi, F.K., Ajayi, O.C., Sileshi, G., Chirwa, P.W. and Chianu, J. (2010) Fertilizer tree systems for sustainable food security in the maize-based production systems of East and Southern

Africa Region: a review. *Agronomy for Sustainable Development* 30, 615–629. doi: 10.1051/agron/2009058.

Barnes, R.D. and Fagg, C.W. (2003) *Faidherbia albida. Monograph and Annotated Bibliography.* Tropical Forestry Papers No 41. Oxford Forestry Institute, Oxford, UK.

Chirwa, P.W., Ong, C.K., Maghembe, J. and Black, C.R. (2007) Soil water dynamics in intercropping systems containing *Gliricidia sepium*, pigeon pea and maize in southern Malawi. *Agroforestry Systems* 69, 29–43.

de Leeuw, J., Njenga, M., Wagner, B. and Iiyama, M. (2014) *Treesilience: An Assessment of the Resilience Provided by Trees in the Drylands of Eastern Africa.* World Agroforestry Centre, Nairobi, 166 p.

Dewees, P.A. (1995) Trees on farms in Malawi: private investment, public policy, and farmer choice. *World Development* 23(7), 1085–1102.

Dossa, E.L., Diedhiou, I., Khouma, M., Sene, M., Lufafa, A., Kizito, F., *et al.* (2012) Crop productivity and nutrient dynamics in a shrub (*Guiera Senegalensis*)–based farming system of the sahel. *Agronomy Journal* 104, 1255–64.

FAO (2010) *An International Consultation on Integrated Crop-Livestock Systems for Development. The Way Forward for Sustainable Production Intensification.* Integrated Crop Management Vol. 13-2010. FAO, Rome, 79pp.

FAO (2011) *Save and Grow. A Policymaker's Guide to the Sustainable Intensification of Smallholder Crop Production.* FAO, Rome, 103 pp.

FAO (2013) *Climate-Smart Agriculture Sourcebook.* FAO, Rome.

Garrity, D., Akinnifesi, F., Ajayi, O., Sileshi, G.W., Mowo, J.G., Kalinganire, A., *et al.* (2010) Evergreen Agriculture: a robust approach to sustainable food security in Africa. *Food Security* 2(3), 197–214.

Giller, K.E., Witter, E., Corbeels, M. and Tittonell, P. (2009) Conservation agriculture and smallholder farming in Africa: the heretics' view. *Field Crops Research* 114, 23–34.

Haggblade, S. and Tembo, G. (2003) *Early Evidence on Conservation Farming in Zambia.* EPTD Discussion Paper 108. International Food Policy Research Institute, Washington DC.

ICRAF (2012) *Creating an EverGreen Agriculture in Africa.* World Agroforestry Centre, Nairobi.

IFPRI (2013) Recent evidence on input subsidy programs in Africa. Highlights from a special issue of *Agricultural Economics.* Policy Brief. International Food Policy Research Institute, Washington, DC.

Kaonga, M. and Bayliss-Smith, T.P. (2008) Carbon pools in tree biomass and the soil in improved fallows in eastern Zambia. *Agroforestry Systems* 76, 37–51.

Kassam, A., Friedrich, T., Derpsch, R. and Kienzle, J. (2015) Overview of the worldwide spread of Conservation Agriculture. *Field Actions Science Reports.* Available at: http://factsreports.revues.org/3966 (accessed 23 September 2016).

Lipper, L., Thornton, P., Campbell, B.M., Baedeker, T., Braimoh, A., Bwalya, M., *et al.* (2014) Climate-smart agriculture for food security. *Nature Climate Change* 4, 1068–1072. doi: 10.1038/NCLIMATE2437.

Makumba, W., Janssen, B., Oenema, O., Akinnifesi, F.K., Mweta, D. and Kwesiga, F. (2006) The long-term effects of a gliricidia-maize intercropping system in southern Malawi, on gliricidia and maize yields, and soil properties. *Agriculture, Ecosystems and Environment* 116, 85–92.

Makumba, W., Akinnifesi, F.K., Janssen, B. and Oenema, O. (2007) Long-term impact of a *Gliricidia*-maize intercropping system on carbon sequestration in southern Malawi. *Agriculture, Ecosystems & Environment* 118, 237–243.

Place, F. and Binam, J.N. (2013) *Economic Impacts of Farmer-Managed Natural Regeneration in the Sahel.* End of technical project report for Free University Amsterdam and IFAD. World Agroforestry Center, Nairobi.

Place, F., Adato, M., Hebinck, P. and Omosa, M. (2005) *The Impact of Agroforestry-based Soil Fertility Replenishment Practices on the Poor in Western Kenya*. Research Report 142. International Food Policy Research Institute & World Agroforestry Centre, Washington, D.C.

Pye-Smith, C. (2008) *Farming Trees, Banishing Hunger: How an Agroforestry Programme is Helping Smallholders in Malawi to Grow More Food & Improve their Livelihoods*. World Agroforestry Centre, Nairobi, 27 pp.

Reij, C. and Winterbottom, R. (2015) *Scaling Up Regreening: Six Steps to Success*. World Resources Institute, Washington, DC, 66 p.

Reij, C., Tappan, G. and Smale, M. (2009) *Agroenvironmental Transformation in the Sahel: Another Kind of "Green Revolution"*. IFPRI Discussion Paper 00914. International Food Policy Research Institute, Washington, DC.

Rockstrom, J., Kaumbutho, P., Mwalley, J., Nzabi, A.W.M., Temesgen, M.L., Mawenya, L., *et al.* (2009) Conservation farming strategies in East and Southern Africa: yields and rain water productivity from on-farm action research. *Soil & Tillage Research* 103, 23–32.

Saka, A.R., Bunderson, W.T., Itimu, O.A., Phombeya, H.S.K. and Mbekeani, Y. (1994) The effects of *Acacia albida* on soils and maize grain yields under smallholder farm conditions in Malawi. *Forest Ecology and Management* 64, 217–230.

Sileshi, G. and Mafongoya, P.L. (2006) Long-term effect of legume-improved fallows on soil invertebrates and maize yield in eastern Zambia. *Agriculture, Ecosystems & Environment* 115, 69–78.

Sileshi, G., Akinnifesi, F.K., Ajayi, O.C. and Place, F. (2008) Meta-analysis of maize yield response to woody & herbaceous legumes in the sub-Saharan Africa. *Plant and Soil* 307, 1–19.

Sileshi, G., Akinnifesi, F.K., Debusho, L.K., Beedy, T., Ajayi, O.C. and Mng'omba, S. (2010) Variation in maize yield gaps with plant nutrient inputs, soil type and climate across sub-Saharan Africa. *Field Crops Research* 116, 1–13.

Snapp, S.S., Mafongoya, P.L. and Waddington, S. (1998) Organic matter technologies for integrated nutrient management in smallholder cropping systems of southern Africa. *Agriculture, Ecosystems & Environment* 71, 185–200.

Winterbottom, R. (2015) Beating famine. Presentation at the *Beating Famine Conference*, Lilongwe, Malawi, April, 2015.

Winterbottom, R., Reij, C., Garrity, D., Glover, J., Hellums, D., McGahuey, M. and Scherr, S. (2013) Improving land and water management: creating a sustainable food future. In: *Instalment Four of Creating a Sustainable Food Future*. World Resources Institute, Washington, DC, 44 pp.

11 Mechanization of Smallholder Conservation Agriculture in Africa: Contributing Resilience to Precarious Systems

Brian G. Sims,[1]* Josef Kienzle,[2] Saidi Mkomwa,[3] Theodor Friedrich[4] and Amir H. Kassam[5]

[1]*Agricultural Engineering Consultant, Bedford, UK;* [2]*Food and Agriculture Organization, Rome, Italy;* [3]*African Conservation Tillage Network, Nairobi, Kenya;* [4]*Food and Agriculture Organization of the United Nations, Havana, Cuba;* [5]*University of Reading, UK*

11.1 Introduction

11.1.1 Why mechanization?

The desire of all involved in smallholder farming (especially women, children and the elderly) to reduce by any means their drudgery and arduous, often painful, struggle to produce food for life and subsistence has been driving the development of tools and implements for many generations. It has encouraged farmers to use their livestock animals not only to produce meat and dairy products but also for draught power that can be applied to crop production, transport, water lifting and crop processing. The invention of the agricultural tractor was driven by the need to curtail dramatically the hard physical work that farming entails for farmers and this situation holds true today, especially in Africa.

The lack of availability (and the non-affordability) of agricultural tools, implements and machines has motivated the youth to turn away from farming and rural dwelling as a way of living and income generation, as able young people are no longer willing to live a life of hard work and insecurity. 'The last thing a young person from rural Africa would like to do is farming', is a commonly heard refrain among youth in Africa. This has led to massive

*briangsims@aol.com

and increasing migration of youth (and fit men) to urban areas in search of alternative and more lucrative lifestyles and jobs. This in turn has increased the demand for mechanized farm operations in order to get the jobs done with a labour force comprising women, the elderly and children.

Here is where the challenge begins. The need to increase food production for a growing population in Africa is clear; yet, at the same time, the rural population is expected to decline substantially in the coming decades due to migration from the rural sector to urban centres. A way to deal with this apparent conundrum is through agricultural mechanization. However, in the 21st century, with a rapidly degrading natural resource base, the mechanization options required will need to be environmentally friendly and sustainable. They will also need to be climate-smart in the face of more frequent and extreme drought spells or floods resulting from climate change, adapted to local conditions and affordable. Moreover, the mechanization options on offer should be promoted by progressive farmers, service providers and private machinery suppliers – not, as so often in the past, by government-led procurement and distribution schemes that do not reach entrepreneurs at village level nor the farming communities that desperately await these services for crop establishment, weeding, harvesting and transport of produce to market.

11.1.2 The importance of timeliness

The issue of timeliness of operations is frequently ignored or forgotten by development planners, policy makers and local authorities (though rarely by farmers). Early crop establishment is key to achieving a good harvest; each day that planting is delayed from the optimum date has a direct negative impact on crop growth and development and on yields at harvest. Systems based on hand labour are not only very arduous but also, as delays will be inevitable, likely to produce biomass and yields that are well below potential levels. The same can apply with mechanized services for crop establishment if the service is not supplied at the correct time for whatever reason (remote, small fields, not prioritized in the community, i.e. households headed by women and orphans, and dependence on equipment from larger farmers). This is one of the reasons why draught animal power is still important and popular (where feasible) among rural dwellers, as animal traction services are often within easy reach of a village or community and can be readily accessed for timely services. The issue of timeliness is explored further in Section 11.3.4.

11.1.3 Current status of farm power use

The principal power sources used in smallholder farming systems are human and draught animal muscles, internal combustion engines and electric motors. The relative use of the different sources varies across regions (Table 11.1). It is remarkable that the percentage of human muscle power is as high as 65% in sub-Saharan Africa (SSA) with a significant percentage (25%) for draught

Table 11.1. Sources of power for land preparation (% of total). From FAO, 2006.

	Human muscle power	Draught animal power	Engine power
Sub-Saharan Africa	65	25	10
East Asia	40	40	20
South Asia	30	30	40
Latin America and the Caribbean	25	25	50

animal power and only 10% engine-driven power. Generally, engine power is on the increase (even in Africa), whilst draught animals are tending to decline in numbers, though locally they can still be very important.

11.1.4 How to make mechanization sustainable

There are many challenges to achieving a sustainable approach to mechanization. Smallholders in Africa are mostly subsistence farmers or farmers who have very limited access to markets and financial capital; their landholdings are fragmented and small, and some argue that they are too small to be considered for mechanization at all (Mrema *et al.*, 2008). The capital value of land is not available as collateral for loans as it is in the developed world and this virtually rules out the availability of credit from banks and input suppliers.

A new approach to mechanization is needed that primarily acknowledges the diversity and complexity of smallholder systems in Africa as well as the scattered holdings, with small fields often integrated with trees and bushland. Mechanization innovations need to be shown to provide solutions that are appropriate, in demand and feasible. They need to be profitable and clearly labour-saving to avoid being rejected but, rather, adopted and further adapted to local systems in the various agroecological regions of SSA.

11.1.5 Using market access and commercialization of services as entry point

Experience has shown that operations that are stationary tend to be easier to mechanize and hence adopted earlier. Therefore small diesel pumps for irrigation and small post-harvest mills tend to be first to appear on the market. In addition, small carts for transport (typically drawn by small tractors or draught animals) are among the first implements to be taken up. The desire to mechanize land preparation (in plough- or hoe-based systems) may be higher from a drudgery alleviation point of view, but this can be more complicated to organize in a way that is workable in the context of smallholder farming systems in Africa.

There are current initiatives to promote the adoption of smaller-horsepower tractors that are simple but robust, mostly emanating from China and India.

Other efforts focus specifically on the introduction of two-wheel tractors (2WTs) and their associated equipment. These, and other attempts, should always draw on lessons learned from previous experiences in market development, which can be distilled into a set of six principles (Baudron *et al.*, 2015):

1. Supporting the emergence of private rural service providers.
2. Considering the need for an honest facilitating broker (this will frequently be an international donor-funded project).
3. Linking mechanization and other input business models to output business models, i.e. viewing mechanization as part of a value chain.
4. Broadening the range of services offered in order to extend the demand for services throughout the year.
5. Bundling hiring services.
6. Providing kick-start subsidies for private sector investment in mechanization service provision.

Additionally, mechanization interventions should be in line with the principles of sustainable production intensification that FAO has summarized in its 'Save and Grow' guidelines. These have environmentally friendly and natural resource conserving and enhancing Conservation Agriculture (CA) mechanization at their heart (FAO, 2011a).

Smallholders in Africa are capital constrained and may not be in a position to purchase any equipment individually. Alternative forms of service provision need to be offered, with mechanization services provided either through independent service providers or through group ownership of mechanization assets (Landers, 2000; Sims *et al.*, 2011; Wongtschowski *et al.*, 2013). Service hiring is not new to Africa; hiring labour and/or draught power is quite common all over the continent. The challenge now is to provide incentives for specific types of mechanization services that are in line with the principles of sustainable intensification. They must also help to make farming not only less arduous but also more resilient to climatic extremes and the impacts of climate change. On the other hand, mechanization (such as ploughing services) that is not in line with such principles should be penalized, or at least not subsidized or promoted. Business models that support these principles will have to be developed and tested. They should be driven by private sector service providers at local level who must have strong links to other key value chain stakeholders.

However, the private sector may be reluctant to invest in mechanization in areas where agricultural markets are weak, awareness and demand for machinery and their services are absent and farming systems are vulnerable to shocks and stresses. Under these circumstances, intermediaries – such as non-governmental organizations (NGOs) and government- or donor-supported projects – may be needed to facilitate linkages between the private sector and smallholder farmers and thus develop viable business models (Kahan, 2007; London and Hart, 2010). Once demand can be assured, intermediaries would be expected to 'exit', allowing the private sector to step in and scale-out the service provision market. Increasingly attention is being paid to identifying such 'scalability' from the outset when designing interventions in market systems (Cooley and Kohl, 2006). In contrast, where markets are stronger with better rewards for investment, the private sector is more likely to emerge as the initial and main driver of the chain, eliminating the need for intermediaries. The existence of market-oriented farm enterprises and their associated cash flows is a prerequisite to affording either investment in, or rental of, farm machinery (Mrema *et al.*, 2008).

The following sections examine many facets of the complexities of mechanizing CA in the smallholder sector of SSA. Section 11.2 describes the hardware requirements for land preparation, seeding and planting, weed and cover crop management and the use of the range of possible power sources. Section 11.3 looks at the ways in which CA adds resilience to agricultural production systems. It does this by rehabilitating and conserving natural resources, sequestering carbon in the soil, improving water and nutrient use efficiency, improving timeliness and extending the agricultural frontier. Section 11.4 analyses the constraints to CA mechanization, which include: affordability; availability; suitability of the available equipment; poorly implemented CA policies; the seasonality of the machinery distribution business; limited technical knowledge; and fragmented land holdings. The section concludes by noting the gender implications of CA mechanization. Section 11.5 examines the supply of mechanization inputs, including through import, local manufacture and specialist CA mechanization service provision. Finally Section 11.6 draws the main conclusions.

11.2 Mechanization Needs for Conservation Agriculture

The most significant differences in cropping practices between conventional tillage-based agriculture and CA are in land preparation and seeding, which require very specific mechanization technologies. Additionally, weed management and harvest operations change, since the management of crop residues and soil cover becomes part of the land preparation for the next crop (Li, 2005; Baker *et al.*, 2006).

11.2.1 Land preparation

The change from tillage-based farming to CA implies a very different approach to land preparation and seeding practices. The use of tillage as a standard periodic operation is completely eliminated in a fully functioning CA system and remains only for very specific tasks, such as creating the conditions for changing over to CA by breaking up compacted soil or forming the soil surface (e.g. creating permanent raised beds, or constructing irrigation canals).

In the early years of CA adoption, and particularly on severely degraded and hard-setting soils, it might be necessary to rip the planting lines in order to break compaction horizons and open the soil until crop roots and soil life have established a more amenable soil structure. Since animal traction rippers are very limited as to their working depth, it is recommended to use tractor rippers for removing compaction created by tractor use and associated tillage operations. Animal traction rippers can be used in situations with less severe soil compaction from the use of animal traction or hand tools (so-called 'hoe pans'). Depending on the amount and condition of the crop residues in the field, rippers might have to be equipped with cutting discs to facilitate the passage of the ripper through a residue-covered field.

Once the soil structure has improved, ripping can be discontinued. At this point the harvest operation under CA becomes the most important part of land preparation for seeding the next crop. It is advisable to adopt harvest techniques that leave the residues in place and that facilitate a quick turnaround between crops, particularly in irrigated farming systems. The management of crop residues during harvest has direct influence on the ease, problems and quality of the subsequent planting operation and on the weed population. Standing residues, anchored in the soil (and, better, tall standing stubble), can facilitate mechanical seeding particularly in heavy or difficult residues, such as rice straw. Another important aspect is the treatment of the residues (see Box 11.1 for views on crop residue retention in SSA smallholder systems that include livestock enterprises). Mechanized farmers practising CA and using combine harvesters often use simple residue spreaders, rather than choppers, which have the additional benefit of a lower energy requirement. One of the most important tools specific to CA, and which is used

Box 11.1. A note on livestock (James Breen, personal communication, 2016).

The current systems of livestock management in Africa are generally inimical to the introduction of CA, in that virtually all residues are appropriated by livestock owners at harvest, with or without payment for them. There is no obligation on livestock owners to provide dry-season/winter feeding for their stock and this increases dependence on crop residues. Moreover, there is no bar on the expansion of livestock numbers when extra stock can be accommodated on the commons. This system of livestock management is leading to overgrazing and consequent soil erosion on a large scale. It is also a cause in some countries of serious conflicts with crop farmers.

The most important aspect of CA is the maintenance of soil cover and this is rendered virtually impossible with the current livestock management system in SSA. That is, unless fences are placed around CA plots and this is often politically impossible. However, the current system is patently unsustainable, inefficient, low yielding and a misuse of land resources. If CA is to be introduced on a large scale, there must be a negotiation with livestock owners and local authorities to allow for enclosure of lands, much as what happened in Europe in the 18th and 19th centuries. In time, the land tenure systems in use in OECD countries will be introduced in SSA, giving farmers a capital asset that can be used as collateral for harvest and longer-term credit. The rapid development of Japan, South Korea and Taiwan since the 1940s was financed by farmers who could leverage the capital value of their land. Agricultural production per hectare also increased as farmers could obtain credit for necessary improved seed and fertilizers. Low-input, low-output farming, both in crops and livestock, cannot be an option for much longer, given the need to increase world agricultural production by 50% by 2050 (De Soto, 2000; Linklater, 2014).

for residue, cover crop and weed management especially in sub-humid or humid climates, is the knife roller, crop crimper or vegetation crusher. Knife rollers are available for animal traction and for tractor use, as trailed or tractor-mounted equipment (Fig. 11.1).

For the harvest operation of grain crops under CA the shift from a two-step harvest, with stationary threshing at the field side or in the farmyard, towards a one-stage harvest with combine harvesting is preferred, as this facilitates the retention, return and spreading of as much crop residue as possible.

In surface-irrigated systems the maintenance of the irrigation furrows between permanent raised beds is a regular operation, usually done once a year to prepare the land for the next crop. Also, care has to be taken in the treatment and distribution of crop residues and the choice of furrow openers, to avoid blockage of the equipment or of the furrows and to ensure that the residues perform their function on the beds.

11.2.2 Seeding and planting – technical features

Probably the most significant requirement for mechanization and change of practices from tillage-based farming to CA is the seeding (continuous flow) and planting (precision spaced). Traditionally this operation in small-scale farming is often still done manually by broadcasting, or combining manual seeding with tillage and land preparation. In CA the seeding and planting, with some exceptions, require specific equipment.

The challenge for seeding and planting in CA is to bring the seed into soil contact in an undisturbed soil with residue cover. With the exception of the use of rippers, where seeding can be done manually into the rip lines,

Fig. 11.1. An animal-drawn knife roller crushing a forage radish cover crop (photo courtesy of Brian G. Sims).

or in cases where planting holes are opened with a dibble stick or hoe, special equipment is required to insert the seed under these conditions (Baker *et al.*, 2006).

At manual technology level, the most common planting tool for row crops is the jab planter. This equipment inserts a pre-set amount of seed and fertilizer into the soil. Besides the jab planter, which is like a mechanized planting stick, there is also a no-till manual planter in the form of a two-pronged hoe, replacing the hoe blade with the points for seed and fertilizer release (Fig. 11.2).

Equipment for animal traction and tractors basically follows the same principles. There are specially designed furrow openers that can penetrate the mulch, without accumulating it or pushing it into the soil, and deliver the seed into the undisturbed soil at the desired depth and position. In order to do this, no-till direct seeding equipment is usually robust enough to resist higher soil forces and is heavier than a conventional seed drill. There are different types of furrow opener available for different conditions and budgets, creating more or less soil disturbance. Chisel or tine furrow openers are cheap and robust and can be used on lightweight equipment but create more soil disturbance. Disc openers, particularly double-disc openers, produce less soil disturbance and have a better capacity to manage unchopped crop residues but require higher equipment weight and come at a higher cost, due to the specific requirement for steel quality. Double-disc openers can require weights as much as 250 kg/row to facilitate the cutting of residues and penetration into the soil. Better steel quality allows for thinner but still durable discs. If those discs are mounted in an offset position, one slightly behind the other, and in addition have different diameters, the weight requirements can be reduced to about 100 kg/row (Fig. 11.3), making this technology eligible for animal traction and single-axle tractor planters (Murray *et al.*, 2006; Ashworth *et al.*, 2010).

As no-tillage systems such as CA mature, the trend goes clearly towards minimum-disturbance no-till furrow openers, such as double-disc, cross-slot

Fig. 11.2. Detail of the fertilizer and seed placement prongs of the Chinese Yunfan jab planter (photo courtesy of Brian G. Sims).

Fig. 11.3. Offset discs of different diameter and, additionally, inclined to the vertical (photo courtesy of Brian G. Sims).

or star-wheel type tools, which also facilitate weed management as they mobilize less of the soil weed-seed bank (Ashworth *et al.* 2010).

However, for low-cost no-till planters, low available draught power and hard-setting soils, simple chisel or tine furrow openers are still popular. For example, rippers used to open the soil can be fitted with seed boxes, converting them into no-till planters, such as the Magoye ripper-planter or similar models developed in different countries in SSA. Simple row-crop precision planters exist for animal traction, often with fertilizer application, using ripper tines or double discs as furrow openers. The designs range from simple lightweight long-beam type ripper-planters to self-suspended one- or two-row animal-drawn planters, eventually even with a ride-on option (Sims *et al.*, 2012).

For single-axle tractors (2WTs) one- or two-row precision planter attachments are available, similar to the ones used on four-wheel tractor (4WT) no-till planters (Fig. 11.4). Particularly in Asia (but also in Brazil) many different options for low-cost no-till planters and seeders for 2WTs have been developed. These developments have spread in recent years to several African countries where local adaptations and developments are being produced. Recognizing the conditions of mature CA systems, the residue-handling capacity of these 2WT seeders and planters has been improved considerably, although particularly the seeders for small-grain crops and narrow row spacing (for example, for rice or wheat) often use high-disturbance furrow openers (Li, 2005). No-till seeders and planters for 2WTs are available with chisel openers and cutting discs or special straw choppers/cutters in front of the chisels, with double-disc openers or with strip-till rotary hoes (Friedrich, 2013).

For 4WTs the full range of no-till seeders and planters, starting with small tractors of 23 kW up to the large tractors of 300 kW, is available. Chisel-type furrow openers prevail in the early years of adoption under arid or semi-arid

Fig. 11.4. A Brazilian chisel-tined direct planter adapted for use with a two-wheel tractor (photo courtesy of Brian G. Sims).

conditions with little residue cover, while disc-type furrow openers are more popular in humid areas with higher residue covers and in mature CA systems. For extreme residues, such as rice straw, no-till planters with a built-in straw chopper cleaning the planting rows ahead of the furrow openers have been developed in India and China and are also being tested in Africa (Sims *et al.*, 2012).

11.2.3 Weed and cover crop management – technical features

A third area of change in the mechanization requirements under CA is the management of weeds and soil cover. Soil tillage has traditionally been seen to be a tool for weed management, since it can provide a clean seed bed. The absence of tillage requires a different approach to weed management. This can be done with widely available tools but there is also specific equipment available for non-soil-engaging mechanical weed management, and finally herbicide weed management is also an option that can be considered.

The main difference in weed management is the approach: weed management under CA is done without soil engagement, i.e. no hoes or cultivators. The purpose of weed management is to avoid competition with the crop, not to clean the field of all traces of weeds. It is important to avoid the propagation of weeds through seeds by not allowing them to flower and to keep them low until the crop can exert its competition. Pulled or chopped weeds should be retained in the field as soil cover and mulch. With this in mind, weeds can be manually pulled, slashed with a cutlass or scraped at the soil surface, rolled down by a roller crimper or simply be suppressed by the crop or a cover crop.

For starting CA in conditions of high weed infestation and limited availability of soil cover, the use of herbicides to control weeds is a useful option for reducing labour requirements. Applying them correctly under CA conditions comes with a much lower environmental footprint than soil disturbance.

For small-scale farmers, simple weed wipers such as the Zamwipe may be an option. They do not require specific knowledge in the application of pesticides and carry a lower risk of drift and crop damage than spraying the herbicide (Sims *et al.*, 2012).

The most popular equipment for herbicide application is still the lever-operated knapsack sprayer. This type of equipment, despite its popularity, is difficult to use properly and it should only be employed after the operator has been trained. As for the technologies available, manual sprayers can be adapted to herbicide application under CA conditions by, for example, attaching multi-nozzle spray booms for wider area coverage, or a spray shield to allow inter-row weed management in row crops with total herbicides without damaging the crops (AAB, 2006).

Apart from that, boom sprayers for herbicide application in CA covering boom widths from 2 m to 40 m are available for manual traction (as manual sprayers mounted on carts), or for animal traction following the same principles but allowing for higher tank loads and boom width up to 8 m, or for 2WTs or 4WTs.

While these more common sprayers all use hydraulic nozzles as the atomizer, there is also equipment available for herbicide application using rotary nozzles. This allows controlled droplet application at much lower volumes than hydraulic nozzles. Such sprayers are, for example, used in CA under cotton out-grower schemes in Cameroon, mainly as manually carried sprayers (AAB, 2006).

11.2.4 Power sources

Another significant change that takes place under CA cropping systems is in farm power requirements. Without the energy-intensive tillage operation, required peak farm power is roughly halved. Small-scale farmers using only manual labour can cover considerable areas without the need to hire additional labour or oxen for ploughing. Farmers using animal traction, who in tillage-based systems would hire 4WTs for ploughing or consider their actual mechanization level as transition to a tractor, can continue farming under CA systems without hiring tractors. They can even start to provide services with animal traction planters, considering the cost and capacity of such equipment. In mechanized systems the energy required per hectare is reduced by 50% when plough-based tillage is abandoned. At the same time, tractor size can be reduced on the farm and this is typically in the region of a 40% power reduction. In Africa this has led to a serious reconsideration of the 2WT as a suitable power source for smallholder farmers and CA service providers. This tractor type has in the past mainly failed in Africa, since it did not have the tractive capacity to plough and hence the only major use for it, apart from transport tasks, was in flooded rice farming for ploughing and puddling. Under CA this power source, with the now available no-till planting and seeding equipment and appropriate trailers for local transport, could become a viable option for African small-scale farmers as it has in many Asian countries.

11.3 Adding Resilience to Agricultural Production Systems

Resilience is normally taken to mean the ability of the production system to withstand or tolerate biotic and abiotic stresses and shocks, especially those generated by climate change (Conway, 2012). Normal small-scale variations in weather may not affect production in areas of adequate rainfall but in dry environments they can have a significant effect on crop performance. Other kinds of shock arise from extreme weather events and include sudden drought or flood, or a major pest or disease outbreak. Sometimes the damage is caused not by a sudden shock but by an accumulated stress, such as a gradual decrease in soil organic matter and destruction of soil structure over time due to mechanical soil tillage, or rising soil salinity due to poor water management. Thus, stresses and shocks can have their origins in the type of prevailing climate and weather (not enough rainfall or high rainfall variability), and/or in the condition of the soil (poor infiltration and moisture retention capacity). They can also have their origin in the way farming is practised. Tillage-based farming, whether practised manually or using animal or mechanical power, has been shown globally to cause extensive damage and degradation to all soil qualities and to the soil's productive capacity (Montgomery, 2007).

Mechanization in CA systems helps to build resilience into production systems in a number of ways: (i) rehabilitating and conserving natural resources, including reduction in soil erosion and compaction, and enhancing soil health; (ii) sequestering carbon; (iii) improving water cycling and water and nutrient use efficiency; (iv) timeliness in operation; and (v) improved cropping ratio. All these aspects operate together and synergistically in mechanized CA systems to build and maintain system resilience from ecological and socio-economic viewpoints. These are expressed in a number of ways, such as: improved and more stable yields and profit; reduced production inputs of seeds, water, nutrients, biocides and energy; and greater farm output and return to investment (Kassam *et al.*, 2013).

11.3.1 Rehabilitating and conserving natural resources

The ecological foundation of resilience in CA systems stems from minimizing or avoiding mechanical soil disturbance. This allows the soil to develop and function biologically, which is necessary for building soil structure and porosity, as well as aggregate stability. This is possible when soil microorganisms, especially bacteria, fungi and mesofauna such as earthworms, termites and other arthropods, are allowed to establish themselves as part of the soil food web and perform their functions. Microorganisms transform soil organic matter and crop residue into humus and eventually into sequestered carbon. During the process of humification and decomposition, there is mineralization of organic matter that releases nutrients for plant uptake. Plant nutrients that are anchored biologically are less likely to be leached or lost through volatilization. Fungi such as mycorrhiza produce glomalin, a carbon compound that binds soil particles and imparts aggregate stability to the body of the soil, which then behaves like a sponge. Mesofauna are responsible for mixing soil with organic matter and creating a network of biopores, permitting water percolation and deep drainage of excess water. Earthworms are responsible for mixing soil particles with organic matter and mucus to form worm casts. Termites also incorporate organic matter into the soil, and both earthworms and termites are known to host nitrogen-fixing bacteria in their guts (French *et al.*, 1976).

For microorganisms and mesofauna to perform their biological functions, they require access to organic matter to serve as substrate which, in CA systems, is provided by soil mulch cover from crop residues and other sources. Also, because of minimum soil disturbance, soil organic matter builds up over time, which serves to create a soil that can capture and retain increased amounts of water from rainfall or irrigation. Thus, in soils under established CA systems, there is minimum runoff and soil erosion and maximum water retention. This also applies to plant nutrients, which are better conserved and retained, and along with water and carbon there is better cycling of nutrients, all contributing to building physical, chemical, hydrological and biological resilience. The result is that risks of floods and droughts, and impacts of insect pests, pathogens and weeds, are reduced. When biotic and abiotic stresses and shocks occur, CA cropping systems are better able to cope. In CA systems, crops with different root systems and depths are used, including legumes as food and cover crops and for biological nitrogen fixation, establishing nutrient cycling and expanding the soil biopore network. So it can be seen that, with CA, there is a reduction in water and wind erosion, due to much lower rainfall runoff and increased topsoil stability. Also, due to CA mechanization, there is far less soil compaction (especially if practised in conjunction with controlled-traffic farming). Both these benefits add to ecological and economic resilience through reduced biophysical and economic risks.

Such diversified cropping systems, along with minimum soil disturbance and soil mulch cover, impart resilience, including against insect pests, pathogens and weeds; in addition they serve to improve and rehabilitate soil health and productivity and conserve natural resources. Also, under CA systems, there is a reduction in input required to achieve or maintain a certain yield level. Generally, in motorized mechanized CA systems, seed rates and the use of fertilizer and other agrochemicals, water, fossil energy and time are reduced and so is the capital and maintenance cost of tractors, equipment and machinery. With smallholders, there are increasing opportunities in Africa for 2WT-mounted equipment and machinery, offering the above benefits. Similar benefits can be seen in manually operated CA systems with simple equipment such as no-till hand planters and hand-operated herbicide appliers, as well as with animal-drawn equipment and machinery. Overall, there is an increase in farm output and crop yields as well as profit; and because of increased resilience, there is greater stability in crop performance over time (Farooq and Siddique, 2014; Jat *et al.*, 2014; Thierfelder *et al.*, 2015).

One major advantage of CA mechanization is the reduction in water and wind erosion due to much lower rainfall runoff and increase in topsoil stability. A second advantage is that there is far less soil compaction. Both these benefits add to ecological and economic resilience through reduced biophysical and economic risks (Kassam *et al.*, 2013).

The role of mechanization in managing CA cropping systems is important for soil rehabilitation and natural resource conservation. This is because through CA systems it is possible to reverse the soil degradation that occurs in conventional tillage systems. Appropriate no-till seeders are able to perform seeding and fertilizer application in a single pass, with minimum soil disturbance, while handling soil surface mulch in different ways, depending on the type of soil openers for seeding (see Section 11.2). Mulch and cropping system diversity along with minimum soil disturbance allows soil to recover biologically and topsoil to be built up over time as soil micro-organisms, mesofauna and crop root systems help to create a biologically active soil. With CA systems, less equipment and machinery are required to manage all farm operations compared with mechanized systems for conventional tillage.

11.3.2 Carbon sequestration

It is difficult to sequester carbon in conventional tillage agriculture, due to constant mechanical soil disturbance leading to greater oxidation and decomposition of the soil organic matter, compounded by an inadequate return of biomass (crop residues) to the soil (Shaxson *et al.*, 2008). Consequently, soil health and productivity potential in conventional tillage agriculture will degrade over time and there is a loss in resilience or adaptability of cropping systems to climate change. The opposite is true for CA systems, in which soil organic matter increases over time (Corsi *et al.*, 2012; Sá *et al.*, 2013). This improves the ecological as well as the economical basis of resilience, the latter because of better productivity and relatively more stable yields from year to year.

Overall, tillage agriculture is a net emitter of greenhouse gases (GHG) whereas CA is a net mitigator of climate change due to reduced GHG (CO_2, CH_4, N_2O) emissions from the soil, reduced fossil fuel use in production operations and thus reduced CO_2 emissions, as well as lower energy use and lower emissions of CO_2 in the input supply chain. Thus, mitigation characteristics of CA systems contribute to overall adaptability – ecological, economic and social – to climate change.

11.3.3 Improved water and nutrient use efficiency and productivity

A major consequence of improved soil biological health and greater soil organic matter is the improved water cycling and soil moisture conditions for crop growth and production, leading to improved water use efficiency and water productivity. Similarly, improved soil organic matter levels help to conserve and retain plant nutrients, and improve nutrient use efficiency and nutrient productivity. These improved water and nutrient responses under mechanized CA add ecological and economic resilience to production systems (Shaxson *et al.*, 2008).

11.3.4 Improved timeliness of farm operations and labour productivity

In many regions there is a growing labour shortage for production in commercial farming. This is also increasingly true in smallholder family farms where alternative opportunities are sought in the urban sector (leading to increasing rural–urban migration). Where family labour is the main source of farm power, there is a limit to the area of land that can be managed and how much production can be handled. Where production intensification is involved, mechanization is seen to be an important input as it improves the timeliness of operations for crop establishment and for weed control, offers opportunity for area expansion where extra land is available and also increases labour productivity. Of course, mechanization generally means greater financial investment and this can be a constraint unless service providers can offer affordable services or where there is government support to purchase equipment and machinery for sharing. In the case of smallholders in Africa, there is an increasing demand and opportunity for appropriate and affordable mechanization options.

Improved timeliness of farm operations such as early or timely direct no-till sowing leads to mobilizing greater land and crop potentials. Often two to three weeks can be lost in the case of tillage agriculture in preparing the land for crop establishment by ploughing and harrowing, which leads to loss of soil moisture and delayed sowing. Similarly, there is improvement in yield and a reduction in labour requirement that can lead to greater labour productivity.

11.3.5 Improved cropping in marginal environments and increasing cropping ratios

In semi-arid areas of Africa where the rainfed growing period is short (90–150 days with 400–800 mm rainfall) and where rainfall variability is high, mechanized CA can improve the quality and length of the growing period. This improvement is possible because the loss of soil moisture before sowing is reduced to a minimum, early sowing can be achieved with mechanized CA and maximum rainfall water can be captured and retained in the soil. Also, with an improved length of growing period and moisture condition, cover crops can be undersown a few weeks before the main crop is harvested. This leads to production of extra biomass, adding additional carbon and biologically fixed nitrogen to the soil (if legumes are used) and improving soil cover for longer. All this imparts greater resilience to production systems, significantly reduces the chances of crop failure and minimizes yield reductions during difficult years.

An important consequence of mechanized CA is that greater cropping ratios are possible, partly because the length of the rainfed growing period can be extended due to greater capture and retention of rainfall, and the maximum amount of time available during the growing season can be utilized for crop growth. For example, in the moist sub-humid areas of Africa, mechanized CA can permit double cropping where only one crop could be grown under conventional tillage agriculture. This increase in cropping ratio and in cropping system diversity contributes to system resilience, leading to improved food and nutrition security.

11.4 Constraints to CA Mechanization in Africa

The constraints to CA mechanization in Africa are discussed in this section in a tentative descending order of severity. Mechanization is generally underdeveloped for smallholders in Africa. The challenge is greater for CA (compared with conventional tillage-based farming) since it is relatively new, having been introduced into the continent only in the last 20 years. It is also not automatically supported by the existing supply chain (see Section 11.5) given that it requires different types of equipment, particularly for planting, in comparison with traditional soil tillage equipment.

11.4.1 Unaffordable CA equipment, spares and repair services

There is sufficient agricultural equipment for conventional farming available in the market but it may not be accessible to many farmers. Major constraints are the high costs, poor geographical penetration of distribution and dealership networks, low purchasing power of the majority of the farmers and poor access to financial services (especially credit and loans) by farmers. CA equipment is at an even greater disadvantage as it is relatively unknown and is generally not yet distributed to stores close to farmers.

Low purchasing power of farmers

However good a technology might be, it must be affordable to the intended end users. The overriding issue in developing commercially successful machines is meeting the market demands within acceptable price levels. Agricultural mechanization requires a relatively heavy initial investment which might not be possible for the majority of farmers in Africa who have low farm incomes. The main causes of farmers' low purchasing power are lack of capital, low volumes of production, low producer prices and the dominance of subsistence agriculture.

LACK OF CAPITAL Most farmers in SSA do not have title deeds to their property and hence lack the capital necessary as collateral for investment loans from banks and other financial institutions.

LOW VOLUMES OF PRODUCTION The average yield of all major cereal crops in SSA is very low (1433 kg/ha compared with a worldwide average of 3851 kg/ha in 2013 (World Bank, 2013)), a result of poor crop husbandry practices, lack of working capital for improved seeds and fertilizers, degraded soils and unreliable weather. In some cases, too, the cropped area is restricted by limitations in available farm power, especially where hand-hoe cultivation dominates.

LOW PRODUCER PRICES In some cases, despite farmers' efforts to adopt improved crop husbandry practices and increase their output, poor marketing facilities and infrastructure result in low producer prices, reducing the revenue that could have been obtained. Food aid inflows have sometimes been a disincentive for farmers in Africa to produce more, because they have had to compete with such subsidized commodities. It has been argued that relatively higher food prices would have a positive effect on the development of African agriculture (MacMillan and Beeden, 2014).

DOMINANCE OF SUBSISTENCE AGRICULTURE It is very difficult to transfer improved technology or practices to typical smallholder farmers, because they often operate in a non-market economy. In near-subsistence economies, maximizing the security of mediocre yields is more important than risking the possibility of little or no yield because of the failure of some innovation that, over time, might result in higher yields (Hyden, 1980). There are cases of farmers committing financial resources to the inputs (e.g. subsoiling, herbicide, additional fertilizer, cover-crop seed) and machinery (e.g. jab planter) required for CA in the expectation of increasing yields through targeted investment, only to find themselves worse off because of natural calamities (especially drought).

When crop production is based on minimizing household food security risks as opposed to commercial orientation, low-value food crops are preferred for marginal and less productive agroecological zones. This subsistence farming model, whereby there is little surplus (if any) for the market, leads to poor profitability in the sector. Subsequently, farmers are unable to pay for mechanization services, and their low purchasing power results in a disincentive for all other players in the mechanization supply value chain. It is no wonder, therefore, that agricultural mechanization in Africa is positively correlated with production of the major field cash crops of cotton, wheat, maize, sugarcane and rice.

There is still a small proportion of rural farmers who do not consider farming as a business but as a way of life. However, widespread mobile phone communication and an inability for a family to survive with mere subsistence farming are heavily impacted by the rapidly changing trends in Africa. These trends oblige most farmers to operate within market economies and be willing to take risks such as adopting mechanized CA in order to increase their profits by decreasing their labour costs and increasing their crop yields.

Poor access to capital and credit

With low purchasing power, access to credit would have been an option where agricultural mechanization services are, in essence, profitable. Credit lines for smallholder farmers have been difficult because of stringent conditions from financial institutions. The majority of smallholder farmers do not have the collateral or the capital needed to guarantee their loans.

Machinery prices have been on the increase in SSA following the collapse of the local manufacturing industry and depreciation of local currencies compared with exporting countries. As an example, the Tanzania Agricultural Mechanization Strategy (URT, 2006) stated that while in 1985 a farmer could purchase a 70 hp tractor with implements by selling 870 bags (of 100 kg each) of maize, in 2006 he had to part with 3000 bags to acquire the same tractor and implements.

With three commercial bank branches for every 100,000 adults compared with eight and 38 for Asia and Europe, respectively (World Bank, 2013), SSA lacks adequate access to financial services (credit, savings and insurance). Moreover, over two-thirds of the adult population have no bank account. For the few who are able to access financial services, the cost of borrowing is very high, with interest rate spreads (lending rate minus deposit rate) in SSA among the largest in the world (World Bank, 2013).

Experience from elsewhere in the world has shown that the provision of credit by crop marketing organizations and financial institutions has led to wide-scale adoption of mechanization technologies; in most cases credit can stimulate demand for mechanization.

11.4.2 Unavailability of equipment, spares and repair services

The industry must be able to provide marketable machines that could meet farmers' operational needs at an affordable price. The majority of locally produced agricultural tools and equipment available to farmers are characterized by low quality and relatively high cost due to the underdevelopment of the manufacturing and supply sub-sector, which in turn is due largely to low demand and resultant poor economies of scale in manufacturing.

Availability of mechanized implements, spares and services has a fundamental influence on whether a farmer adopts mechanized CA or not. Functional equipment will need to match the available power sources (hand, animal or tractor), the different soils and crops, terrain and ergonomic needs. Mechanization equipment and machinery are in many cases not available close to the end users due to a number of causes, including in particular: poor distribution and dealership systems; inadequate local manufacturing; weak artisan, blacksmith and craft workshops; and marketing difficulties.

Poor distribution and dealership systems

The development of an efficient distribution and dealership system is affected by the cost of capital and management effort tied up in stocks due to the high seasonality of sales of mechanization equipment. The demand is also very unpredictable, as it is usually correlated with agricultural performance in the previous year. Agricultural machines are heavy, bulky and hence costly to transport, a problem that is perpetuated further by the poor transport infrastructure (railways, waterways and rural roads) for most of Africa. These constraints limit the scope of business in agricultural machinery and many traders opt for fast-moving consumer goods with more regular sales.

Inadequate local manufacturing

Inadequate local manufacturing is attributed to low demand for the mechanization equipment. This, in turn, is attributed to inadequate knowledge by farmers of mechanized CA technologies and lack of comprehensive demand forecasts, obliging manufacturers to rely on often ill-informed estimates. Economies of scale demand a minimum number of units to be manufactured to justify the invested capital.

Lack of trade financing is another important factor. Manufacturing and trading in agricultural machinery and spares is highly seasonal, creating cash-flow problems for most of the year. About 70% of annual machinery sales are realized in three to four peak farming months of a year. In order to ensure timely delivery at the start of the farming season, suppliers need to build up their stocks off-season when their sales are at the lowest level. This is difficult to manage when credit is not easily accessible and is very expensive.

Lack of an encouraging business environment, including high customs tariffs, import duty and sales tax applied to raw materials such as steel, have at times made it more profitable to import finished products free of such charges, putting local manufacturing at a competitive disadvantage.

In Kenya, although agricultural machines are exempted from import duties and sales taxes, they are still subjected to value added tax and other levies. Ndume Ltd of Kenya reports that in the case of importing a small 25 hp 4WT from China, while the unit is cheap enough in China, the cost more than doubles when port charges and other affiliated import charges are factored in (R. Brendon, Kenya, personal communication, 2014).

Weak artisan, blacksmith and craft workshops

In many parts of Africa, village-level repair services are inadequate. Contractors and others in need of spares have to spend a great deal of money or time (in addition to machine downtime) as they travel to major towns in search of parts and/or repair services. The few available blacksmiths and local craftsmen have limited skills and usually work with inferior tools. Government-based initiatives to support rural workshops and traditional blacksmiths have declined following the imposition of trade liberalization in the 1980s.

The working life of CA equipment is reduced by poor maintenance, making the offered services more costly or unprofitable and simultaneously discouraging investors. Poorly serviced equipment is unlikely to perform its intended functions with precision, thus compromising the quality of work.

In parts of Africa, the adoption of simpler smallholder mechanization technologies (e.g. animal traction) depends to a large extent on village artisans who can repair and even manufacture implements. Such craftsmen are more in touch with the needs of farmers.

Marketing difficulties

Pilot CA initiatives have generally not created enough equipment demand, only small quantities being required by the projects and early adopters. Unit production and distribution costs are therefore high and as a consequence reduce profit margins for value chain stakeholders.

The use of CA equipment saves time and inputs (seeds, fertilizers, fuel, machinery wear and tear) making CA more competitive. In the transitional 'shortage' of CA services, the economies of scale are not attained and the services can be more expensive compared with mechanized tillage-based services.

The failure of farmers to express their concerns openly may result in the marketing of unsolicited CA equipment. The African smallholder farmer may be reluctant even to express disagreement when an extension worker attempts to introduce a new practice or piece of machinery the farmer does not like. The farmer may be even more courteous when government-funded or foreign development workers are involved (as handouts or subsidies could be on offer). It is important to establish 'true CA equipment demands' prior to contemplating the production or importation and marketing of large numbers of machines.

11.4.3 Unsuitable equipment

The quality of available CA equipment varies greatly depending on price and manufacturer. The market for simple hand and animal traction equipment (weed-control scrapers, rippers and sub-soilers) is satisfactorily catered for by local workshops and manufacturers. Hand jab planters and precision animal-drawn seeders are also manufactured by some but the quality is often unsatisfactory and more development work is required.

Experiences from Africa identify import-based technology transfer strategies for the introduction of mechanization technologies (and CA in particular) as a significant cause of failures, due to the inappropriateness of technologies developed in and for countries with larger farm sizes. On the other hand, machinery development efforts at public research institutions are geared towards satisfying the farmers' functional needs, rather than meeting the market demand for new machines. Unlike private-sector local manufacturers adept at developing commercially viable machines, the responsiveness of the public sector is not propelled by clearly perceived market demands, it being more preoccupied with innovations.

The technology institutes operating in this field for Africa are primarily the universities and national government agricultural engineering institutions. Despite the good work carried out by some of these institutions, relatively few of their designs have been taken up by manufacturers, primarily due to unproven end-user demand. Furthermore, the quality of the products produced by them is highly variable, with good work produced by some and very poor work from others. The lack of suitable CA seeders for small to medium-sized land-holdings has been identified as the most prominent factor limiting the widespread adoption of CA in Asia (He *et al.*, 2014).

11.4.4 Poor implementation of mechanized CA policies and strategies

Few African countries have mechanization strategies but those that do emphasize the development and promotion of the whole range of manual, animal traction and tractor-based mechanization to cater for variations in farm sizes, topographies and field conditions. CA has also come to be included in such strategies in the last 20 years to address labour and land productivity, climate change mitigation and adaptation and gender issues. However, the tractor is viewed as a symbol of modernization in Africa and, almost invariably, tractors are supplied with a destructive arsenal of tillage implements. Inconsistencies between declared policy goals and programme implementation plans are derailing adoption of mechanized CA.

11.4.5 Seasonality of mechanization equipment businesses

The mechanized CA equipment service provision business is highly seasonal with key services being demanded for a mere 3–4 months in a year in some cases. Innovations are required to spread the ownership costs to make the business profitable. Such enterprise models include: farm contractors moving across geographical areas synchronized with the farming seasons; and mechanization of other value chain operations, including irrigation, transport and postharvest processes in addition to direct seeding.

11.4.6 Limited knowledge and technical skills

Limited knowledge of farmers

Limited access to knowledge by many smallholder farmers hinders the potential for them to make use of opportunities to improve crop and livestock production and is considered to be a priority constraint. The main causes of the limited knowledge of farmers about mechanized CA are: inadequate and ineffective extension services; unskilled extension workers in mechanized CA; weak linkages between research, extension, training, manufacturing and farmers; and mixed messages on the role and effects of herbicides.

INADEQUATE AND INEFFECTIVE EXTENSION SERVICES The ratio of extension workers to farmers varies greatly in Africa; nonetheless some typical figures are 1:2500 for Uganda, 1:3000 for Nigeria and 1:3500 for Gambia. The high numbers of client farmers, coupled with inadequate facilities (such as transport), mean that the number of farmers who can access adequate extension services is very limited.

UNSKILLED EXTENSION WORKERS IN MECHANIZED CA Extension workers are key people in technology transfer. They need interpersonal communication skills and technical qualifications as well. The majority of extension workers have been trained at college in conventional farming methods and lack the background to integrate CA mechanization technology in the total farming system.

Although the academic levels of universities, technical colleges and vocational training centres are acceptable, there is a misalignment in some countries between the type of curricula and skill endowments of graduates from universities and the needs of industry. CA is yet to be included in most university and training institutions' curricula and they are also poorly equipped with field and laboratory equipment.

WEAK LINKAGES BETWEEN RESEARCH, EXTENSION, TRAINING, MANUFACTURING AND FARMERS Many farmers are unaware of the available technologies like CA, while on the other hand manufacturers are unaware of the demand or farmer requirements and preferences.

MIXED MESSAGES ON THE ROLE AND EFFECTS OF HERBICIDES Casão-Junior *et al.* (2012) described the expansion of CA in southern Brazil as being attributed primarily to: (i) the discovery and reduction in price of locally produced and effective herbicides; (ii) the market availability of good quality precision planters and seed drill models; (iii) economic and energy crises; and (iv) availability of no-till system technology for all the main annual crops. The development and promotion of mechanized CA in Africa faces the particular difficulty of having to deal with a majority of smallholders (generally with < 2 ha/family) being exposed to mechanization and use of external inputs for the first time. While herbicides can be part of CA, they are in many cases presented as a luxury option, as manual weeding and biological control are logical alternatives. Concerns about safety and environmental issues associated with their improper use has meant that many extension workers do not to want to be associated with herbicides.

Low and inadequate technical skills of machine operators, repair and maintenance artisans

The shortage of skilled operators for tractors (for example) is considered to be one of the main reasons why government-supported mechanization schemes in Africa have failed. Motorized mechanization of agriculture for predominantly hand-hoe based farmers can be a difficult transformation.

11.4.7 Fragmented small land holdings

According to FAO (2012), 80% of all farms in Africa have an agricultural area of < 2 ha. This has dual impacts on machinery usage: small farm sizes contribute not only to low family income, but also limits the effectiveness of large machinery in fieldwork. While some government-run schemes exist for renting machinery to farms, these are neither widespread nor particularly effective, as cooperative membership remains low. In land preparation and harvesting operations, mechanizing small and non-contiguous parcels of land will usually be inefficient.

11.4.8 Gender issues in mechanization of CA

In SSA, women contribute between 60% and 80% of the labour for food production. Moreover, agriculture is becoming predominantly feminized as a consequence of increasing male out-migration to urban centres (FAO, 1994). Women now constitute the majority of smallholder farmers and provide most of the labour, managing a large part of the farming activities on a daily basis. On the other hand, it is often the men who control the resources required to invest in mechanization (especially capital). Studies show that increasing women's access to productive resources to be at par with those of men would increase farm yields by 20–30% (FAO, 2011b). It would be sensible to consider how women can access or have control of resources invested in mechanization for outscaling CA.

11.4.9 The way forward

It is crucial to recognize the inter-linkage between input markets (including mechanization) and output markets (Lundy *et al.*, 2012), with smallholders lying at the interface between the two. The viability of a mechanization service-provision business also depends on the rates of use of equipment and machinery, which can be extended if there is potential for providing off-farm services such as transport and, perhaps, the maintenance of rural road networks (Petts, 2012). This allows expensive mechanization equipment to be in productive use for more days of the year and to reduce the unit cost of custom work, but of course it is very site-specific and not yet a widespread occurrence. The cost of a unit (service or product) may also be reduced through 'bundling' mechanization services with complementary packages of inputs and materials, assistance in marketing produce, financial services and general advisory service support. A form of bundling that is likely to appeal to customers is the integrated provision of services through support centres that offer a range of goods and services as a 'one-stop shop' where constraints can be addressed in an orchestrated manner (Downing, 2001).

11.5 Supply of Mechanization Inputs

11.5.1 Imports and support networks

As explained in the previous section, one of the major obstacles to the adoption of CA by smallholder farmers in SSA is their gaining access to the most appropriate mechanization options; and more detail on this is provided below. These options include power sources, be they draught animal power or two- or four-wheel tractors (human muscle power is always available but represents the yoke of drudgery and should be alleviated whenever possible).

Generally, organizations that are promoting CA, including government extension services, international donors and NGOs, solve this challenge through the import of machinery that is the most suitable for the conditions being encountered. Although this is a quick and easy option it can result in problems. Firstly, technical backup is unlikely to be available after the project or programme comes to an end, as there is no network of importers and dealers with an interest in providing this service. This situation is slowly (but surely) being overcome as international donors understand the importance of engaging local stakeholders in the process, so that machinery is imported via established dealers who are then motivated to invest in the personnel and infrastructure to be able to supply quality backup services. Of course machinery dealers are only one of the many stakeholder groups comprising the support network for smallholder CA mechanization (Fig. 11.5).

The stakeholders indicated in Fig. 11.5 may often be combined, so that a machinery supplier can also offer a financial package that facilitates machinery acquisition. This could be on the model of lease-buy, in which a farmer leases the machine and becomes the owner when all repayments have been made. What is important, however, is that the support network is recognized and not bypassed by development projects hoping to save costs by cutting corners through importing unsupported equipment and supplying it directly to farmers, often at subsidized rates. This approach, while being

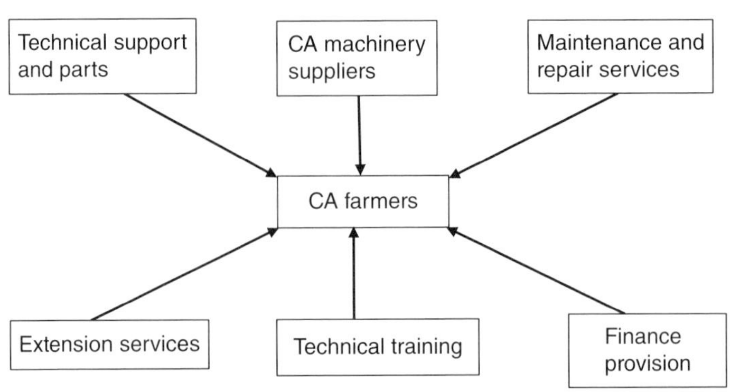

Fig. 11.5. Stakeholders in the CA farm machinery support network for smallholder farmers.

initially attractive to farmers who thereby obtain equipment quickly and cheaply, is unsustainable if the necessary support backup is not engaged.

11.5.2 Local manufacture

CA equipment can also be made locally by national manufacturers. This is a fledgling situation at the moment in SSA, but there is wide potential for development in the area. Although, as has been noted, import of CA equipment is a relatively easy option, it is expensive. Prices will typically double or more from factory to user when the costs of transport, import duties, exporting and importing agents' fees and local retail mark-up are taken into account. Furthermore, imported equipment may not be well suited to local conditions. For example, compacted and stony terrain may be difficult to sow using no-till planters with twin offset disc openers; and moving equipment between scattered smallholder plots may require a transport mode to be supplied with the planter.

Local manufacture should be encouraged whenever possible, as machinery can be produced that is appropriate for local circumstances. Some of the needs and obstacles to be overcome will now be discussed, though it is clear that there is a wide range of inter-country differences, competencies and capabilities.

Creation of a nurturing environment for the incubation of CA manufacturing capability

There are lessons to be learned in this regard from the Brazilian experience. From the 1980s the Brazilian government has been concerned to develop and disseminate CA equipment for smallholder farmers and this public sector support has been crucial to the success of CA for this farming sector. Appropriate machinery has been developed by consortia of private and public sector expertise and the farming community. Finance for machinery acquisition has been made available to smallholder farmers at very advantageous rates and this has stimulated the manufacturing industry to become self-sufficient. This is in stark contrast to the situation in many SSA countries where public sector investment in agricultural sector development is often very low compared with other developing regions.

Materials supply

Manufacturers in SSA frequently point to the uneven playing field on which they are expected to perform and compete. There may be import duty on steel and machinery parts but not on imported tractors and agricultural machinery (e.g. in Kenya). This will disadvantage local manufacturers as it might be difficult for them to compete on price with their local products being made with materials subject to tax whilst imported agricultural equipment is tax free. Addressing this apparent injustice is one way that governments can work towards creating the necessary nurturing environment. Scrap steel may be appropriate for small-scale blacksmiths, but not larger-scale manufacturers.

Lack of appreciation of the importance of critical part design and materials considerations

Local manufacturers making an effort to break into the CA machinery market may often be at a disadvantage due to their limited knowledge of the details of design parameters for such machinery. Examples are: jab planter beaks; tine dimensions, attack angles and materials (use of high-carbon spring steel or easier-to-work alternatives such as Bennox steel); vertical loading on residue cutting discs; seed metering, placement and covering; seed plates for different crops, fertilizer placement, depth control.

Creation of demand

Manufacturers are reluctant to manufacture before having firm orders; at the same time farmers complain that CA technologies are not in the marketplace. This highlights the importance of development projects that, as happened in Brazil, can create the demand and maintain the interest of manufacturers until effective farmer demand can take over. Other actions have been seen to be important and these include the creation of dealer networks and their role in promotion through demonstrations and field days. The public sector's role in providing subsidies and credit in the initial stages is also important. A specific targeted subsidy system (e.g. with vouchers) that supports the use of CA equipment would help.

Need for training to improve the skills base

Skills need to be improved at both the user and manufacturer levels in order for equipment to be effectively designed and properly operated and maintained.

Current situation

The current situation with regard to local manufacture of smallholder CA equipment in SSA is that the simpler tools (such as chisel-tined rippers) are being successfully manufactured but other equipment, such as direct planters and sprayers, is generally not widely available on the local market (Sims *et al.*, 2012).

11.5.3 Specialist CA service provision

Almost by definition the demand for equipment directly by smallholders will be very low in SSA, as the majority of these farmers can be described as resource-poor. The investment cost will generally be too high and there is a high opportunity cost for available capital. As a result, CA mechanization services will have to be supplied from elsewhere and it would seem that the most attractive option to improve access by smallholders to CA mechanization would be to offer the service from well equipped and well trained local service providers.

Service providers do not necessarily have to operate on a large scale. Neither do they necessarily need to be practising farmers, though that would be an advantage, as discussed below. It seems that with smallholder agricultural mechanization and sustainable intensification in rural areas, the range or availability of reliable, skilled and appropriately equipped service providers is very limited or non-existent, especially in SSA. Yet the need for services related to farm power-intensive and time-sensitive agricultural tasks is very high. Timely, good quality and affordable services are needed in many places in rural areas.

For farmers who already own implements, and maybe draught animals or a tractor, the step to becoming a service provider is not so great. However, a potential service provider should preferably be a good farmer who is familiar with, and concerned about, agricultural practices and methods that are in line with sustainable intensification and with efficient use and care of natural resources and external inputs. Yet for service providers themselves, the most important challenge is to make a living and income out of the offered services.

For farmers to make the transition to service provision they need to be part of a network and have access to knowledge and training (see Fig. 11.5). They need to be aware of, or should be equipped with, the principles of good agricultural practices and what it means in the context of the local neighbourhood, where most if not all of their potential clients will be located. Potential service providers should: (i) have a reputation as reliable and knowledgeable farmers; (ii) be open to innovations in farming such as CA, and its specific approach to mechanization, and the use of alternative implements such as rippers and direct seeders; and (iii) be able to operate, repair and maintain agricultural equipment.

In order to become a good business person and service provider, an entrepreneur needs to develop the skills needed to make the business financially sustainable. As a result, the following challenges may have to be overcome: (i) it may be important to focus on services that have the highest demand or that are not yet provided by another service provider; and (ii) it may be necessary to acquire additional or new equipment, and so the need for credit becomes clear and with it the need for a detailed business plan.

If equipment is purchased and if the investment is large, e.g. a tractor, it is important to devise an annual plan for services that can be provided at different times of the year and for services that may be independent of agricultural seasons. Such additional services, e.g. transport or power provision for irrigation or milling, can make the difference between a service business being sustainable or not.

The specifications of the equipment needed will depend on the requirements of the particular site landscape, the power source available and the quality of the supply chain. The minimum requirement is likely to comprise a ripper, no-till planter and sprayer, plus perhaps a trailer. The skills that are needed, and alluded to above, will often require specific training on both technical and management aspects and such training can be (and in practice is) provided by the public sector or international donors, or both in combination. The content of any training course will be very site-specific and should concentrate on examining and solving local problems and deficiencies.

In general terms the technical content would include sessions on machinery field operation, maintenance and calibration. Management training would cover feasibility studies, record keeping, accounting, costing of CA equipment use, income assessment and business planning and marketing.

During practical examination of market research, business planning and marketing training, the trainees and trainers may conclude that CA service providers depend on very short agricultural seasons with maybe one or two months of working time. This will not usually be sufficient for a valid business plan and may not justify the investment that has to be made in the equipment. Therefore, potential service providers need to consider broadening their business into other areas for which the power source can be used. Such potential business expansion could include, for example: a tractor-drawn trailer for all-year-round transport services of all types of goods; a tractor-powered maize sheller or harvester; or possibly a tractor-driven pump for water lifting and irrigation.

By examining the business holistically and creating business opportunities for otherwise slack periods, potential service providers will be able to develop an inclusive all-year business model that would include the CA planting and crop care and management services.

11.6 Conclusions

This chapter has shown the crucial importance of mechanization for smallholder farming in SSA. The alleviation of drudgery, increase in labour productivity, better timeliness of farming operations, greater area under cultivation and the adoption of innovations are all facilitated by well-focused mechanization.

The mechanization of CA is best done by private sector entrepreneurs offering a range of services according to a robust business model that addresses the needs of the community, while providing a livelihood for the entrepreneur. Mechanization interventions should have the principles of sustainable production intensification at their heart and should achieve increased and more stable production whilst conserving natural resources, especially soil and water. It is important for smallholders to have access to CA mechanization without exposing themselves to high levels of risk. To that end the use of qualified and competent CA service providers is an appropriate vehicle to widen CA adoption.

The mechanization requirements for successful CA expansion are already available. Implements exist for land preparation (e.g. rippers and subsoilers), residue, weed and cover crop management (e.g. straw spreaders, knife rollers, sprayers, slashers and scrapers) and direct seeding and planting (with chisel-point and disc openers). Equipment is also available for manual, draught-animal or tractor-power (both 2WTs and 4WTs) systems.

CA and its associated mechanization confers resilience to smallholder farming systems in the sense that there is better toleration of environmental and economic stresses and shocks. Soil erosion and compaction are dramatically reduced and carbon is added to the system, resulting in improved soil fertility, structure and water-holding capacity. This results in improved and more stable yields and profits, reduced use of inputs (including energy) and better returns to investment.

CA, when sustainably mechanized, is an invaluable tool for rehabilitating degraded landscapes and conserving natural resources. It also encourages carbon sequestration through the return of biomass to the soil and the reduced oxidation of soil organic matter. The greater resilience of soils managed under a CA regime means that cropping is made possible in marginal environments, cropping ratios can be increased in favourable environments and the timeliness of crop establishment can be greatly improved. CA also has the potential to improve rural livelihoods and stem rural–urban migration by creating new business opportunities in the agricultural sector.

However, there are constraints to the adoption of CA mechanization in SSA due to the resource-poor condition of many smallholder farmers. Constraints range from unaffordable equipment to unavailability of equipment in local markets, poor extension services and access to knowledge, and poor and conflicting government policies and strategies. The first step is to identify the constraints in a particular region and then to develop a plan to conquer them.

The best way to provide smallholder farmers with CA mechanization technology is through private sector mechanization service providers. Such providers are part of the supply chain that includes machinery manufacturers and suppliers, maintenance and repair services, extension support, access to finance and, crucially, technical and business skills training. Well trained and knowledgeable service providers well versed in the sustainable crop production intensification paradigm are ideally placed to accelerate the expansion of CA mechanization for smallholder farmers.

Local manufacture of CA equipment should be encouraged and supported and will help to ensure that equipment is best suited to local conditions. The public sector has a vital role to play in creating a level playing field that is not disadvantageous to local manufacture and in fostering an enabling environment for facilitating acquisition of CA equipment by farmers and service providers.

References

AAB (Association of Applied Biologists) (2006) International advances in pesticide application 2006. *Aspects of Applied Biology* 77, 524.

Ashworth, M., Desbiolles, J. and Tola, E.K. (2010) *Disc Seeding in Zero-till Farming Systems – A Review of Technology and Paddock Issues*. Western Australian No-Tillage Farmers' Association (WANTFA), Northam, Western Australia.

Baker, C.J., Saxton, K.E., Ritchie, W.R., Chamen, W.C.T., Reicosky, D.C., Ribeiro, M.F.S., *et al.* (eds) (2006) *No-tillage Seeding in Conservation Agriculture*, 2nd edition. FAO, Rome, and CAB International, Wallingford, UK.

Baudron, F., Sims, B., Justice, S., Kahan, D.G., Rose, R., Mkomwa, S., *et al.* (2015) Re-examining appropriate mechanization in Africa: two-wheel tractors, conservation agriculture, and private sector involvement. *Food Security* 7(3), 16.

Casão Junior, R., Guilherme de Araújo, A. and Fuentes-Llanillo, R. (2012) *No-Till Agriculture in Southern Brazil*. Food and Agriculture Organization of the United Nations, Rome.

Conway, G. (2012) *One Billion Hungry: Can We Feed the World?* Cornell University Press, Ithaca and London. p. 439.

Cooley, L. and Kohl, R. (2006) *Scaling Up – From Vision to Large Scale Change: A Management Framework for Practitioners*. Management Systems International, Washington DC.

Corsi, S., Friedrich, T., Kassam, A., Pisante, M. and de Moraes Sá, J.C. (2012) Soil organic carbon accumulation and greenhouse gas emission reductions from Conservation Agriculture: a literature review. *Integrated Crop Management* 16, 89.

de Moraes Sá, J.C., Bürkner dos Santos, J., Lal, R., de Moraes, A., Tivet, F., Freire Machado Sá, M., *et al.* (2013) Soil-specific inventories of landscape Carbon and Nitrogen stocks under no-till and native vegetation to estimate Carbon offset in a subtropical ecosystem. *Soil Science Society of America Journal* 77(6), 2094–2110. doi:10.2136/sssaj2013.01.0007.

De Soto, H. (2000) *The Mystery of Capital: Why Capitalism Triumphs in the West and Fails Everywhere Else*. Black Swan-Random House, London.

Downing, J. (2001) Introduction: business services for small enterprises in Asia. In: Levitsky, J. (ed.) *Small Business Services in Asian Countries*. Intermediate Technology Development Group, Rugby, UK.

FAO (1994) *Women, Agriculture and Rural Development, a Synthesis Report of the Africa Region*. Food and Agriculture Organization of the United Nations, Rome.

FAO (2006) *Farm Power and Mechanization for Small Farms in Sub-Saharan Africa*. Agriculture and Food Engineering Technical Report 3. Food and Agriculture Organization of the United Nations, Rome.

FAO (2011a) *Save and Grow. A Policymaker's Guide to the Sustainable Intensification of Smallholder Crop Production*. Food and Agriculture Organization of the United Nations, Rome.

FAO (2011b) *The State of Food and Agriculture 2010–2011: Women in Agriculture – Closing the Gender Gap for Development*. Food and Agriculture Organization of the United Nations, Rome. Available at: http://www.fao.org/docrep/013/i2050e/i2050e.pdf (accessed 23 September 2016).

FAO (2012) *The State of Food and Agriculture 2012. Investing in Agriculture for a better Future*. Food and Agriculture Organization of the United Nations, Rome. Available at: http://www.fao.org/docrep/017/i3028e/i3028e.pdf (accessed 15 November 2016).

Farooq, M. and Siddique, K.H.M. (eds) (2014) *Conservation Agriculture*. Springer International Publishing, Cham, Switzerland.

French, J.R.J., Turner, G.L. and Bradbury, J.F. (1976) Nitrogen fixation by bacteria from the hindgut of termites. *Journal of General Microbiology* 95, 202–206. Available at: http://mic.microbiologyresearch.org/content/journal/micro/10.1099/00221287-95-2-202 (accessed 15 November 2016).

Friedrich, T. (2013) Agricultural mechanization and the environment. In: Kienzle, J., Ashburner, J.E. and Sims, B. (eds), Mechanization for rural development: a review of patterns and progress from around the world. *Integrated Crop Management* 20, 181–204.

He, J., Zhang, Z.Q., Li, H.W. and Wang, Q.J. (2014) Development of small/medium size no-till and minimum-till seeders in Asia: a review. *International Journal of Agricultural and Biological Engineering* 7(4), 1–12.

Hyden, G. (1980) *Beyond Ujamaa in Tanzania. Underdevelopment and An Uncaptured Peasantry*. Heinemann Educational Books, London.

Jat, R.A., Sahrawat, K.L. and Kassam, A.H. (2014) *Conservation Agriculture: Global Prospects and Challenges*. CAB International, Wallingford, UK.

Kahan, D.G. (2007) *Business Services in Support of Farm Enterprise Development: A Review of Relevant Experiences (Vol. 18)*. Food and Agriculture Organization of the United Nations, Rome.

Kassam, A.H., Basch, G., Friedrich, T., Shaxson, F., Goddard, T., Amado, T., *et al.* (2013) Sustainable soil management is more than what and how crops are grown. In: Lal, R. and Stewart, R.A. (eds)

Principles of Soil Management in Agro-ecosystems, Advances in Soil Science. CRC Press, Boca Raton, Florida, USA.

Landers, A. (2000) *Farm Machinery: Selection, Investment and Management*. Farming Press, United Business Media, London.

Li, H.W. (2005) *The Current Status of Conservation Tillage in China*. Conservation Tillage Research Centre of the Ministry of Agriculture, Beijing, Peoples' Republic of China.

Linklater, A. (2014) *Owning the Earth, the Transforming History of Land Ownership*. Bloomsbury, London.

London, T. and Hart, S. (2010) *Next Generation Business Strategies for the Base of the Pyramid: New Approaches for Building Mutual Value*. FT Press, New Jersey, USA.

Lundy, M., Becx, G., Zamierowski, N., Amrein, A., Hurtado, J.J., Mosquera, E.E. and Rodríguez, F. (2012) *LINK Methodology: A Participatory Guide to Business Models That Link Smallholders to Markets*. CIAT Publication No. 380. Centro Internacional de Agricultura Tropical, Cali, Colombia.

MacMillan, A. and Beeden, P. (2014) Perhaps we should pay more for our food. *Agriculture for Development* (UK) 23, 10–14.

Montgomery, D. (2007) *Dirt: The Erosion of Civilizations*. University California Press, Berkeley and Los Angeles, California, USA.

Mrema, G.C., Baker, D. and Kahan, D. (2008) *Agricultural Mechanization in Sub-Saharan Africa: Time for a New Look*. Agricultural Management, Marketing and Finance Occasional Paper 22. Food and Agriculture Organization of the United Nations, Rome.

Murray, J.R., Tullberg, J.N. and Basnet, B.B. (2006) *Planters and Their Components: Types, Attributes, Functional Requirements, Classification and Description*. The University of Queensland and Australian Centre for International Agricultural Research, Canberra.

Petts, R. (2012) *Handbook of Intermediate Equipment for Road Works in Emerging Economies*. Intech Associates, Leatherhead, UK.

Shaxson, F., Kassam, A., Friedrich, T. Boddey, B. and Adenkunle, A. (2008) Underpinning conservation agriculture's benefits: the roots of soil health and functions. In: An international technical workshop on investing in sustainable crop intensification: the case for improving soil health. *Integrated Crop Management* 6, 73–111.

Sims, B., Röttger, A. and Mkomwa, S. (2011) *Hire Services by Farmers for Farmers*. Diversification Booklet 19. Food and Agriculture Organization of the United Nations, Rome.

Sims, B.G., Thierfelder, C., Kienzle, J., Friedrich, T. and Kassam, A. (2012) Development of the Conservation Agriculture Equipment Industry in Sub-Saharan Africa. *Applied Engineering in Agriculture* 28(6), 813–823.

Thierfelder, C., Bunderson, W.T. and Mupangwa, W. (2015) Evidence and lessons learned from long-term on-farm research on Conservation Agriculture systems in communities in Malawi and Zimbabwe. *Environments* 2, 317–337.

URT (2006) *Tanzania Agricultural Mechanization Strategy (TAMS)*. Ministry of Agriculture Food Security and Cooperatives, Irrigation and Technical Services Department, Agricultural Mechanization Section, Dar es Salaam, United Republic of Tanzania.

Wongtschowski, M., Belt, J., Heemskerk, W. and Kahan, K. (eds) (2013) *The Business of Agricultural Business Services: Working with Smallholders in Africa*. Royal Tropical Institute, Amsterdam; Food and Agriculture Organization of the United Nations, Rome; and Agri-ProFocus, Arnhem, The Netherlands.

World Bank (2013) *World Bank Indicators – World – Agricultural production*. Available at: http://www.tradingeconomics.com/sub-saharan-africa/cereal-yield-kg-per-hectare-wb-data.html and at: http://www.tradingeconomics.com/world/cereal-yield-kg-per-hectare-wb-data.html (accessed 23 September 2016).

12 Conservation Agriculture in South Africa: Lessons from Case Studies

Hendrik J. Smith,[1*] Erna Kruger,[2] Jaap Knot[3] and James N. Blignaut[4]

[1]*Grain SA, Pretoria, South Africa;* [2]*Mahlathini Organics, Pietermaritzburg, South Africa;* [3]*KEL Growing Nations Trust, Ladybrand, South Africa;* [4]*University of Pretoria, South Africa*

12.1 Introduction

Mainstreaming sustainable agriculture systems in South Africa has become imperative. Severe environmental degradation, low farm profitability and poverty associated with current conventional production systems have brought the agricultural sector to a crossroads. If farmers in South Africa are offered a better chance to survive on the farm and if sustainable and economically viable agriculture is to be achieved, then the paradigms of agricultural production and management must be changed.

Conservation Agriculture (CA) is an approach to managing agroecosystems for improved and sustained productivity, increased profits and food security while preserving and enhancing the resource base and the environment. CA is characterized by three linked principles (FAO, 2001; Lal, 2010), namely: (i) continuous no or minimum mechanical soil disturbance; (ii) permanent organic soil cover; and (iii) diversification of crop species grown in sequences and/or associations, including the use of cover crops.

Since sustainable agriculture systems, such as CA, are social constructs (Coughenour and Chamla, 2000), this 'paradigm change' or innovation process can only be undertaken on farms and by farmers and has in fact already been happening for decades. This process thereby critically functions as on-farm, farmer-centred innovation systems (ISs), embracing not only the science suppliers but also the totality and interaction of actors involved in innovation. This IS approach brought the understanding that innovation emerges from the interaction of multiple stakeholders, i.e. researchers, advisory service providers, non-governmental organizations, farmers' organizations and private sector actors (Röling and Wagemakers, 1998; Hall *et al.*, 2006;

*hendrik.smith@grainsa.co.za

Waters-Bayer *et al.*, 2009). Under this paradigm of 'constructivism', sustainable agriculture is not seen as a simple model or package to be imposed; it is more a process for learning or a self-perpetuating learning system (Smith, 2006; Putter *et al.*, 2011). Unfortunately, to date the introduction of CA to farmers (commercial and smallholder systems) has largely excluded IS approaches and models, more notably in the government programmes with larger impact potential.

In this chapter, the diverse and risky South African environment, the major agrarian production systems, land degradation and the adoption of, and some initiatives on, CA mainstreaming are described. The latter emphasizes lessons learned from two recent case studies in South Africa. A review of a modelling approach for commercial grain production is provided in the first case. How smallholder structures are better defined and transformed through CA IS, whereby not only soil health and crop productivity are improved but also access to other components of the agricultural system, such as land, finance, inputs, technology, knowledge and diverse markets, are discussed in the second case.

12.2 Background

12.2.1 South Africa's risk-prone environment and food security

South Africa's landscapes and agroecology are highly complex and diverse with large variation, especially in terms of soil properties (Fey, 2010). Depending on factors intrinsic to arability, 16.7 million hectares (12–14%) of agricultural land is rated, soil-wise, as suitable for dryland cropping and 83.9 million hectares (68.6%) for grazing land. Only 3% of agricultural land is of high agricultural potential (Smith, 1998). Nearly 91% of the country is sub-arid, semi-arid or dry-humid, while 8% is hyper-arid. A 500 mm rainfall line running from north to south divides the country into more or less equal parts, with the drier western part being theoretically unsuitable for regular sustainable crop production. Even in moister areas receiving more than 500 mm rain per year, production risk is high because of poor seasonal distribution and erratic rainfall with moisture stress at critical plant development stages (Barnard *et al.*, 2000; De Villiers *et al.*, 2004). The country is divided according to rainfall into three seasonal zones: summer, year-round and winter rainfall areas, allowing virtually all field and horticultural crops to be produced.

High temperatures, relatively low humidity and high vapour saturation deficits result in comparatively high evapotranspiration rates, often exceeding precipitation. This is exacerbated by unproductive soil water loss through runoff, *in situ* evaporation and deep drainage. The rule rather than the exception of weather is high-energy rainstorms (resulting in soil sealing and crusting), high runoff, flash and prolonged floods, droughts of varying intensity and duration, hail storms and frost (Bennie and Hensley, 2004).

Placing the above picture into a global context: South Africa's rainfall is half the global average, with challenges related to water availability exacerbated by rainfall variability and a decline in water quality. These challenges are fast becoming major constraints for South Africa's vision of sustainable agriculture and rural development to foster major macro-economic objectives that are aimed at accelerating poverty alleviation and its multitude of consequences as well as reaching its sustainable development goals (Blignaut and Van Heerden, 2009; Turton, 2015).

The environmental conditions described above are inextricably linked to food production and security. The changes in such conditions and the pressures as a result of the extreme conditions are a matter of national importance if (and to what extent) food security is to be achieved under rapidly mounting population pressure. With agricultural resources that are limited in quantity and/or quality, South Africa will find it difficult to reduce poverty and maintain or make adequate progress in food security in the future (Blignaut and van Heerden, 2009). This warning is in contrast to the view that South Africa is largely a food-secure nation (DAFF, 2011; FAO, 2015) producing enough staple foods or having the capacity to import food, if needed, in order to meet the basic nutritional requirements of its population. Despite this national view on food security, one in four people currently suffers hunger on a regular basis and more than half of the population live in such precarious circumstances that they are at risk of going hungry (Demetre *et al.*, 2009). The unemployed urban poor, the landless rural poor and unemployed youth are particularly vulnerable to food insecurity. Furthermore, rural people (smallholders) do not appear more food-secure because they engage in subsistence farming. On the contrary, most people in rural areas of South Africa buy their food from shops (Oxfam, 2014). Seventy per cent of the country's poorest households live in these rural areas. A common problem is that smallholders have limited or no resources to invest in improved agricultural production practices (Baiphethi and Jacobs, 2009). Exaggerated or inaccurate views on national food security are therefore dangerous from a national security perspective as they signal that 'all is good', while the situation is dire from a grassroots perspective.

The national security concern that is tied to food (in)security is further affected by the impact of erratic weather conditions (exacerbated by a combination of the El Niño effect and climate change). The late start of seasonal rains in November 2015 resulted in South Africa's maize harvest decreasing by some 30% compared with the 2013/14 crop (a bumper crop of 14.3 million tonnes), a situation that triggered food (maize-meal, samp, etc.) price increases of around 18% compared with 2013/14 and adversely affected the food security situation, especially among poor people (W. Sihlobo, Pretoria, personal communication, 2015).

12.2.2 Major agrarian production systems: an overview

The major field crops, planted across South Africa according to their specific agroecological suitability, are maize, wheat, grain sorghum, groundnuts, sugarcane, sunflower, soybeans, oats, barley, rye, dry beans, cowpeas, chicory, cotton, wattle bark, hay (including lucerne) and tobacco. These crops are produced in a highly dualistic production system comprising a well developed commercial sector (about 35,000 farm units) and a small-scale farming (smallholder) sector (about 3 million small-scale units) (Cousins, 2015). This skewed agrarian structure of South Africa should be the main focus of policy; and understanding its underlying dynamics is key for designing effective interventions (Table 12.1).

Ideally, sustainable dryland farming systems should be conducive to soil resilience, soil and water conservation, sequestering carbon (C), reducing greenhouse gas emissions, curbing land degradation and pollution and minimizing inputs while sustaining soil fertility and soil productivity (Smith *et al.*, 2010). Cultivation methods in South Africa are influenced by the agricultural sector, agroecological regions and the choice of the farmer as land owner/user. Current conventional (mainstream) commercial farming is associated with high external input (HEI) systems in close cooperation with various input supplying companies. There is enough evidence that this HEI system becomes increasingly risky (financially) and environmentally

Table 12.1. The agrarian structure of South Africa in 2014 (from Cousins, 2015).

Farmers	Numbers	Key features
Top 20% of large-scale commercial farmers on private land	7,000	Sophisticated, specialized, capital-intensive farmers, producing for export or for agroprocessing and large retailers; produce bulk of produce, perhaps as much as 80%
Medium- to large-scale commercial farmers on private land	9,000	Some farmers succeed, some struggle, some are unable to earn a living from farming alone
Small- to medium-scale commercial farmers on private land	19,000	Many cannot survive from farming alone; includes hobby farmers
Small-scale commercial farmers in communal areas and in land reform contexts	5,000–10,000	Many farmers earn income from off-farm incomes and businesses in addition to farming
Market-oriented smallholder farmers in communal areas and land reform contexts, supplying tight value chains (e.g. under contract)	5,000–10,000	Many grow fresh produce under irrigation, others are livestock producers, and a few engage in dryland cropping
Market-oriented smallholder farmers in communal areas and land reform contexts, supplying loose value chains	200,000–250,000	Many grow fresh produce under irrigation, others are livestock producers. Few depend wholly on farming
Subsistence-oriented smallholder farmers growing food for themselves and selling occasionally	2–2.5 million	Most crop production takes place in homestead gardens, some of which are quite large. Occasional livestock sales by some

unsustainable (Blignaut *et al.*, 2015). For the past 50–100 years, conventional cropping practices in South Africa have been based on aggressive tillage involving primary and secondary soil preparation for weeding and seedbed preparation before planting. A range of tractor-drawn implements, such as mouldboard ploughs, discs, chisels, rippers, rotary tillers, tine cultivators and disc harrows, are used in this system. After planting, weeding is done through a combination of mechanical and chemical control practices. Conventional HEI practices are dependent on high levels of external inputs, such as synthetic fertilizers, pesticides, fungicides and herbicides. Due to the negative effects of these conventional practices on the environment and farmers' profitability, there is a great need for alternative low external input sustainable agriculture (LEISA) systems, based on the rigorous application of CA principles at farm level to conserve and improve the natural resources soil and water and to contribute to farm productivity and profitability.

12.2.3 Land degradation

South Africa's soil mantle is complex and diverse, with a large variation in soil properties. Low soil organic matter content in combination with other poor soil qualities (e.g. shallow depth, coarse texture and degree of leaching), in addition to topography, result in not only natural limitations for agricultural production, but also high susceptibility to biological, chemical and physical soil degradation. Up to 70% of the fertility in South African topsoil is associated with organic matter (Du Preez *et al.*, 2011b). Agricultural activities on these soils that lead to a depletion of organic matter will cause a significant reduction in the fertility status of the soils. It has been shown (Du Preez *et al.*, 2011a) that South African soils have low organic matter levels. About 58% of soils contain less than 0.5% organic carbon and only 4% contain more than 2% organic carbon. It has been indicated that in South Africa between 46% (Swanepoel *et al.*, 2015), 50% (Du Preez, 2003) and 65% (Lobe *et al.*, 2001) of soil organic matter has been lost in arable soils due to conventional tillage-based cropping practices. In soils cultivated for approximately 90 years, carbon concentrations have reduced to 34% of the concentrations in native grasslands (Lobe *et al.*, 2001). Depletion of soil organic carbon (SOC) has not yet stabilized, because of the progressive losses of silt-size particles and its associated organic matter caused by wind (Lobe *et al.*, 2001) and water erosion (Le Roux and Smith, 2014).

Gully erosion is widespread across the country, as indicated by the area affected by gullies in Table 12.2.

The percentages mentioned above as a fraction of total land in the province seem low but have permanently affected 592,003 ha. To put it in perspective, it implies feed for approximately 100,000 livestock units/year, or potentially more than 1 million tonnes of maize grain, enough to feed up to 1 million households per year with maize meal. The average predicted

Table 12.2. Area affected by gully erosion in South Africa (Mararakanye and Le Roux, 2012).

Province	Area (ha)	Percentage of province
Eastern Cape	161,517	0.9
Northern Cape	160,885	0.4
KwaZulu-Natal	92,543	0.9
Free State	64,674	0.5
Limpopo	58,669	0.4
Western Cape	25,403	0.2
Mpumalanga	17,420	0.2
North West	10,782	0.1
Gauteng	110	< 0.1
Total	**592,003**	

annual soil loss rate for all land in South Africa is 12.6 t/ha, while the average rate under annual cultivated cropland (grain crops) is 13 t/ha (Le Roux *et al.*, 2008), which is much higher than the annual natural soil formation rate of < 5 t/ha. This simply means that much more topsoil is lost than what is gained as mineral particles due to weathering of the parent material in the subsoil, or losing 3 t of topsoil for every tonne of maize produced every year. Additionally there are various other soil degradation threats, such as soil compaction and crusting, soil acidity, alkalization and waterlogging, desertification and a resultant decline in biological productivity and soil pollution. The serious need for a soil protection strategy is underlined by the state of soil degradation in South Africa.

12.3 Conservation Agriculture in South Africa: A Reflection

12.3.1 Why CA?

CA is more than a viable alternative to HEI conventional agriculture (CV) as it provides potential solutions to a number of wide-ranging challenges, including economic viability, ecological sustainability and the social acceptability of farming (see Box 12.1). It is also a viable method for smallholder and commercial farming at all scales, addressing both food and job security – both issues of high importance from a national security perspective to mitigate against the risk of domestic social revolution.

The success of CA under diverse agroecological conditions is now being documented in South Africa and appears to justify investment of human and financial resources, whenever and wherever conditions permit it (Smith *et al.*, 2008, 2010; Knot, 2014; Blignaut *et al.*, 2015; Modiselle *et al.*, 2015). Multiple benefits are delivered by CA in terms of yield, factor productivity (efficiency), sustainability of land use, income, timeliness of cropping practices, ease of farming and ecosystem services.

Box 12.1. Why CA? A movitation.

- The increasing cost-pressure and declining gross margins of farming enterprises using conventional tillage.
- The decline and collapse of soil quality and soil ecosystem services. At this stage competitive yields are not feasible without the use of inorganic fertilizer. Declining yield trends in some areas show that the effect of this practice is reaching its limit and that soil ecosystem services should be restored to regain soil productivity, reduce risk and increase profitability. Soils can be rebuilt or rehabilitated with CA through rigorous application of all its principles.
- The impact of climate change on weather patterns, water regimes, biodiversity and ecosystems services will put pressure on farmers to adapt their farming systems and management styles to increase their resilience and sustainability.
- A growing awareness, knowledge and self-organization among farmers (as stewards of the land and natural resources), scientists and agribusiness to use and promote sustainable agricultural practices. The networking of these key actors creates so-called innovation platforms, which are ideal structures to promote and scale out CA.
- A need to improve the resource use efficiency and competitiveness of farming practices relies on healthy soils, healthy biodiversity and innovative farmers.
- The need to rebuild the status and image of farming, which has been severely damaged by a negative environmental footprint and poor socio-economic conditions. CA innovation platforms have the ability to generate or contribute to social capital in rural societies to the benefit of the society as a whole.

12.3.2 Adoption of CA in South Africa

Large-scale commercial production sector

There are many agroecological zones in South Africa within which CA systems are currently scaled out by farmers within their own socio-economic realities, as well as soil, climate and other environmental conditions. CA systems have been adapted in climates ranging from semi-arid (250–300 mm per annum) to sub-humid (1000 mm per annum), soils from sandy (< 5% clay) to clay (> 60% clay), landscapes from flat (< 3% slope) to steep (> 15% slope), and crops ranging from maize, soybean and sunflower to pastures such as kikuyu, lucerne and clover (Fowler, 2004).

The adoption of CA principles and practices has increased considerably under commercial agriculture in the major grain-producing regions of South Africa, especially in the past decade. Initially, after sporadic adoption approximately 40 years ago, it was slow, primarily due to a lack of awareness, knowledge and skills among key stakeholders, such as farmers, scientists and extension services. As experience grew and conventional methods proved limiting, especially in addressing land degradation and input cost pressures, the increasing exposure to successful CA practices and the benefits thereof helped in the transition process. According to the most recent assessment, 40% of commercial farmers across all grain-producing areas of South

Africa have adopted all CA principles (Findlater, 2015) and it is expected that adoption trends will increase sharply over the next decade. The primary reasons for commercial farmers to adopt CA are financial benefits (profit margins), positive impact on water use efficiency, the reduction of water runoff and soil erosion and positive environmental impacts. It can be argued that it should be easy for commercial farmers to adopt CA, especially those who can justify the expensive CA equipment and move to more minimal labour requirements.

The adoption patterns between the agroecological regions or provinces differ considerably. According to Findlater (2015) and Modiselle *et al.* (2015), the most successful CA adoption has been in the winter rainfall regions (Swartland and Ruens area on undulating topography) of the Western Cape with around 80% adoption. The major cropping system in this region is wheat in rotation with legume pastures (medic and lucerne), which has improved the integration and management of livestock and soil health status. Farmers are also using canola in rotation but are mostly reluctant to use cover crops, due to the limited seasonal rainfall. However, a growing number of farmers, supported by researchers from the Western Cape Department of Agriculture (J. Strauss, Stellenbosch, personal communication, 2015), are starting to incorporate cover crops in their annual rotation systems to improve their soil health status.

In the relatively high summer rainfall area (> 750 mm per annum) of KwaZulu-Natal, adoption rates are between 60% and 70%. In the mist-belt region, many farmers continue to monocrop maize, and the maize–soybean system has been described as 'repetition, not rotation'. In addition, the removal of maize stover for silage has resulted in low soil cover and producers have been experimenting with winter cover crops (Fowler, 2004). In the higher-altitude regions closer to the Drakensberg, a maize–soybean–wheat rotation system (especially under irrigation) has been adopted and annual cover crops have started to be included. Winter cover crops used in a relay intercropping system seem to be more popular, with the use of cover crops such as black and white oats, (stooling) rye, grazing vetch and radishes (Japanese, tillage).

Considerable interest in CA adoption in the dry, sandier areas of the North West Province (20%) and in higher rainfall pockets across the Highveld of South Africa, such as Mpumalanga, has been experienced in the last decade. Although crop rotation has room for improvement in these regions, their CA systems are increasingly characterized by improved crop rotations (maize–sunflower–soybean–sorghum combinations) and crop residue retention. In the Free State Province, however, the adoption level is still very low (< 20%), though there are areas with higher adoption rates within the province (Findlater, 2015).

In the very sandy soils of the drier western regions, such as the north-western Free State and North West Province, deep ripping, or rip-on-row, is still practised widely. Although contradictory to established CA principles, there is currently little evidence to show that CA could be done successfully in these conditions without using deep ripping to alleviate soil compaction, or without significant soil health improvements through a fairly long period (a decade or more) of rigorous CA application. Improved soil health and more farmer-led research initiatives are required to continually manage and adapt CA principles in these situations to maximum advantage.

Crop rotation with three or more crops is used by only 40% of the commercial farmers practising CA, with monoculture being done by 10% and two crops, such as a maize and soybean rotation (Findlater, 2015), being used by 50%. Certainly there is a need to improve crop diversity in rotation systems, as well as for better management of crop residues during the fallow period. Annual cover crops are used by around 20% of commercial farmers, while (perennial) ley cropping for 3–5 years, such as weeping lovegrass (*Eragrostis curvula*), is used by 25% of the commercial farmers (Findlater, 2015).

No-till planters with tines are used by 36% of commercial farmers in SA and disc planters are used by 13% (Findlater, 2015). Initial unavailability of suitable tractor-drawn planters severely limited the introduction of CA in South Africa during the 1970s and 1980s (Fowler, 2004). However, over the past two decades the interest in and consequent influx of planters from Brazil, Australia, USA and Argentina into South Africa, combined with the ingenuity of local engineers, has greatly eased the shortage of planters and most requirements can now be met.

Smallholder production sector

According to Putter *et al.* (2014), a real opportunity for smallholders to escape from a vicious poverty cycle, which is exacerbated by extremely low levels of production, degraded and infertile soils, high input costs and a lack of access to markets, is presented by CA. Smallholders are motivated to adopt CA in order to realize the increase in crop yields and food diversity, a reduction in production costs (mainly due to a cut in mechanization costs) and an immediate increase in disposable family income.

Although no formal CA adoption figures exist, all indications are that the adoption of CA within the smallholder production sector in South Africa is still very low (< 5%). Smallholder CA promotion interventions are mostly done through projects funded by various agencies, mainly government. Peaks of adoption seem to be reached with some of these projects, while unfortunately dis-adoption tends to follow most project completions and the drying up of external funds, with only small pockets of enthusiasm and participation remaining. Major constraints experienced by smallholders include availability of resources, such as land, production inputs, labour, information, financial means, markets and access to infrastructure.

Smallholders are often older persons, pensioners and mostly women. Agriculture is often perceived as an inferior lifestyle by younger people and they move to the cities in search of more lucrative employment. Unfortunately the HIV-AIDS pandemic still plays a devastating role, especially amongst the normally economically productive sector. Smallholder CA initiatives have many active women, compared with conventional systems where the men are responsible for ploughing and land preparation.

Where CA-related activities have been rejected, it was largely due to lack of commitment, awareness of its advantages, knowledge, skills and finances to make the necessary adaptations and the inability to implement all the principles simultaneously and properly. In this respect there is little difference between community and commercial farms, though the latter probably have a greater potential financial advantage in making the conversion.

Although the manufacturing industry targeting smallholder CA farmers is still in its infancy in South Africa (Sims *et al.*, 2012), the availability of animal- and hand-drawn or operated equipment suitable for smallholders has improved slightly during the past few years. Brazilian equipment is being imported by one manufacturer and commercially available adaptations to the South African Farm Implement Manufacturers (SAFIM) planter are being furthered and made by another. Also, a prototype roller planter has been developed and there are plans to introduce locally manufactured animal-drawn planters to smallholders, for which there is an immense market opportunity.

12.4 Making Conservation Agriculture Work: Lessons and Challenges

12.4.1 Major CA initiatives among large-scale commercial enterprises in South Africa

Most CA activities among commercial farmers are driven by farmers themselves. Following similar international trends, innovative and successful local CA farmers tend to act as role models, inspiring those who witness and experience their success. In most cases these farmers organize themselves into groups (innovation platforms, or IPs), leading towards enthusiastic and successful CA promotion and adoption initiatives, characterized by activities such as on-farm trials, conferences, field days, workshops and other awareness-creating activities. Several of these initiatives have sprung up among farmers in the past decade.

The No-Till Club in KwaZulu-Natal has been the best example of a successful CA farmer group for a long period. Officially established in 1997 when six farmers met on Gourton Farm, Anthony Muirhead was elected as the first chairman and Bill Russell, Chief of the Soil Conservation Service of the provincial Department of Agriculture, as secretary. Currently the club has approximately 100 members and primary objectives include: providing a forum for the exchange of knowledge; establishing the needs of its members; disseminating CA-related knowledge and information; and encouraging research into problem areas. Initially a number of farmers' days were organized and, with inputs from researchers at Cedara College of Agriculture, newsletters were circulated. An annual conference is held and in September 2015 more than 300 delegates attended, including visitors from the rest of Africa.

A few new CA structures have been established in the past couple of years. The CA Western Cape and the Ottosdal No-Till Club in the North West Province are two of the most prominent new structures created to serve the information need of farmers in those regions. In collaboration with Grain SA, the national grain producers' organization, and various other stakeholders, the Ottosdal No-Till Club has been conducting on-farm trials and an annual conference, which has raised a significant level of awareness in the drier western grain-producing areas of the country.

For the past decade or so the Maize Trust of South Africa has invested in CA research projects financially, through the Agricultural Research Council (ARC) and other stakeholders. From 2013, the Maize Trust, in cooperation with Grain SA, supported the development of the CA Farmer Innovation Programme (FIP), primarily using farmer-led IPs to launch activities such as on-farm trials and awareness events. The main aim of the CA-FIP is to mainstream CA to and through grain farmers and hence follows a farmer-centred IS philosophy (Smith and Visser, 2014). Four major CA IPs or projects with commercial farmers and two with smallholder farmers in different regions have been established and coordinated by the CA-FIP. The commercial farming project sites are at Ottosdal (North West Province), in the sandy soils region of the north-western Free State and at Vrede and Reitz (both in north-eastern Free State).

12.4.2 The impact of CA: a modelling outcome

There is a range of indicators that can be used, either individually or in combination, to measure CA success and adoption (Box 12.2). While each of the measures is important, and ideally a combination should be used on an ongoing basis, one is limited in measuring the impacts of potential (future) CA projects. An attempt has therefore been made to estimate the potential impact of CA on the South African dryland maize industry using a system dynamics model. System dynamics modelling is one approach for modelling complexity and entails understanding how systems change over time with feedback loops, time lags and nonlinearity characterizing the relationships among the building blocks of a system dynamics model. This is facilitated through the use of user-friendly graphical simulation packages in Vensim® that helped to formulate a model using the stock flow components based on differential equations. A model was constructed to model the transition from conventional (CV) to CA systems in four maize-producing regions in South Africa, namely Western Free State (WFS), Eastern Free State (EFS), KwaZulu-Natal (KZN) and North West (NW), over a 20-year period (De Wit *et al.*, 2015). Four region-framed production and environmental sub-models were therefore constructed that made provisions for the unique farming characteristics of both CV and CA systems. Data was obtained from a number of sources (e.g. farmer interviews, Department of Agriculture, Forestry and Fisheries, Orange Free State Cooperative (OVK), Grain SA, Novon, Pannar and Profert) and was verified by experts through Grain SA networks.

In modelling the transition from CV to CA systems the relationships, as established in literature, between soil organic matter (SOM), soil organic carbon (SOC) and water-holding capacity were used to inform changes in yield. In addition, the production data was used to model CA systems' gradual yield increases (due to improved soil health) over a 20-year period, whereby cost reductions are phased in over a 10-year period.

Box 12.2. Measuring Conservation Agriculture.

The following list of indicators can be used either individually or in combination to measure CA success and adoption:

1. Return on investment with respect to yield (t/ha).
2. Levels of (reduced) external production inputs: measured in R/ha and/or kg/ha/year for fertilizers, herbicides, pesticides, and l/ha/year for fuel use.
3. Soil health measurements – chemical:
  a. balanced ratio of certain micro- and macro-nutrients, pH, acid level, etc. (see also Soil Health Tool, below).
4. Soil health measurements – biological:
  a. Soil Health Tool (SHT Index by crop systems), and/or
  b. microbial genetic diversity (DNA sequencing), microbial functional diversity (BIOLOG assay), carbon cycling (Solvita CO_2 respiration, soil enzymes), nitrogen cycling (part of SHT), soil biomass (microbial biomass, earthworm populations) and key species (mycorrhiza, pathogens).
5. Soil health measurements – physical:
  a. soil organic matter (SOM) and soil organic carbon (SOC) build-up with respect to an appropriate baseline. Consider different Soil C fractions, e.g. active or labile fractions, and
  b. aggregate stability.
6. Water use efficiency (WUE) measured as kg/mm/ha.
7. Reduced riskiness (combination of yields, WUE and return on investment linked to knowledge and management levels).
8. Soil loss (t/ha/year) through soil loss modelling.
9. Number of CA farmer groups, such as study groups, clubs, etc. (measured by impact survey).
10. Number of CA awareness events, such as farmers' days, conferences and cross-visits.
11. Number of farmers adopting CA per region (adoption rate).
12. Number of no-till planters sold per region per year.
13. Number of infestations by pests or other forms of invasive alien organisms per season per region.

A very large monetary benefit was found by the study (De Wit *et al.*, 2015), as indicated by the net present value (NPV) of adopting CA systems, with or without the incorporation of positive externalities, such as improved ecosystems services. The viability of maize production is improved in all regions with the adoption of CA systems but the potential is more so in the Eastern and Western Free State. Large-scale experience-based research in this matter is justified by the theoretic conclusion. This outcome is the result of cost reduction owing to lower input use, increases in yields, less emissions into the environment and carbon sequestration. While there are improvements in the financial viability of CA systems versus CV in the North West, CA systems remain negative indicating that the investment was not yet economical even with adopting CA.

12.4.3 Other major CA initiatives in South Africa, including smallholders

Most CA activities among smallholders are done through projects funded by government and other implementing agents, such as LandCare and Grain SA.

The National LandCare Programme (NLP) is still seen by the Department of Agriculture, Forestry and Fisheries (DAFF) as the ideal vehicle for CA. This idea has merit and it is potentially a perfect synergistic combination, with LandCare being a bottom-up, community-based approach to the sustainable management and use of agricultural natural resources (DAFF, 1999) and CA a social construct that 'happens' by farmers on their farms (Coughenour and Chamla, 2000). Hence, in 2014 DAFF launched a strategic framework for mainstreaming CA into the NLP in South Africa. Mainstreaming CA into the NLP is seen as one way of ensuring that South Africa's food security responses duly reflect long-term natural resource sustainability (DAFF, 2014). This CA Mainstreaming Framework, developed in collaboration with key stakeholder groups, will be utilized to raise political will and commitment and spur action for CA mainstreaming into the NLP. Approaches and tools for CA mainstreaming into the NLP at multiple levels of authority and types of responsibility are provided by the Framework. Prompted by the NLP, which is implemented at provincial level, the provincial governments also appear to invest in CA-related projects, with the Western Cape leading by example. The Department of Rural Development and Land Reform (DRDLR) has been supportive with their conservation agriculture technologies (CATs) projects for a number of years already (Smith *et al.*, 2010). There are also limited contributions by some international organizations, cooperatives and universities. Since 2007, interested stakeholders have been involved in the National CA Task Force (NCATF), supported by the regional FAO office and chaired by DAFF, aiming to promote and coordinate CA activities through a national level IP; this initiative has not yet been able to make a significant impact.

The ARC itself has continued to support CA research initiatives since the 1980s (initially working mostly with commercial farmers, such as in KwaZulu-Natal), and from 1997 mostly with smallholders in LandCare-related projects. Around 20 LandCare-related CA projects have been implemented by the ARC in the last 17 years, particularly by their Institute for Soil, Climate and Water (Smith *et al.*, 2008, 2010), directly involving hundreds of farmers and reaching thousands more through awareness activities.

On the Grain SA CA-FIP study sites a series of selected on-farm farmer-led trials were established where farmers are the lead or equal partners (in identifying research needs and designing, implementing and evaluating experiments). The experience is that this independence and control over trials and other activities gave farmers much more ownership of the data and results. Experiments are well designed with appropriate treatments and sufficient replications spread over the entire agroecological region and/or on a sufficient number of farms. Data from properly designed experiments provide a much stronger starting point for discussion and investigation of a farmer's claims or problems. Furthermore, scientifically valid data are generated providing incentives for bringing people together in groups or networks across boundaries and on different levels. It is the potential for getting real answers that makes the group(s) hopeful that working together will be worthwhile; it makes all parties interested in how the experiment is conducted. Involving agricultural scientists (e.g. from the ARC) and other stakeholders in group problem solving and on-farm research (through a typical IP) becomes easier if valid data and results are important outcomes. The smallholder projects implemented by the CA-FIP are in the Bergville (KwaZulu-Natal) and Matatiele (Eastern Cape Province) districts.

12.4.4 Introduction of CA in smallholder systems, models and approaches

As mentioned above, approaches and technologies tailored to the constraints, needs and priorities of smallholders need to be introduced in a conscious manner. Approaches that have worked with scaling down commercial farming systems have only been marginally successful. Smallholder farming systems are not just smaller commercial farms; they are also a complex web of social, economic, environmental and agronomic practices specific to this farming system. According to Braun and Duveskog (2008), one of the biggest problems with smallholder interventions 'has been the tendency to generalise and make recommendations for farmers across large and highly heterogeneous areas' (Braun and Duveskog, 2008). All of these practical issues vary on a small spatial scale, within the farm boundary. This local specificity requires implementers to accept and use farmers as experts and innovators in their own right and reality, aiming to build on and improve their innovation capacity. Therefore, the development of innovation approaches and processes over the last few decades 'reveals a steady progression in the accommodation of ecological heterogeneity and farmer control of agro-ecosystem management' (Braun and Duveskog, 2008).

Introduction of CA into any farming system requires continuous innovation, learning and change in a number of different areas, including social, economic, environmental and agronomic considerations. In the smallholder context it requires the design, introduction and facilitation of a reasonably complex IS approach by the implementers, and of practice, labour and resources (including natural and financial resources) by the farmer that has system-wide implications. In most of these complex adaptive systems it has not been nearly enough just to use mainstream conventional (linear) research and extension approaches, nor just to introduce the new idea, try it out and compare it with the present local practices. There is an interplay of a number of different factors, all of which need to be integrated, thus requiring a well designed and facilitated IS approach.

Challenges with the introduction of CA into complex communal-tenure smallholder farming systems can be summarized as follows:

- Smallholders are not a homogenous grouping.
- Smallholders are not linked into information systems.
- Smallholders lack finances and financing options.
- Smallholders have limited access to resources, both natural and financial.
- Smallholders are risk averse and more vulnerable to impacts.

- Smallholders have extremely limited access to agribusiness and marketing support.
- Smallholders do not have access to improved agricultural technology, such as CA equipment.

The first two points above will be discussed in more detail below.

Smallholders are not homogenous

Smallholder farmers in South Africa are predominantly situated in communal tenure areas and live in communities within villages and in their homesteads, which consist of their dwellings and gardens. Permission is given to occupation-type arrangements through traditional and local authorities for use of crop lands which typically range in size from around 0.1 ha to 2.5 ha. Grazing for livestock is almost universally set up as an open-access system, where livestock can graze land anywhere outside of homesteads that is not designated for another purpose. Local grazing rules are set and enforced by the local and traditional authorities and may include movement of livestock to different areas in different seasons, typically closer to and into the villages in winter, when fodder availability in the grazing areas becomes very limited. Cropping fields are often extremely vulnerable to grazing pressure and damage even if they are fenced, which in many cases they are not. Grazing of crop residues in the fields during the off-season (winter months of June to September) is practised throughout.

It is becoming clearer that smallholder farmers themselves fall within different categories of resource availability, capabilities and aims for their farming. The concepts of subsistence and emerging commercial farmers have not been particularly useful in providing appropriate models of support to assist individuals to produce at scale or to include the myriad of important factors into the thinking process. These would be, for example, including the imperative of food and good nutrition and the imperative of resource conservation and sustainability into prospective interventions.

Smallholder farmer segmentation or typology approaches have become more popular of late. Internationally, the Consultative Group to Assist the Poor (CGAP) (Christen and Anderson, 2013) has worked on a segmentation approach. In South Africa, the Abalimi Bezekhaya ('planters of the home') initiative and the Institute for Poverty, Land and Agrarian Studies (PLAAS) have developed a similar model (Abalimi Bezekhaya, 2010; Cousins, 2015).

A pragmatic approach to responsible market development is combined by CGAP with an evidence-based advocacy platform to increase access to financial services for the poor. The challenge of providing financial services that support the multiple goals of rural households, including those related to their more universal, general household needs and those linked to their agricultural activities, is addressed by CGAP's segmentation approach. Three broad segments have been proposed: (i) non-commercial smallholders; (ii) commercial smallholders in loose value chains; and (iii) commercial smallholders in tight value chains. The farmer segments are differentiated by what they grow, how they engage with markets as buyers and/or sellers and how those markets are organized. These segments are meant to be broad categories based on common traits that can begin to illuminate the financial mechanisms that might best fit the given financial goals and cash flows. Gender, size of land holding, crop mixes, engagement with markets, access to improved agricultural technologies and access to financial services are some of the criteria they include.

For the purposes of the Grain SA Smallholder Farmer Innovation Programme (SFIP), a combination of the above three farmer segmentation processes is being developed to provide the concepts for a development continuum in this context and the needed sustainability criteria to design a specific sustainability index for this application.

The initial outline of the smallholder farming systems and farmer categories involved in the Bergville CA project are shown in Table 12.3. Information has also been drawn from an assessment conducted by the SaveAct Trust (a micro-finance initiative), where an in-depth livelihoods analysis was undertaken with 150 participants in the Grain SA SFIP (Kruger and Lewis, 2014).

Through this analysis, the basis for providing targeted support to the different segments of the farming community is provided. The aim is to move or develop interested or volunteering farmers from the first to the last categories.

Smallholders are not linked into information systems

Most smallholder farmers in South Africa, in this case the KwaZulu-Natal region, generally still glean most of the information that they use through word of mouth and through personal experience. Communication is frequently done through informal associations and very rarely, if at all, through formal organizations. The vast majority of smallholders are not very literate and do not make decisions about their farming practices based on written information.

They are thus functionally excluded from the vast array of ICT (information and communication technologies) and published information available. Conventional awareness-raising campaigns and strategies have not had a great impact on social and agricultural change in the developing world, unless farmers are directly involved in a more participatory empowerment experience.

The challenge is thus twofold, with different and innovative communication strategies being required as well as providing for meaningful and direct farmer involvement and learning.

More recently, information and communication innovations have focused around wireless and mobile technology, given the universal distribution of mobile phones even among the rural poor. The use of radio programmes is also popular.

Farmer information and advisory systems and other forms of extension augmentation, such as through participatory radio and video, entertainment media, interactive voice response and text-based systems, are being piloted and supported in developing countries (Global Centre for Food Systems Innovations, 2013).

Table 12.3. Farmer segmentation in the Bergville smallholder farming system.

Category	Non-commercial smallholders	Semi-commercial smallholders	Commercial smallholders in loose value chains	Commercial smallholders in tight value chains
People in each category (%)	72	23	5	–
Farmer priorities	Most production consumed by the household and additional food is bought in	Production is intensified. Selling becomes more significant and supplements household income	Consumption and sale in various percentage mixes but moving to more sales	Primarily for sale – working within existing well defined commodity value chains
Gender	Mostly women (89%)	Mostly women (96%)	Women (40%), men (60%)	Mostly men
Resources	Low external input systems are used with a minimum of bought inputs	Mixed (low and external) input systems are used with a minimum of bought inputs	Mixed (low and external) input systems are used with greater reliance on bought inputs	Mostly high external input systems
Level of mechanization	Hand cultivation	Hand cultivation, animal traction	Animal traction, tractors	Tractors
Land size	≤ 0.1 ha	0.1–1 ha	1–2.5 ha	> 2 ha
Farm productivity, including labour access	Extremely low	Low to high	Low to high	Low to high
Access to improved agricultural tech and information	Very limited	Limited	Limited	Good
Access to financial services	Very limited if at all	Very limited if at all	Very limited	Informal and some formal through buyers
Local organization	Almost non-existent	Almost non-existent	Informal farmers groups	Farmers associations and cooperatives
Agribusiness support	Very limited	Very limited	Informal but growing	Reasonable
Engagements with markets	Very little; entirely informal	Limited and still informal for the most part	Both informal and formal	Can be good due to value chain farming bundles

Continued

Table 12.3. Continued.

Category	Non-commercial smallholders	Semi-commercial smallholders	Commercial smallholders in loose value chains	Commercial smallholders in tight value chains
Environmental performance	Generally not considered	Generally not considered, some adoption of conservation and sustainable practices	Generally not considered, some adoption of conservation and sustainable practices	Some adoption of conservation and sustainable practices
Crop mix	Staple crops, crop livestock mixes – focusing on 4–5 commodities	Staple crops, some cash crops, crop livestock mixes – focusing on 3–4 commodities	Staple crops, some cash crops, crop livestock mixes – focusing on 2–3 commodities	Mostly cash crops – focusing on 1, maybe 2 commodities
Livelihood (food security, total monthly income, assets, poverty likelihood)	Food security: low Monthly income: R0–2000 Assets: minimal Poverty likelihood: high	Food security: low– medium Monthly income: R2001–4000 Assets: minimal – starting to build Poverty likelihood: medium	Food security: medium-high Monthly income: > R4000 Assets: reasonable Poverty likelihood: low	Food security: high Monthly income: ... Assets: ... Poverty likelihood: ...

12.4.5 A model for implementation

As mentioned in Section 12.1, sustainable agriculture is not seen as a simple model or package to be imposed; it is more a process for learning or a self-perpetuating learning system. This has alerted some CA practitioners to the now growing family of IS approaches, processes and tools that enable participants to share, act, observe, reflect, plan and learn, creating a culture of learning that would allow people to be innovative and interactive in managing natural resources in a sustainable manner (Röling and Wagemakers, 1998; Smith, 2006, 2014; Braun and Duveskog, 2008; Kruger and Gilles, 2014). Intensive interactions with smallholders are needed in all approaches throughout the life cycle of a project and have a strong emphasis on co-learning and using farmers as key innovators.

The CA-FIP approach followed by Grain SA has been using a model built on observations and experiences of CA IS processes employed for smallholder farmers to date to incorporate all lessons learned and minimize previous challenges (Kruger and Gilles, 2014; Smith and Visser, 2014). This farmer-centred IS model has been particularly complemented by participatory innovation development (PID) (Kruger and Gilles, 2014) and farmer field school (FFS) (Braun and Duveskog, 2008) approaches and has been employed in the Bergville study since 2013. The model strongly relies on interactive in-field discovery learning and using a systems approach that includes the whole value chain.

A number of elements within the IS process are considered crucial for this model:

1. Stakeholder analysis and mobilization of community members must lead to practical implementation, direct observation and experiential learning having the following major components:
   a. working with organized and voluntary farmer groups as ideal innovation platforms for learning and sharing (these would include learning groups, local savings and credit groups, farmers' associations and cooperatives);
   b. farmer-led trials as opposed to researcher-managed trials or demonstrations; and
   c. adaptive, informed decision making within a continuous cycle of planning, action and observation.

2. Use of an interactive on-farm learning approach:
   a. based on FFS and using a farmer-led 'trial-and-error' approach to adapt CA principles and practices to local or individual situations. This provides for:
   - farmer-led trials as the main learning tool;
   - dealing with complexity in learning in a step-wise way (testing ideas in a progressive way, integrating complexity over time);
   - timely and responsive actions and operations in the fields;
   - ongoing adaptive management and dealing with issues that arise;
   - season-long learning based on the crop calendar, involving farmers and researchers/extension; and
   - in-depth and critical reflection on implementing new ideas.

   b. facilitation of social learning in small groups (five people) and larger groups (20–30 people); and
   c. inclusion of awareness-raising events, such as cross-visits, learning materials and farmers days.

3. Working within the whole value chain using a livelihoods framework.

4. Ensuring access to appropriate implements and inputs.

5. Vigorously facilitating scaling-out (horizontal) and scaling-up/down (vertical) processes.

6. Including participatory monitoring and evaluation (PM&E) using agronomic, environmental, economic and social indicators.

7. Action research processes whereby farmers work as key innovators or researchers.

12.4.6 Targeted support to the different segments of the farming community

The function of the IS approach and model described above is to: (i) introduce different types and sets of interventions for each different category of smallholders, within the development and sustainability categories defined in Table 12.4, to achieve the greatest impact at a local farm level; and (ii) move or migrate farmers from lower to higher categories, or levels of innovation

Table 12.4. Appropriate interventions for non-commercial and semi-commercial smallholders in the Bergville study area.

Category	Semi- and non-commercial smallholders	Appropriate interventions	Gaps and complementary options
People in each category (%)	95	For all semi- and non-commercial smallholders	Very few smallholders are commercial level
Farmer priorities	Household consumption and sales	Focus on household consumption first – household storage and value adding at household level (*small mechanical mills, storage drums, tablets for weevil control, rat bait*)	Improve harvesting efficiency for non-mechanized systems (like introduction of hand threshers)
Gender	Mostly women (92%)	Technology and systems that are appropriate for women (*hand tools that women can easily use; saving and credit groups, working with support groups*)	Men are mostly absent due to migrant labour practice or lack of interest
Resources	Mixed (low and external) input systems with a minimum of bought inputs	Focus on localized natural systems as much as possible (*reduced herbicide use (only pre-spray of herbicide) and fertilizer usage (micro-dosing), increase cover crops, intercropping, mulch*)	Introduce rainwater harvesting in minimum tillage systems, especially during initial stages when sufficient crop residues are lacking Use more local resources for compost making, apply kraal manure
Traction	Hand cultivation, animal traction	Provision of hand planters, animal drawn planters and two-row tractor-drawn planters for CA to try out, rent and purchase	Access to tractors for ploughing services through government programmes is a dis-incentive for CA due to reliance on provincial government Formalize farmer centres
Land size	> 0.1 ha – 1 ha	Plan for increased production (*expansion of activities to incorporate more unused land; long-term allocation of certain plots to CA*)	Access to unused land is not easy and requires linkages with traditional leadership and/or land reform processes in the area
Farm productivity, including labour access	Low	Minimize labour (*farmer experimentation, working groups, use of pre-plant herbicides, close spacing, intercropping, soil cover, appropriate implements*) Maximize soil fertility (*CA and cover crops, legumes*)	Introduction of labour saving technologies and implements

Continued

Table 12.4. Continued.

Category	Semi- and non-commercial smallholders	Appropriate interventions	Gaps and complementary options
Access to improved agricultural technology and information	Limited	Partnerships with service providers, NGOs, government and agribusiness for provision of information, skills development and appropriate technical support (*working partnerships where organizations consciously work together at farmer level*)	Introduce appropriate technology for hand cultivation, animal traction and small implements for tractors Establish farmer centres in different localities to supply and support Use incentive scheme such as Payments for Ecosystem Services (PES) for alternative and ongoing funding and scaling out and up
Access to financial services	Very limited if at all	Arrangements for savings need to be made. External credit provision is not appropriate at this scale (*locally based saving and credit groups linked to group accounts at local banks (in partnership with banks)*)	Suite of services available in formal system – e.g. mobile and cell phone banking, branchless banking, savings, insurance and loan products for smallholders
Local organization	Almost non-existent	Focus on local organization (*saving and credit groups for debt management, consumption smoothing and input supply, farmer learning groups and cooperatives*) Organize across villages and areas (*digital systems for communication, ordering, payment, etc., to organize large groups of smallholders efficiently – being developed*)	Improved group organization for local storage and processing (milling) Set up and strengthen input buying groups Poor access to software and service providers for digital systems, internet and cell phone reception and access for smallholders
Agribusiness support	Very limited	Focus on local organization for numbers and efficiency (*commodity interest groups, farmer platforms for bulk buying, negotiation of 'deals' with businesses and suppliers in the area*)	Need agribusiness systems to support smaller quantities, 'packages', cash on delivery, pre-ordering, paying off over time, save storage of inputs on site (insect/pest control in case of purchasing early, theft control)

Continued

Table 12.4. Continued.

Category	Semi- and non-commercial smallholders	Appropriate interventions	Gaps and complementary options
Engagements with markets	Limited and still informal for the most part	Arrangements with milling companies in local towns and support local marketing options (*local marketing of maize and beans – market is readily available; sale of beans to shops in local towns is generally easy*)	Lack of formal arrangements with local markets
Environmental performance	Generally not considered, some adoption of CA practices	Embed environmental concerns in the programme design (*FFS sessions to underpin activities in CA into environment and environmental concerns; focus on increased soil cover to gain runoff control, water infiltration soil health*) *VSA - Visual Soil Assessment methodology piloted for local monitoring*	Absence of environmental stakeholders and potential long-term funders of incentive schemes
Crop and seed diversity	Produce mostly staple crops	Work within food and nutrition diversification (*focus on inclusion of legumes for nutrition, environmental and economic benefit*) *Cover crop mixes as fodder crops for livestock. Crop–livestock mixes focusing on 4–5 crops and commodities*	Set up local and participatory breeding, production and multiplication of seeds
Livelihood (food security, monthly household income, assets, poverty likelihood)	Food security: low–medium; monthly income: R0–4000; assets: minimal, starting to build assets; poverty likelihood: medium to high	Increased food security and access to nutritious food Increased sales to augment household income	Quantify the economic impact and potential of such small systems Set up local seed stores, local mills, local sales, local fodder mixes and production

and farming, in order to improve the overall sustainability of the farming system. Table 12.4 is a summary of interventions seen to be most appropriate and required for non-commercial and semi-commercial smallholders. The appropriate interventions are based primarily on present implementation within the Bergville study through Grains SA SFIP. The gaps are those processes and interventions that emerged from the continuous planning, action, learning and adaptation, towards creating a much more coherent process. It should be clear that this is a complex, long-term multi-stakeholder approach to change and improvement.

12.5 Lessons Learned

12.5.1 Lessons from modelling the potential of large-scale commercial CA

The results from the modelling exercise are of such a convincing nature that it justifies large-scale testing and in-field verification – more so since maize production is one of the backbone industries in the country and hence the long-term profitability and sustainability thereof are essential to maintain and develop the resilience of rural societies and the economy in South Africa at large. Important issues to consider (Blignaut *et al.*, 2015) are as follows.

- The better the soil quality, the higher are the yields that are sustained and the lesser the need for using (high) external inputs.
- The better the soil quality, the better is the water use efficiency in a drought-prone country.
- The lesser the disturbance of the soil, the lower is the environmental demand per unit of maize produced and the more sustainable the system.
- The lesser the use of external inputs like fertilizer, pesticides and herbicides, the lower is the environmental demand per unit of maize produced and the more profitable and sustainable is the system.

To rectify the situation requires the following key outcomes to be achieved, among others:

- improved access to suitable maize and rotational crop varieties, as well as a range of cover crop species, to be able to build resilient cropping systems in area-specific environments;
- improved crop and livestock management to allow soil quality recovery and enhanced effectiveness and productivity;
- more plant (crop) diversity for disease, pest and weed control methods to avoid high rates of pesticides and herbicides; and
- the build-up of organic matter in the soils by having a rigorous application of minimal mechanical soil disturbance, permanent organic soil cover and diverse cropping systems.

The above can be achieved through the following actions.

1. Facilitate the formation and operation of farmer innovation platforms, for sharing, learning, implementation and scaling out of CA practices.
2. Facilitate farmer-driven research where different stakeholders (i.e. scientists/researchers, extension officers, farmers, agribusiness) share responsibilities in support of **1** above.
3. Involve and enhance extension officers to learn, participate and facilitate in **1** and **2**.
4. Improve the general awareness and understanding, through, among others, social media, publications, conferences and farmers' days, of the impact and the sustainability of the various farming systems.
5. Identify and/or strengthen the various rural institutional arrangements, especially under smallholder farmers, as platforms to improve local crop production systems through CA.
6. Raise awareness of the main (market) stakeholders of the qualities, role and success of CA.
7. Restore the soils to improve net primary production.
8. Introduce incentive and market-based mechanisms to facilitate CA on a broader scale across the country.

12.5.2 Lessons from Bergville smallholder model

From the above the following lessons can be identified.

Farmer-led experimentation is invaluable in the learning, adoption and adaptation process

As part of the IS approach and on-farm experimentation protocol, participating farmers set up their CA treatments or plots alongside their traditional farming plots that are used as a control for visual and measured comparison of differences and improvements over time. An example of such plots is shown in Fig. 12.1.

Fig. 12.1. Comparison of CA treatments or plots alongside traditional farming plots. (a) A control plot planted by Mr Dlezakhe Hlongwane from Bergville in KwaZulu-Natal, South Africa, on 15 November 2014. He used CA practices for his control but tested single cropping and wide spacing. Weeds are a problem and germination is somewhat patchy. (b) Mr Hlongwane's CA trial plot planted on 13 November 2014. This is an intercropped plot of maize and sugar beans with a test of close inter- and in-row spacing. Here the soil cover and growth of the crop are a lot more impressive.

The guiding principles for farmer-led experimentation being formulated thus far are as follows:

- Some resources need to be provided for experimentation with new ideas, so that a learning platform is created in a low-risk environment.
- New ideas need to build upon existing practices and priorities of the smallholder farmers, including their need to produce food for themselves.
- Environmental concerns need to be embedded into the process design as it is usually not an explicitly recognized outcome (goal) for smallholders.
- Changes in the farming system need to reduce labour and costs to be adoptable.
- Access to inputs, materials and tools is a prerequisite.
- Introduction of fodder crops for livestock as an incentive to use cover crops works well as a progressive idea (start small and increase incrementally) to include the time needed to reorient grazing arrangements in the community.
- Farmer segmentation (grouping) is important in the experimentation process to provide appropriate support for increasing the scale of production.

The farmer-led experimentation (on-farm trial) process starts with an introduction by the project development facilitators (on project level and farm level) through an intensive, season-long and practical learning-by-doing process. This includes, for example:

- choice of seed types and varieties;
- fertilizer micro-dosing practices based on soil sample analysis;
- choice and use of herbicides and safe use of knapsack sprayers;
- use of no-till planters (including changing seed plates, calibration and maintenance);
- plant spacing;
- intercropping with legumes;
- relay cropping with fodder-based cover crops in summer and winter rotations; and
- integrated pest and disease management programmes.

The trial and control plots are monitored intensively and jointly between the farmers and development facilitators during the season using both agronomic and environmental indicators. At the end of the season a joint review (or social learning) session for the learning group based on these observations for all the individual farmers serves to clarify lessons learned and serves as a basis for designing the next season's cropping and trials.

Inclusion of the whole value chain is important

The learning process also involves farmers undertaking to measure crop yields of their CA treatments and their control plots. Harvesting and joint and improved storage options are considered and discussed. Discussions are held around marketing and sales and incomes made, as well as availability of food for households. Farmers make decisions around joint saving for input costs and set up savings groups for bulk buying of inputs.

Farmer-led trials, based on testing of different ideas for progressive and continued learning, work well

The level of farmer experimentation continues to grow in complexity and choice into the second and third years of experimentation (3 years is regarded as the minimum length of time for a project cycle), while new farmers are brought into the experimentation cycle every year. The new entrants are assisted by the more experienced farmers in their CA practice. In each area or group, local facilitators or best performers emerge, who then take on a facilitation and linkage role with stakeholders and role players within and outside the community.

Use of visual indicators to assess progress and assist in informed decision making is important

The intention has been to use a number of different types of indicators and methods for PM&E, to be able to monitor, evaluate and compare these and consolidate the process into a 'best bet' framework for monitoring that is both robust enough for the inaccuracies and difficulties of smallholder monitoring processes, but sensitive enough to pick up yearly changes in individual management practices. This kind of PM&E framework and process would need to be tested and streamlined on a case-by-case basis.

The visual CA and agronomic indicators used in the CA Grain SA SFIP are percentages of: ground cover at planting; weed infestation; pest and disease occurrence; and growth (which includes germination, plant height, colour and general vigour). The indicators are assessed at planting, then 6 weeks later at top dressing and again towards the end of the growing season. In addition, the visual soil assessment (VSA) scoring process has been introduced to take soil and environmental indicators into account (Soil Management Initiative, 2011). These indicators include: soil erosion; soil structure, porosity and colour; soil compaction; water infiltration and water-holding capacity; and earthworm counts. Basic soil health and improvement thereof are indicated by these. In addition plant-based indicators (e.g. size and development of root systems, crop height at maturity) and economic indicators (e.g. production costs and cost/benefit analysis) are included.

In the Grain SA SFIP, the above two scoring techniques have been evaluated and compared with each other over two seasons and related also to the yield data as part of a process of refining the monitoring indicators for the programme. The results of using these scores and the yields are shown in Fig. 12.2.

There is a definite relationship between the CA scores, the VSA soil indicator scores and yields (Table 12.5). From this result a score card combining these three sets of indicators into a comprehensive framework of visual indicators has been compiled as a first step towards setting up a Payment for Ecosystems Services (PES) incentive scheme assessment process.

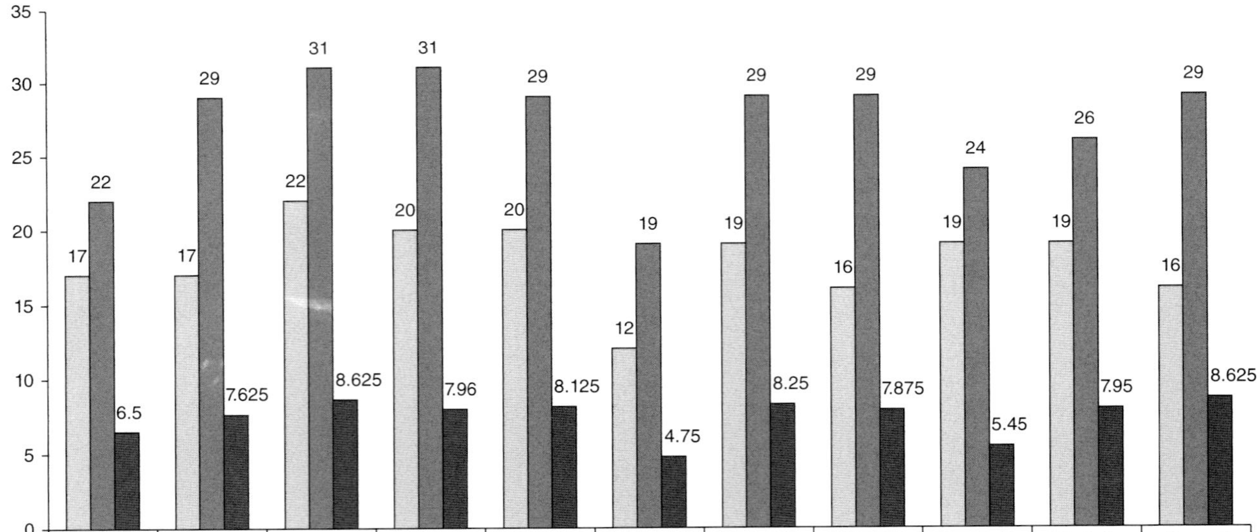

	Bangeni Dlamini Year 2 2,7	Cazile Zimba Year 2 3,6	Cupile Buthelezi Year 2 4,8	Dlezakhe Hlongwane Year 2 6,7	Khulekani Dladla Year 2 5,7	Landile Nsele Year 2 2,2	Makethi Dladla Year 2 3,9	Mtholeni Dlamini Year 2 1	Phasazile Sithebe Year 2 5,1	Thulislie Hlongwane Year 2 3	Zamani Dladla Year 2 1,8
Trial plot – VSA score plant	17	17	22	20	20	12	19	16	19	19	16
Trial plot – VSA score soil	22	29	31	31	29	19	29	29	24	26	29
Trial plot – CA score (10)	6.5	7.625	8.625	7.96	8.125	4.75	8.25	7.875	5.45	7.95	8.625

Fig. 12.2. A comparison of CA scores, visual soil assessment (VSA) soil and plant scores and yields for 11 CA farmer participants in the Stulwane learning group, Bergville study area, South Africa, July 2015.

Table 12.5. A summary of the three sets of indicators used into three broad categories.

	CA monitoring scores	VSA soil scores	VSA plant scores	Yields
Above average	≥ 7	> 28	> 15	3–8.9 t/ha
Average	5–6.9	11–28	7–15	1–2.9 t/ha
Below average	3–4.9	< 11	< 7	≤ 1 t/ha

12.6 Conclusions and Policy Implications of the Case Studies

From the two case studies described above, the following conclusion and policy implications are made:

- Move from programmes that support conventional approaches and technologies to programmes that support CA IS approaches and technologies; provide support for at least 3–5years.
- Set up processes that support and include women.
- Work within the locally defined farming systems and start development from what exists.
- Provide access to CA implements and tools in innovative distribution systems that can reach the farmers effectively.
- Assist farmers to access seed for maize, legumes and cover crops (summer and winter).
- Run projects and processes that are holistic and include the whole grain and legume production value chain.
- Move from CA as projects to CA as innovation systems and processes within a sustainability and food (and hence national security) context linking to existing policies high in the policy hierarchy.
- Move from researcher-managed demonstrations to farmer-led trials for meaningful farmer involvement, learning, adoption and adaptation.
- Provide technical (research, extension, facilitation, etc.) support for farmer CA ISs.
- Provide technical support for the design of appropriate monitoring and evaluation (M&E) systems, sustainability indices and incentive schemes to facilitate the mainstreaming (scaling) of CA among farmers.
- Set up market pull arrangements wherever possible, rather than providing specific markets, and support local marketing initiatives.

12.7 Chapter Summary

The following points summarize the contents of this chapter:

- Conservation agriculture is seen as a sustainable and economically viable agricultural system for South African farmers as it provides potential solutions to a wide-ranging number of challenges.
- South Africa's limited and degraded agricultural resources are inextricably linked to food production and security, i.e. if and to what

extent food security can be achieved under rapidly mounting population pressure.

- Current conventional (mainstream) commercial farming in South Africa becomes increasingly risky (financially) and environmentally unsustainable and there is a great need for alternative low external input sustainable agriculture (LEISA) systems, based on the rigorous application of CA principles at farm level.
- The adoption of CA has increased considerably under commercial agriculture in the major grain-producing regions of South Africa. Most CA activities among commercial farmers are driven by farmers themselves.
- Despite very low levels of adoption among smallholders, CA presents a real opportunity for them to escape from a vicious poverty cycle, which is exacerbated by extremely low levels of production, degraded and infertile soils, high input costs and a lack of access to markets. Farmer segmentation or typology approaches serve as the basis for providing appropriate support to the different segments of the farming community.
- A modelling case study showed a very large monetary benefit of adopting commercial CA systems, illustrating improved viability of maize production in all regions modelled, as a result of cost reduction owing to lower input use, increases in yields, less emission into the environment and carbon sequestration. The results from the modelling exercise are of such a convincing nature that it justifies (and supports) large-scale testing and in-field verification.
- Introduction of CA into any farming system, especially in the smallholder context, is not seen as a simple model or package to be imposed, but requires the design, introduction and facilitation of a reasonably complex IS approach. This approach involves intensive interaction with farmers and has a strong emphasis on co-learning and using farmers as key innovators.

References

Abalimi Bezekhaya (2010) The Sustainable Development Chain. Newsletter No 36. Available at: http://abalimi.org.za/wp-content/uploads/2014/03/nl36.pdf (accessed 15 November 2016).

Baiphethi, M.N. and Jacobs, P.J. (2009) The contribution of subsistence farming to food security in South Africa. *Agrekon* 48(4), 459–482. Available at: http://ageconsearch.umn.edu/bitstream/58216/2/6.%20Baiphethi%20%26%20Jacobs.pdf (accessed 26 September 2016).

Barnard, R.O., Van der Merwe, A.J., De Villiers, M.C., Van der Merwe, G.M.E. and Mulibana, N.E. (2000) Problem soils including degraded soils: extent, present use, management and rehabilitation in South Africa. In: *Proceedings FAO/ISCW Expert Consultation on Management of Degraded Soils in Southern and East Africa (MADS-SEA)*, 18–22 September 2000, Pretoria, South Africa. FAO, Harare, Zimbabwe, pp. 126–140.

Bennie, A.T.P. and Hensley, M. (2004) Advances in soil physics and soil water management research in South Africa, 1979–2003. *South African Journal of Plant and Soil* 21(5), 268–277.

Blignaut, J.N. and Van Heerden, J.N. (2009) The impact of water scarcity on economic development initiatives. *WaterSA* 35(4), 415–420.

Blignaut, J.N., de Wit, M., Knot, J., Smith, H., Nkambule, N., Drimie, S. and Midgley, S. (2015) *Promoting and Advancing the Uptake of Sustainable, Regenerative, Conservation Agricultural in the Maize Production Sector*. Green economy policy brief series. Green Fund, Development Bank of Southern Africa, Midrand, South Africa.

Braun, A. and Duveskog, D. (2008) *The Farmer Field School Approach – History, Global Assessment and Success Stories*. Background Paper for the IFAD Rural Poverty Report 2011, IFAD, Rome.

Christen, R.P. and Anderson, J. (2013) *Segmentation of Smallholder Households: Meeting the Range of Financial Needs in Agricultural Families*. CGAP Focus Note No 85. Available at: http://www.cgap.org/sites/default/files/Focus-Note-Segmentation-of-Smallholder-Households-April-2013.pdf (accessed 26 September 2016).

Coughenour, M. and Chamla, S. (2000) *Conservation Tillage and Cropping Innovation: Constructing the New Culture of Agriculture*. Iowa University Press, Ames, Iowa, USA.

Cousins, B. (2015) Through a glass darkly: towards agrarian reform in South Africa. In: Cousins, B. and Walker, C. (eds) *Land Divided, Land Restored. Land Reform in South Africa for the 21st Century*. Jacana, Auckland Park, Johannesburg, pp. 250–269.

DAFF (1999) *Implementation Framework for the LandCare Programme*. Available at: http://www.nda.agric.za/docs/landcare/landcare.htm (accessed 15 November 2016).

DAFF (2011) *Food Security*. Available at: http://www.nda.agric.za/docs/genreports/foodsecurity.pdf (accessed 15 November 2016).

DAFF (2014) A strategic framework for mainstreaming conservation agriculture into the LandCare programme in South Africa. Unpublished document.

Demetre, L., Yul, D. and Zandile, M. (2009) The assessment of food insecurity in South Africa. Available at: http://www.hsrc.ac.za/en/research-data/view/4359 (accessed 15 November 2016).

De Villiers, M.C., Eloff, J.F., Barnard, R.O., Mulibana, E., Mkhize, S. and Msomi, B. (2004) Land and water management: South African country situational analysis. In: Nhira, C. and Mapiki, A. (eds) *Regional Situational Analysis: Land and Water Management in the SADC Region. SADC Workshop Proceedings* 1, 2004, 93–112.

De Wit, M.P., Blignaut, J.N., Knot, J., Midgley, S., Drimie, S., Crookes, D.J. and Nkambule, N.P. (2015) Sustainable farming as a viable option for enhanced food and nutritional security and a sustainable productive resource base. Synthesis report. *Green Economy Research Report*. Green Fund, Development Bank of Southern Africa, Midrand, South Africa.

Du Preez, C.C. (2003) Volhoubare landgebruik en grondkwaliteit: organiese materiaal as indikator. *SA Tydskrif vir Natuurwetenskap en Tegnologie* 22(4), 106–112.

Du Preez, C.C., Van Huyssteen, C.W. and Mnkeni, P.N.S. (2011a) Land use and soil organic matter in South Africa 1: A review on spatial variability and the influence of rangeland stock production. *South African Journal of Science* 107, 354–362.

Du Preez, C.C., Van Huyssteen, C.W. and Mnkeni, P.N.S. (2011b) Land use and soil organic matter in South Africa 2: A review on the influence of arable crop production. *South African Journal of Science* 107(5/6), Art. #358, 8 pages. doi:10.4102/sajs. v107i5/6.358.

FAO (2001) Conservation Agriculture: Case studies in Latin America and Africa. *FAO Soils Bulletin* 78.

FAO (2015) Regional overview of food insecurity: African food insecurity prospects brighter than ever. FAO, Accra. Available at: http://www.fao.org/3/a-i4635e.pdf (accessed 26 September 2016).

Fey, M. (2010) *Soils of South Africa*. Cambridge University Press, Cambridge, UK.

Findlater, K. (2015) Improving South African grain farming. PhD thesis, University of British Columbia, Vancouver, Canada.

Fowler, R.M. (2004) *Conservation Agriculture in South Africa – Yesterday, Today and Tomorrow*. FAO Report, as part of project No. FAO/NDA TCP 2902. ARC-grain Crops Institute, Pietermaritzburg.

Global Centre for Food Systems Innovations (2013) *Assessing the Role of Information and Communication Technologies to Enhance Food Systems in Developing Countries*. White paper 2013 series. Michigan State University, East Lansing, Michigan, USA.

Hall, A.W., Janssen, W., Pehu, E. and Rajalahti, R. (2006) *Enhancing Agricultural Innovation: How to go Beyond the Strengthening of Research Systems*. World Bank, Washington, DC.

Knot, J. (2014) Conservation agriculture and commercial farmers in the Eastern Free State. PhD thesis, University of the Free State, Bloemfontein, South Africa.

Kruger, E. and Gilles, J.L. (2014) *A Review of Participatory Agricultural Research and Development in South Africa and KwaZulu Natal*. Missouri University of Science and Technology, Rolla, Missouri, USA. Available at: www.communitycommons.org (accessed 10 November 2015).

Kruger, E. and Lewis, E. (2014) *Matatiele Livelihoods Assessment. Emerging Data*. Internal Report. Available at: http://www.mahlathini.org/wp-content/uploads/2016/11/Matatiele-Livelihoods-Assessment-September-2014-Emerging-data_2-December-2014.pdf (accessed 21 November 2016).

Lal, R. (2010) Enhancing eco-efficiency in agro-ecosystems through soil carbon sequestration. *Crop Science Society of America* 50, 120–131.

Le Roux, J.J. and Smith, H.J. (2014) Soil erosion in South Africa – its nature and distribution. *SA Grain* 16(11), 32–35.

Le Roux, J.J., Morgenthal, T.L., Malherbe, J., Pretorius, D.J. and Sumner, P.D. (2008) Water erosion prediction at a national scale for South Africa. *Water SA* 34, 305–314.

Lobe, I., Amelung, W. and Du Preez, C.C. (2001) Losses of carbon and nitrogen with prolonged arable cropping from sandy soils of the South African Highveld. *European Journal of Soil Science* 52, 93–101.

Mararakanye, N. and Le Roux, J.J. (2012) Gully location mapping at a national scale for South Africa. *South African Geographical Journal* 94(2), 208–218.

Modiselle, S., Verschoor, A.J. and Strauss, J. (2015) Western Cape wheat producers and conservation agriculture technology uptake. *SA Grain* 17(5), 50–51.

Oxfam (2014) Hidden hunger in South Africa: the faces of hunger and malnutrition in a food-secure nation. Available at: https://www.oxfam.org/sites/www.oxfam.org/files/file_attachments/hidden_hunger_in_south_africa_0.pdf (accessed 15 November 2016).

Putter, C.A.J., Botha, R., Keuneman, E., Friedrich, T., Kassam, A., Lange, D., *et al.* (2011) Integrating conservation agriculture (CA) into formal education systems. Paper presented at the *World Congress on Conservation Agriculture*, Brisbane, Australia.

Putter, C.A.J., Smith, H.J. and Lange, D. (2014) Transforming the benefits of conservation agriculture into a pro-CA 'manifesto'. *SA Grain* 16(3), 78–81.

Röling, N.G. and Wagemakers, M.A.E. (1998) *Facilitating Sustainable Agriculture: Participatory Learning and Adaptive Management in Times of Environmental Uncertainty*. Cambridge University Press, Cambridge, UK.

Sims, B.G., Thierfelder, C., Kienzle, J., Friedrich, T. and Kassam, A. (2012) Development of the conservation agriculture equipment industry in Sub-Saharan Africa. *Applied Engineering in Agriculture* 28(6), 1–11.

Smith, H.J. (1998) Definition and demarcation of high potential land for rain-fed annual crop production. M Inst. Agrar. thesis, University of Pretoria, Pretoria, South Africa.

Smith, H.J. (2006) Development of a systems model facilitating action research with resource-poor farmers for sustainable management of natural resources. PhD thesis. University of the Free State, Bloemfontein, South Africa.

Smith, H.J. (2014) Strengthening community participation for real stewardship and custodianship of natural agricultural resources. Oral paper presented at the *National LandCare Conference*, Durban, 22–25 September 2014.

Smith, H.J. and Visser, M. (2014) A farmer-centred innovation systems approach to stimulate adoption of conservation agriculture in South Africa. Poster presented at the *World Congress on Conservation Agriculture*, 23–25 June 2014, Winnipeg, Canada.

Smith, H.J., Matlou, M.C., Mulibana, N.E., Kidson, M.V., Maribeng, L., Mashabane, K.W., *et al.* (2008) *Identification, Demonstration and Adoption of Appropriate Eco-technologies to Ensure Sustainable Resource Management in the KwaZulu-Natal, Limpopo and Mpumalanga Provinces*. Eco-technology Contract Final Report. ARC-ISCW Report Number: GW/A/2008/73.

Smith, H.J., Barnard, R.O., De Villiers, M.C., Beukes, D.J., Lange, D. and Putter, T. (2010) *The Establishment and Management of Demonstration Sites and Synthesis of Conservation Agriculture (CA) in South Africa*. ARC-ISCW Report number GW/A/2010/37.

Soil Management Initiative (UK) (2011) Visual Soil Assessment Field Guide. Available at: http://www.smi.org.uk/ (accessed 1 November 2015). Agriculture Document Library. University of Hertfordshire, Hatfield, UK.

Swanepoel, C.M., Van der Laan, M., Weepener, H.L., Du Preez, C.C. and Annandale, J.G. (2015) Review and meta-analysis of organic matter dynamics and associated greenhouse gas emissions in cultivated soils in Southern Africa. Presentation at the *Sustainable Soil Management Symposium,* 5–6 November 2015, Stellenbosch, South Africa.

Turton, A. (2015) *Sitting on the Horns of a Dilemma: Water as a Strategic Resource in South Africa.* The policy bulletin of the IRR, No. 6, 2015/10, Issue 22. ISSN: 2311-7591 1.

Waters-Bayer, A., Sanginga, P.C., Kaaria, S., Njuki, J. and Wettashina, C. (2009) Innovation Africa: An Introduction. In: Sanginga, P. *et al.* (eds) *Innovation Africa: Enriching Farmers' Livelihoods.* Earthscan, London, pp. 1–8.

13 Extending Conservation Agriculture Benefits Through Innovation Platforms

MICHAEL MISIKO*

International Maize and Wheat Improvement Center (CIMMYT), Nairobi, Kenya

13.1 Introduction

African smallholder farmers engage in many attempts to overcome poverty through individual and collective processes that depend on natural resources management (NRM) for agricultural production and livelihood. NRM for land-use development is often seen as a collective process (IFAD, 2006; Di Gregorio *et al.*, 2012; also see World Bank, 2008; Oweis and Hachum, 2009; Pathak *et al.*, 2009; Rockström *et al.*, 2010). Such collective action is common over forest or water resources management (Altieri and Toledo, 2005). However, protected areas known for providing environmental services (forests and communal water resources) continue to deteriorate or disappear; populations are rapidly urbanizing, turning more virgin areas into irrigated lands and into individual farms or seeking alternatives away from NRM-related livelihoods. NRM initiatives in smallholder contexts are therefore increasingly difficult to organize or sustain due to these complex target landscape changes. With decreasing scope for communal action and because of the increasing focus on household farm-level enterprises, there is a need for expanding the scope of NRM in land use and rural development (McCarthy *et al.*, 2004). Smallholder household actions largely comprise agricultural practices, which often form part of the problem, including low investments in erosion control (Muchena *et al.*, 2005). Smallholder agriculture, therefore, must play a multifunctional role beyond food production and fibre. Its functions need to include 'renewable natural resources management, landscape and biodiversity conservation and contribution to the socio-economic viability of rural areas' (Renting *et al.*, 2009, p. 112). To support NRM among smallholders, a clear understanding of their agricultural contexts is needed

*m.misiko@cgiar.org

and it requires supportive policies, institutions (Carmona-Torres *et al.*, 2011), services and investments (IFAD, 2006). In this chapter, the focus is on institutions, through the analysis of agricultural innovation platforms (AIP) as a valuable entry point in NRM at farm level. The rationale is that AIP is a viable approach to multifunctional smallholder agriculture that generates a range of productivity, ecological, economic, social and other benefits necessary for sustainable NRM among smallholders.

13.1.1 A case of Conservation Agriculture for natural resource management

The degradation of natural resources is often linked to high rates of population growth and has cause–effect relationships with poverty (Shiferaw, 2006) and opportunity to innovate and adopt new knowledge or test new practices. Managing agricultural development sustainably amid the growing populations in sub-Saharan Africa requires intensification options that both conserve resources and increase agricultural productivity. This is a formidable challenge. With little opportunity for continued area expansion for agriculture, and with the fragile ecosystems unsuitable for farming (e.g. steep slopes and marginal lands) being degraded, there are not many straightforward remedies. One serious option is Conservation Agriculture (CA). With declining fallow periods, CA can be adaptively applied, amid its challenges (Fowler and Rockstrom, 2001), in progressively stemming NRM constraints on small farm plots. Such challenges commonly relate to sustainable water and soils management (Rockström *et al.*, 2002).

13.1.2 CA as NRM

CA is concerned more with soil and water management. These are key resources in any farming system. Soil cover, especially by residue retention, is important in conservation of soil and water but there are numerous challenges, including the use of cover as fodder and fuel. CA is, however, not all about residue (Baker *et al.*, 1996): it is a more holistic system of agroecosystem management (see www.fao.org/ag/ca for definition and description of CA). Crop production, while at the same time ensuring soil cover, must go hand in hand with the minimization or avoidance of soil mechanical disturbance. Among smallholders, it means the abandonment of the hand-hoe and mouldboard plough for soil tillage, which are the hallmarks of smallholding in Africa as well as elsewhere. CA means that there has to be technological innovation for appropriate, economically viable and socially acceptable means of farming. Progressing beyond the hoe will need more than demonstrated (i.e. research) evidence for CA benefits; it will require serious attention among governments and development partners as well as the education and private sectors. Yet small equipment is currently key in smallholder efforts to conserve soils. In conventional systems, soil tillage is required when preparing ridges and/or planting basins, which are necessary to stop surface runoff or increase water retention in areas of low-intensity rainfall and on slopes that have poor soil cover. About 73% of Africa's drylands (1 billion hectares) are moderately or severely degraded (Convention to Combat Desertification, 1995). The key problem in smallholder agriculture is soil erosion and degradation, aggravated by continuous soil disturbance (Derpsch and Moriya, 1998). In addition, systems such as CA that avoid or minimize soil disturbance need small but different equipment to carry out the field operations

Alternatives to agriculture, including off-farm employment and enterprises, often suffer complex constraints just as in farming. For instance, initiating and sustaining a meaningful business in Kenya requires significant capital investment to cover registration, land or labour, etc. Agriculture on inherited land is still the most followed pathway for livelihoods. These lands have long historical use-restoration gaps and are increasingly overcultivated, with less option for natural regeneration (i.e. fallow methods). In such systems, CA can reduce soil loss to sustainable levels (Vogel, 1994; Nyagumbo, 1999). CA has social benefits too, including increased agricultural productivity, reduced weeds and saved labour. Sometimes these benefits might take longer to manifest or accrue. For instance, weed reduction requires skilful use of herbicides, soil cover and/or consistent removal before seeds can spread. Notwithstanding, CA is a good basis for responding to resource scarcity – a major entry point for sustainable intensification, in which CA technologies are more justifiable among smallholders.

This thinking partly links to institutional innovation (Hayami and Ruttan, 1985) required to ensure short-term payoffs to smallholders, such as yield increase, which are perceived to be a major shortcoming for CA and a preferred talking point among CA critics. However, it is important that yield should be analysed in a broad social and technological context. The broad focus by CA sceptics on challenges relating to crop yield, on residue retention, or on slow improvements of soil properties is important but this focus is often narrow. Debates need to be framed against the backdrop of alarming poverty among smallholders cultivating in extremely degraded soils. Any practice that restores soil or land, however minimally, must be supported. The connections between CA results and benefits must be explained to farmers. For instance, getting more water into the soil profile results in enormous differences in crop yield; reduced soil erosion means nutrient preservation, or lesser need for fertilizer, etc. Also, preparatory de-compaction might enable greater returns per unit input of labour in rain-poor areas (Andrews, 2015). CA benefits to the poor cultivator need to be framed in terms of NRM, labour and cost-saving, as well as profit (financial and social). And compared with what alternatives? If minimal benefits are realized, they may not necessarily encourage rapid adoption of CA but instead underscore the sheer complexity of alternatives for smallholder intensification.

The focus must be on how to innovate and ensure an integrated CA system to be practised among smallholders over extended periods, especially in synchrony with the necessary supportive institutional changes. On small plots there are limited alternatives to supporting smallholder NRM (especially soils, land and water), except through integrated resource-saving intensification that relies on application of CA-based technologies supported through the right institutional arrangements at the landscape level (see also Lestrelin *et al.*, 2012).

13.1.3 Institutional arrangements

Without the necessary institutional arrangements in place, the NRM catastrophes that are building up will reduce smallholder agricultural economies to ruins. Individuals, communities, development organizations, researchers and countries must design NRM strategies that are consistent with realities of future challenges, which are likely to be shaped by the balances between population, politics, policies and poverty. African smallholders are on the periphery, almost never in the picture when key strategies are formulated, yet they bear a substantial burden of environmental goals pursued by governments.

Studies show the difficulty in reconciling poverty reduction and environmental sustainability objectives, yet possibilities exist (Lee and Barrett, 2001; Barrett *et al.*, 2002; Ribot, 2002). Evidence is scant on how to foster 'win–win' sustainable results and minimize trade-offs between NRM and smallholder development goals. Even with the promise of CA, success is difficult to scale given that scientists develop technical portfolios through meticulous controlled experimentation and then recommend wider use based on their own results, without careful regard to local social and institutional conditions. Failure is predictably a common scenario, or at best successes prove difficult to replicate (Ostrom, 1990; Ghimire and Pimbert, 1997; Pritchett and Woolcock, 2003). Even in conditions where social scientists are involved, especially to understand institutional and social contexts, there is often an inappropriate design of interventions that depend heavily on biophysical solutions (Gjertsen and Barrett, 2004). This regularly results in poor local ownership over strategies for tackling NRM at scale. This is partly the result of weak farmer organizational capacity, which limits their role in taking NRM programmes forward and thus compromises the effectiveness of any participatory process. Attention among development and research actors must focus on AIP-based approaches, building grass-roots level institutions and partnership capacities for NRM.

13.1.4 Institutions

The term institutions is used in relation to the rules devised by and/or for groups that shape farmer interaction and which influence expectations and outcomes from planned change. Such rules tend to apply more in organized entities in communities, governments, etc. For effective NRM, these rules must be clear, accepted and enforced consistently. At the local level, such institutions work better when incentives are clear. There remain hurdles in illustrating good institutional arrangements that best foster synergies among agriculture, NRM and broader development efforts. AIP offer the best available institutional framework for having different organizations collaborate to maximize the chances of addressing rural poverty reduction while at the same time enhancing NRM and agricultural goals. Although such goals are similar, they cannot be approached in the same manner everywhere, but rather they have to be adapted from site to site. Designers of CA projects need to embed agricultural innovation system (AIS) approaches. Firstly, researchers and development practitioners have to move beyond wishful thinking, theorizing or recurrent critical reflections and embrace the diversity of institutions. This often means confusion, especially given the poor people-facilitation skills available in NRM-focused projects. When such skills exist, they must always be embedded, armed with common-sense knowledge and 'street' skills to navigate complexities of institutional interactions for NRM.

The ability to achieve 'win–win' outcomes in CA for NRM and poverty reduction depends largely, but not exclusively, on institutional arrangements that influence the incentives and constraints faced by smallholders in rural Africa. AIP is an approach to improve institutional collaboration and to facilitate partnerships necessary for commercial benefits and social incentives, both at individual and collective levels, that are much needed in CA promotion. Extending CA benefits on a large scale is highly dependent on how smallholder groups are organized and how effectively they harness social capital for shared interests. Local-level organization must be linked to district and national governance structures. Stewardship of NRM is more successful when governance systems are adaptive to smallholder contexts, i.e. with the right institutional arrangements in place (Dietz *et al.*, 2003).

There is a fundamental need for a multi-scalar approach. It fits well within a context of AIS (World Bank, 2012), especially in Africa where contexts vary within small geographical distances. It is a good precondition for the functional implementation of community, district and national level priorities in agricultural development through successful CA promotion. However, any multi-scalar approach must integrate strong participatory structures, in which local priorities are indeed followed. This is well illustrated by the current situation in Kenya. Several devolved (county) governments in Kenya are prioritizing use of four-wheel tractors in the effort to 'revolutionize' agriculture. This is a popular, yet disjointed effort. In villages where average parcel sizes are as little as 0.4 ha, sustainability of large machinery is debatable. It shows that the ongoing decentralization of governance is yet to fully tackle underlying constraints of rural development and smallholder agricultural sustainability. A credible mechanism would be to establish basic financial services and human resource skills through an institutional foundation appropriate for sustainable smallholder development.

Empowering local institutions to lead CA promotion can be difficult where political and social expectations are biased. The reality is that only processes that ensure communal approaches to CA through institutional development will address impediments to efforts to develop agriculture. This is through AIPs that effectively bring together informal community groups, community-based organizations (CBOs), local governments, national governments, private businesses and non-governmental organizations (NGOs) accordingly to harness their comparative multi-scalar advantages (see also Barrett *et al.*, 2001; Ribot, 2002). There is also need for multi-sectoral approaches that integrate the complementary roles of public, private and not-for-profit sectors for CA (Misiko, 2014).

13.1.5 Policy, economic incentives and beyond

Policy instruments, not only legislative, may be used to bring about meaningful change (Fowler and Rockstrom, 2001). Agricultural and conservation policies function well when institutions are in place and technologies are acceptable. It is critical for local stakes to be preserved or ensured when CA-related policy is instituted for large-scale application. Local institutions are critical for ensuring that local perspectives or even interests are preserved for rural institutional policy to be most efficiently implemented (Ellis-Jones, 1999). Besides, over the years, many programmes have been rated 'best bet' based on economic benefits. It is often believed that economic incentives backed up by 'good' policy will drive agriculture and conservation but this logic is often disproved in Africa (Kamar, 1998). For instance, such approaches that ignored social, household and institutional requirements in Kenya failed to address the complexity of farming systems. The social landscape was not factored into the drivers of change and sustainability in agriculture. There is also much emphasis on policy formulation, when in reality implementation is the main constraint (Erenstein, 1999). Implementation is only possible when there is the right institutional landscape, including strong local (farmer) institutions to partner with high-level organizations. The right institutional arrangements will enhance positive outcomes when partnerships are organized based on strong alliances, or institutional AIPs.

13.1.6 Agricultural innovation platforms and the African smallholder

Agricultural innovation platforms (AIPs) are a viable form of collective action and a useful approach to turning smallholder farming into multifunctional agriculture for sustainable NRM (Renting *et al.*, 2009). AIP here refers to an affiliation of actors along a value chain with aligned objectives and efforts to sustainably generate and share mutual benefits. Elsewhere, AIPs are defined as alliances of stakeholders formed to diagnose constraints, explore opportunities and analyse solutions along a value chain to generate and facilitate technology applications for sustainable impact (Adekunle and Fatunbi, 2012).

AIPs are necessary because efforts to promote NRM among smallholders are often constrained by limited adoption and unsustainable use of production practices (Cary *et al.*, 2001). Among the many factors that contribute to this, the most important is lack of sufficient incentives (Pagiola, 1998; Antle and Diagana, 2003; Fox *et al.*, 2005). Sufficient incentives, especially those of economic and social nature, affect adoption of NRM practices in a major way

(Aklilu *et al.*, 2007; Woldeamlak, 2007; de Graaff *et al.*, 2008). This makes it necessary to embrace a functional model of AIP, to generate benefits through reduced costs, increased profitability, reduced risks and cumulative investments in resource-conserving practices, especially CA (see also Goldman *et al.*, 2007). AIP is a promising approach, given that many CA initiatives involve actors who often operate singly in few sites with stakeholder platforms that are loosely organized and unable to target landscape-level benefits. Besides, NRM is not a usual value chain commodity; to take off, it needs to offer strong economic, ecological and social benefits.

The AIP approach offers a solid departure point for improving returns from NRM-related investments through institutional innovation suited for the smallholder context (Rosenzweig and Binswanger, 1993). The institutional dimension usually receives limited attention; it can be approached through AIP by innovating for weak and underdeveloped markets for NRM and related service provision. AIPs function to provide a framework (or even rules) for actor interactions targeted to address marketing or market constraints, the problem of availability, costs, and quality of both inputs and information, etc., necessary for critical incentives for NRM-related agricultural investments (Cortner *et al.*, 1998; Darghouth *et al.*, 2008; Hagedorn, 2008; Merrey and Gebreselassie, 2011).

AIPs enable a new model of collective action. Such collective action involves a critical membership level to make it possible for affordable credit, insurance, information or prices that would usually be prohibitive or full of risks when smallholders act individually. When such benefits are realized, investments in NRM among smallholders are more feasible (Doss and Meinzen-Dick, 2015). AIPs therefore have a key role to coordinate resource use and management both at household level and at group level in creating, using and managing common-pool resources necessary for generating the long-term NRM returns that are often realized from consistent CA adoption over extended periods.

13.2 Methods

With information from secondary sources and illustrations from three case studies (Gerring, 2007), this chapter explains how AIP can work to generate multiple benefits at scale among smallholders. These case studies are related to the Sustainable Intensification of Maize–Legume Cropping Systems for Food Security in Eastern and Southern Africa (SIMLESA) project. The Bungoma South Farming Innovation Platform (BUSOFIP) of SIMLESA in western Kenya is supporting the adoption of CA-based practices. The Maize Innovation Platform (MIP) of Nyagatare in northern Rwanda is a multifunctional platform, registered and actively engaging in maize chain development, including marketing, with successful gender and policy dialogue and implementation. The Kieni AIP in central Kenya is a multifunctional alliance of actors, uniquely integrating climate change insurance into smallholder agriculture, local businesses and social savings.

BUSOFIP, MIP (Nyagatare) and Kieni were purposively selected based on existing ethnographic data collected in 2014. These data are ideal for illustrative, retrospective and sequential analyses. These AIPs have significance for many African smallholder contexts. Field visits and participant observation, focus group discussions with AIP actors, and institutional mapping and analyses of local maize value chains were also undertaken to build case studies.

The goal of this chapter is to analyse how longer-term CA benefits such as reduced soil degradation and labour saving can be achieved among many smallholders, through an AIP approach that enables shorter-term benefits such as lower farming costs and early planting to be harnessed.

13.3 Findings and Discussions

13.3.1 Introduction to three case study AIPs

The Bungoma South Farming Innovation Platform of SIMLESA (BUSOFIP) is registered formally. It has a wide range of actors (see also Section 13.3.6 below) and originated through the support of the SIMLESA project. BUSOFIP stakeholders are mostly based in Bungoma, western Kenya. Its key stakeholder is a local CBO. That CBO has ten resident groups that include farmer field schools (FFS), representing about 630 farmers. The most consistent activities include CA demonstrations, field days, information sharing, joint acquisition of inputs and joint mobilization of resources. This AIP has had several capacity enhancement activities, engages in scaling of other systems intensification technologies (especially seed) and catalyses farmer market chain linkages to increase production and income generation. It has active links with county government. Kieni Innovation Platform has numerous similarities with BUSOFIP but is found in Embu, eastern Kenya. It also has active insurance actors (Kilimo Salama, the agricultural insurance programme supported by the Syngenta Foundation) and has an advanced business component that is core to its collective generation of member benefits.

The Maize Innovation Platform (MIP) of Nyagatare in Rwanda has a robust market link. It is involved in the maize value chain, especially collective marketing (Markelova and Mwangi, 2010) and other aspects of maize development. This platform has a commercial subsidiary called Nyagatare Maize Investment Group Ltd (NYAMIG), which secures contracts for the supply of maize at competitive returns for members. One of the AIP's key achievements is the warrantage programme. Besides having many qualities possessed by BUSOFIP and Kieni, MIP has a solid grounding in policy, an advanced gender equity plan and highly advanced participation by commercial banks, input suppliers and high-level government and international institutions.

13.3.2 Why AIPs?

These selected AIPs are alliances of stakeholders. They are meaningful for smallholder agriculturally based NRM development because they consist of supportive organizational partners rather than weak rural groups. Supportive institutional partners are critical for strong CBOs, which is often a missing ingredient in rural NRM transformation. AIPs combine farmer institutions, local partners and higher-level organizations: for example, banks, research and CBOs are a key component in the structured alliances; they ensure that smallholder voices contribute to and shape decisions among stakeholders (Ostrom, 1990; Baland and Platteau, 1996; Kassam *et al.*, 2014). In other words, AIPs are about negotiating benefits, building consensus for action and having each actor contribute to implementing decisions. AIPs solve the ever recurrent conundrum of scale, through interactive partnerships among institutions at the landscape level. This is innovative when different yet complementary organizations partner to add various comparative advantages to the complex challenge of CA on smallholder plots. The selected AIPs have enabled a value chain approach, a cross-level approach (among local and district/national structures) and a cross-sectoral approach to integrate and solve different interests among local farmers and stakeholders. For instance, all these AIPs incorporate public extension agencies and private sector companies, beyond national research institutes. It is the public extension that has been central in reaching the scale, whereas the private sector has played a key role to sustain CA-related practices among adopters through easy access to inputs (especially herbicides), markets (of produce or products) and processing (technology or equipment). This model, based on extension–business collaboration, is incentivizing wider and diverse partner alliances for such functional innovation, beyond traditional participatory agricultural approaches (which principally include research–extension–farmer methods). Traditional participatory approaches often presuppose that complex technologies and research portfolios will eventually get adopted after being preferred during testing.

13.3.3 Ensure CA is not the problem but rather an African smallholder solution

Field visits, direct observation and interviews in northern Rwanda and eastern and western Kenya reinforced common knowledge of the existence of the twin problem of NRM and farm productivity. In these sites, land is frequently tilled, including on slopes, resulting in degradation. In Bungoma, Kieni in Kenya and MIP (Nyagatare, Rwanda), smallholder farming often happens at the expense of NRM. Smallholders in these sites share insignificant resources; they farm on individual plots and prioritize few landscape activities. Due to high poverty rates, smallholders are keener on immediate benefits, including food and incomes. They need high yields in current seasons and can ill afford to wait for long, or invest repeatedly over the years. They have no spare money to invest in equipment for CA, or to spare crop residue for soil cover instead of fodder; they have no fences to keep animals from feeding on the residue. Their current knowledge about herbicides is incomplete; improved germplasm is often unavailable; and traditional intercrops are favoured over rotations, etc. These appear to illustrate CA as an 'inconvenience' or a burden for the smallholder, especially because short-term benefits get overlooked.

The two issues of NRM and farm productivity can be tackled simultaneously, to the advantage of both. Through AIP the many agricultural development initiatives in Africa that yield shorter-term benefits can be conjoined with NRM programmes that target elusive longer-term results. CA must be framed as a development option to be implemented as an entry point to conjoin many initiatives, including businesses and subsidies. These are being innovatively combined to create many incentives for shorter-term and longer-term benefits through the SIMLESA programme in Bungoma and Kieni (Kenya) and through other initiatives in MIP (Nyagatare, Rwanda).

13.3.4 Linking CA benefits across scales

An AIP ensures that benefits do not simply add up, or are only generated, but rather create new ones immediately to ensure that longer-term ones are realized. For instance, besides targeting improved yield on individual plots, CA can reduce conflicts resulting from degradation, etc. In the immediate term, farmers save production costs, reduce drudgery, produce commercially for markets, save money that would be spent on labour, save land from animal degradation, and so on. To realize these quick benefits, different actors must realign (see also World Bank, 2012; Kassam *et al.*, 2013). Entry points may not necessarily be land management, but rather in business models that bring farmers reduced costs, saved money or improved yields. These are critical for sustained CA implementation to realize landscape-level or large-scale benefits among rural cultivators (see Section 13.3.8 below).

13.3.5 Illustrating CA and NRM-related benefits at community level

NRM among smallholders on degraded farm slopes is only achievable through coordinated interventions at a large spatial scale (Rao *et al.*, 2010). As conceptualized in Section 13.3.8, social, economic and ecological objectives across farms in a landscape must be inclusively addressed. This means that, in a typical African smallholder village, individual farmers have to be persuaded to work collaboratively (Fig. 13.1). Unlike in the past, where communal forests, water and grasslands or mutual threats such as invading warriors united residents, AIP must convincingly identify new common pool resources or uniting problems to incentivize engagements beyond farm-scale. Most smallholder livelihood enterprises are at farm level, yet they are interconnected at village level, and beyond.

NRM, especially through CA, can benefit smallholders making a living through intensive cultivation on slopes. CA has potential for both short-term farm-level and long-term communal benefits. To realize landscape results, farmers need to work in a structured manner, to practise CA widely and consistently so that both farm-scale and landscape benefits are realized. A local entity is needed, firstly to ensure that residents synchronize their actions by having a common NRM vision (Nederlof and Pyburn, 2012; Makini *et al.*,

Fig. 13.1. Interconnected livelihoods, disconnected enterprises (Misiko *et al.*, 2013; see also Msuya, 2013; Andrews, 2015).

2013) and secondly to represent and act on behalf of farmers to engage with higher-level institutions. Farmers need to organize their efforts through a strong community-based organization, which can attract or create opportunities that enable partnerships with other actors along the critical value chain.

13.3.6 Targeting landscapes through complementarity of actors and actions

CA innovations are usually developed to improve the sustainable use of soils for agriculture. For CA to result in broader benefits, including at landscape level, there is need for AIP structures that support multifunctional action (Misiko *et al.*, 2013).

Improving crop and livestock yields while at the same time conserving soils, water and ecosystems is a delicate process. These need structured approaches to plan, act and spread benefits. For instance, residents living on rocky sloping farmlands will minimize landslides, gullies or nutrient wash-away when both up-the-hill and down-the-hill cultivators collaborate beyond mere efforts to adopt CA. In many SIMLESA sites where CA is being promoted, other initiatives on different but complementary components such as livestock are being implemented. Residents in these sites have multiple, yet related problems. For instance, Bungoma smallholders have grazing problems and still seek to expand their farmland. BUSOFIP is providing a framework for them to partner and to engage outside institutions to strengthen local innovation. Smallholder representative CBOs are playing a critical role as focal points for these AIP engagements. Complementarity in BUSOFIP is found when CBOs, private companies, input dealers and livestock initiatives, among others (Fig. 13.2), work collaboratively.

The core activities need not involve all actors at once. In reality, each activity in BUSOFIP only engages core actors at a time. Business models are being explored among actors for mutual benefits.

13.3.7 Targeting landscapes means scaling CA to ensure large-scale adoption

It is widely noted that achieving significant yield increases among most farmers adopting CA is not always forthcoming in initial cropping seasons (Kassam *et al.*, 2009). Crop-related benefits of CA initiatives, given their longer-term nature, raise two fundamental issues: (i) how such benefits can be disseminated more widely; and (ii) what justifications can be provided to persuade smallholders to apply CA-based practices consistently for meaningful impact. Low adoption numbers imply that crop yields and ecological targets are not adequate incentives; there have to be better reasons to market CA while targeting NRM benefits. Far-reaching changes must be made in policies (Muchena *et al.*, 2005), institutions (FAO, 2009; Lestrelin *et al.*, 2012; Kassam *et al.*, 2013, 2014), and research and development agendas (Misiko 2014) to enable marginalized smallholder women and men to access the technologies and gain benefits equitably, consistently and broadly. Support for CA must hinge upon security of land tenure and increased public investments in participatory approaches, among others, for collective action.

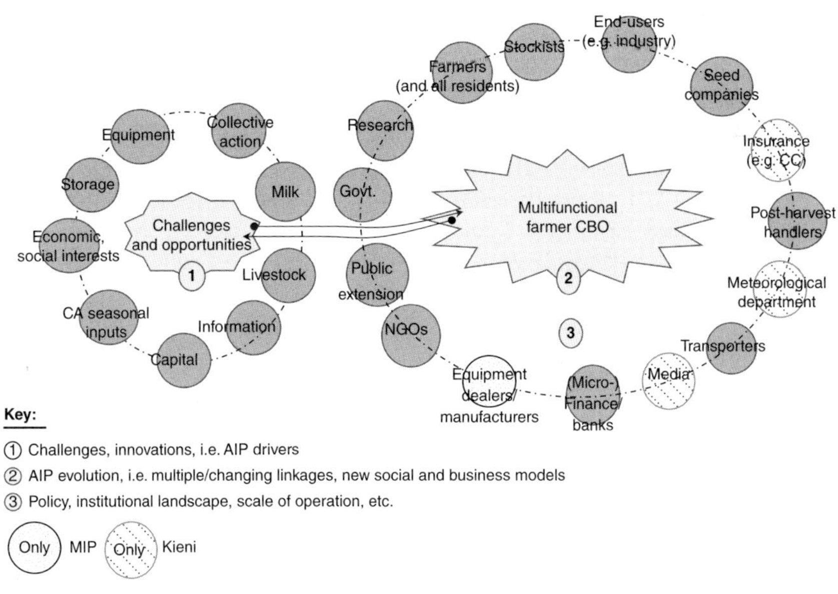

Fig. 13.2. A generalized representation of divergent agricultural innovation platforms (AIPs) for Bungoma South Farming Innovation Platform (BUSOFIP), Kieni and Nyagatare maize innovation platform (MIP).

A good case in point is MIP of Nyagatare (Rwanda), which is anchored in public policy and is fully supported by other programmes of the government of Rwanda. The main priority must now be to systematize this and other current successes, however rare, and apply the underlying success factors widely for rapid yet lasting impact under highly varying African social, agroecological and economic contexts.

In the SIMLESA programme, these factors must be brought to bear on use of new germplasm (maize and legumes), use of herbicides, soil fertility practices and cropping systems (i.e. smart intercrops and crop rotations). The SIMLESA programme is applying opportunities for integrated scaling of options that have met expectations of local smallholder householders, amid numerous constraints. A fundamental aspect has been to promote CA through AIP-based approaches, including by expanding capacities of public extension through training, incorporating NGOs, church-based organizations, CBOs, and applying business models through external facilitation (e.g. Fig. 13.2). This approach is ensuring sustainable intensification, widely applied over many seasons, necessary to harness agroecological benefits.

13.3.8 An innovation platform framework for supporting smallholder NRM through CA

Based on the case studies presented here, a progressive model of collective action can be built, beyond traditional participatory approaches that are often led by international or national institutes' researchers. Such a model (Fig. 13.3) must rely on research findings (McCarthy *et al.*, 2004), but needs to build upon non-research input and go beyond mere extension goals, targeted to galvanize collective action for cheaper credit, insurance, information or prices that would usually be prohibitive or full of risks when smallholders act alone without stronger institutional linkages.

The proposed AIP and CA for NRM (CA4NRM) smallholder framework comprises five major components. The first component relates to CA principles and practices. CA proponents will, however, have to move beyond strict prescription. By principles, it means that end results matter and variations in CA practice need to be acceptable if outcomes meet the goals. The second and third components are farm-scale shorter-term rewards and outcomes, respectively. Outcomes include assessment of longer-term ecological environmental changes due to CA. This third component also refers to the essential social attributes that influence continued CA application and include enhanced adaptation to climate change. These are possible as a result of the fourth component, which is actor activities under AIP, and the fifth one, which refers to peculiarities of innovations suitable to contexts. It is the contexts that inform AIP organizational arrangements necessary to generate economic, social and ecological benefits and incentives to invest in CA for NRM. These are key for farm-level investments, necessary for adoption decisions. Landscape benefits, which are elusive, can benefit from AIP-enabled investments. In turn, these investments result in short-term (farm-level) benefits that catalyse continued CA practice for long-term (landscape) benefits. The

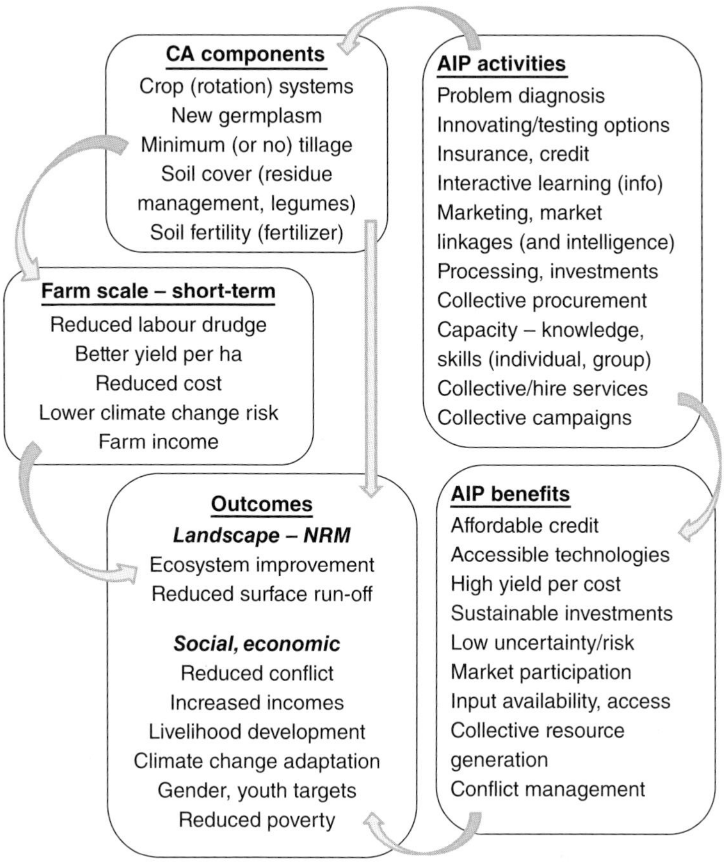

Fig. 13.3. An integrated agricultural innovation platform (AIP) and CA for natural resources management (NRM) smallholder framework. Smallholder CA is about balancing short-term and long-term incentives.

fourth component in the CA4NRM framework requires relevant institutional actors in place, necessary to address issues related to the kind of services that smallholders need. It also requires the identification of rules (provisions) and policy to guide the process (Makini *et al.*, 2013).

AIP approach must always be shaped for each context to enhance livelihood diversification and resilience. This is the way African smallholders can practise and sustain CA and ensure wider benefits for NRM. It is not possible to prescribe and predict outcomes of such innovation but rather it is required that actors work to shape them. AIP must be a learning process and adaptive to changing conditions to ensure continued progress toward the guiding vision. In this process, therefore, various actors (farmer CBOs or cooperatives, governments, experts and non-farmer actors) need a mutual arrangement guided by vision and good leadership. With good leadership, and multifunctional AIPs as shown in SIMLESA (Kenya and Rwanda) experiences, institutional innovations deliver better impacts when they address the multiple dimensions of smallholder needs. Integrating different AIP activities (fourth component) fosters positive synergies for effective benefits that contribute to livelihood resilience. CA for NRM must therefore be linked to credit, market, information, insurance, training, policy dialogue, conflict management, etc. This is only possible through a landscape-based institutional approach to foster a coordinated problem-solving agenda beyond individual households or a village. Such an approach must embrace multiple collectively agreed objectives at a scale necessary to manage trade-offs and reduce conflicts.

13.3.9 Coordinated action across the landscape

For landscape-level results to be realized, synchronized actions across the landscape are needed. For instance, residents on slopes need a collective, coordinated approach that sees the land protected through changed use or abandoned activities. Farmers living up the hill need to plant more cover crops, protect their soils through no tillage and soil mulch cover, reduce or eliminate grazing livestock on slopes, etc. With shorter-term benefits, both up-the-hill and down-the-hill farmers can widely practise CA, with benefits being realized at landscape and at farm level. The up-the-hill farmers can avoid losing their crops or homes to erosion. The down-the-hill farmers can avoid losses resulting from mud-floods, etc. When their actions are synchronized, and when AIP improves actor complementarity, then CA benefits are manifested in the short term (such as reduced drudgery and fertilizer costs) and at landscape level (such as reduced landslides). In their totality, such benefits outweigh any opportunity costs relating to abandoned crops and livestock. There is scope for more social inclusion and harmony (Andrews, 2015) and a much larger-scale practice of NRM through CA (Kassam *et al.*, 2013). For instance, through the AIP approach, nomadic herders need to be incorporated in CA platforms to mitigate animal destruction of maize residue or legume cover crops. As Msuya (2013) explains, the overlapping interests of farmers and animal herdsmen on crop fields used for grazing cannot be settled through policy or legislation alone.

13.4 Conclusion and Way Forward

Smallholders can improve livelihoods through diversification that assists in the conservation of the natural resource base and agrobiodiversity. Whether the potential and spread of CA is realized depends on several factors and actions.

Firstly, proposed CA4NRM strategies have to be adapted for smallholder contexts, beyond mere yield increments and natural resource conservation. Strategies must create immediate incomes, ease access to critical farm inputs and improve output markets. MIP (Nyagatare) in Rwanda illustrates how smallholders can have complementary marketing and value addition activities. It also illustrates how traditional gender skills and cultural trends can be harnessed to ensure that AIPs succeed to foster collective smallholder self-reliance. MIP (Nyagatare) gender-equitable market opportunities and its other activities are ensuring that farmers and consumers are linked more directly and sustainably. The ultimate challenge that SIMLESA is researching is to unlock larger investments that take such successes to scale across Africa.

Secondly, AIP is unveiling processes that enhance smallholder organization and is enabling participation and focus on interactive learning that is key for innovation. BUSOFIP is a classic AIP example of how to grapple with challenges of diversification in an area where a single cash commodity has for decades defined poverty pathways. The drive for BUSOFIP must be to rely on collective sugarcane-related discontent to rally alternative action.

Thirdly, researchers and relevant development partners will have to translate generic CA concepts and on-station results into practical actionable practices that directly address the needs and fit the contexts of smallholders. These include demonstrating how CA is resource conserving, saves labour, increases farming efficiency, supports diversification of smallholder systems and improves incomes. CA as a technological solution is not cheap and it therefore requires well thought-out AIP business models. There must be opportunities to explore existing smallholder skills and knowledge through strong CBOs.

Fourthly, there must be dialogue for policies and institutional backup. CA adoption must bring forth equitable, broadly accessible and multiple benefits. These are key reasons for investing in AIP and the search for incentives for several value chain actors to stay engaged with the smallholder to unlock the full range of CA benefits.

13.5 Acknowledgements

I am grateful to the African Conservation Tillage Network (ACT) for the stimulating conversations on the topic, especially during the Conservation Agriculture Conference in Zambia. I thank the reviewers of this chapter and Dr Saidi Mkomwa for their illustrious work. I deeply appreciate the immeasurable support from International Maize and Wheat Improvement Center (CIMMYT), my colleagues at the Rwanda Agricultural Board, and SIMLESA. Finally, my sincere gratitude to the Australian Centre for International Agricultural Research (ACIAR) for funding this research.

References

Adekunle, A.A. and Fatunbi, A.O. (2012) Approaches for setting-up multi-stakeholder platforms for agricultural research and development. *World Applied Sciences Journal* 16(7), 981–988.

Aklilu, A., Stroosnijder, L. and de Graaff, J. (2007) Long-term dynamics in land resource use and the driving forces in the Beressa watershed, highlands of Ethiopia. *Journal of Environmental Management* 83(4), 448–459.

Altieri, M.A. and Toledo, V.M. (2005) Natural resource management among small-scale farmers in semi-arid lands: building on traditional knowledge and agroecology. *Annals of Arid Zone* 44(3&4), 365–385.

Andrews, C. (2015) Promoting sustainable farming methods in Malawi: the Tiyeni method. In: *Tropical Agriculture Association. Agriculture for Development. Special Issue on Soils*. No. 24, Spring 2015.
Antle, J.M. and Diagana, B. (2003) Creating incentives for the adoption of sustainable agricultural practices in developing countries: the role of soil carbon sequestration. *American Journal of Agricultural Economics* 85(5), 1178–1184.
Baker, C.J., Saxton, K.E. and Ritchie, W.R. (1996) *No-Tillage Seeding: Science and Practice*. CAB International, Wallingford, UK.
Baland, J.M. and Platteau, J.-P. (1996) *Halting Degradation of Natural Resources: Is There a Role for Rural Communities?* Clarendon Press, Oxford, UK.
Barrett, C.B., Brandon, C.B.K., Gibson, C. and Gjertsen, H. (2001) Conserving tropical biodiversity amid weak institutions. *BioScience* 51(6), 497–502.
Barrett, C.B., Place, F. and Aboud, A.A. (Eds). (2002) *Natural Resources Management in African Agriculture: Understanding and Improving Current Practices*. CAB International, Wallingford, UK.
Carmona-Torres, C., Parra-López, C., Groot, J.C.J. and Rossing, W.A.H. (2011) Collective action for multi-scale environmental management: achieving landscape policy objectives through cooperation of local resource managers. *Landscape and Urban Planning* 103, 24–33.
Cary, J.W., Webb, T. and Barr, N.F. (2001) *The Adoption of Sustainable Practices: Some New Insights. An Analysis of Drivers and Constraints for the Adoption of Sustainable Practices Derived From Research*. Land and Water, Canberra, Australia.
Convention to Combat Desertification. (1995) *Convention to Combat Desertification*. Interim Secretariat, Geneva, Switzerland.
Cortner, H.J., Wallace, M.G., Burke, S. and Moote, M.A. (1998) Institutions matter: the need to address the institutional challenges of ecosystem management. *Landscape and Urban Planning* 40(1), 159–166.
Darghouth, S., Ward, C., Gambarelli, G., Styger, E. and Roux, J. (2008) *Watershed Management Approaches, Policies, and Operations: Lessons for Scaling Up*. Water Sector Board Discussion. Paper Series, Paper No. 11. The World Bank, Washington, DC.
De Graaff, J., Amsalu, A., Bodnar, F., Kessler, A., Posthumus, H. and Tenge, A. (2008) Factors influencing adoption and continued use of long-term soil and water conservation measures in five developing countries. *Applied Geography* 28(4), 271–280.
Derpsch, R. and Moriya, K. (1998) Implications of no-tillage versus soil preparation on sustainability of agricultural production. *Advances in GeoEcology* 31, 1179–1186.
Di Gregorio, M., Hagedorn, K., Kirk, M., Korf, B., McCarthy, N., Meinzen-Dick, R., *et al.* (2012) Property rights and collective action for poverty reduction: A framework for analysis. In: Mwangi, E., Markelova, H. and Meinzen-Dick, R. (eds) *Collective Action and Property Rights for Poverty Reduction: Insights from Africa and Asia*. University of Pennsylvania Press, Philadelphia, Pennsylvania, pp. 25–48.
Dietz, T., Ostrom, E.P. and Stern, C. (2003) The struggle to govern the commons. *Science* 302, 1907–1912.
Doss, C.R and Meinzen-Dick, R. (2015) Collective action within the household: insights from natural resource management. *World Development* 74, 171–183.
Ellis-Jones, J. (1999) Poverty, land care and sustainable livelihoods in hillside and mountain regions. *Mountain Research and Development* 19, 179–190.
Erenstein, O.C.A. (1999) The economics of soil conservation in developing countries: the case of crop residue mulching. Thesis. Wageningen University, The Netherlands.
FAO (2009) *Scaling-up Conservation Agriculture in Africa: Strategy and Approaches*. The FAO sub-Regional Office for Eastern Africa, Addis Ababa.
Fowler, R. and Rockstrom, J. (2001) Conservation Agriculture for sustainable agriculture: an agrarian revolution gathers momentum in Africa. *Soil and Tillage Research* 61, 93–107.

Fox, P., Rockstrom, J. and Barron, J.P. (2005) Risk analysis and economic viability of water harvesting for supplemental irrigation in semi-arid Burkina Faso and Kenya. *Agricultural Systems* 83(3), 231–250.
Gerring, J. (2007) *Case Study Research: Principles and Practices*. Cambridge University Press, Cambridge, UK.
Ghimire, K. and Pimbert, M. (1997) *Social Change and Conservation*. Earthscan, London.
Gjertsen, H.E. and Barrett, C.B. (2004) Context-dependent biodiversity conservation management regimes. *Theory and Simulations Land Economics* 80(3), 323–341.
Goldman, R.L., Thompson, B.H. and Daily, G.C. (2007) Institutional incentives for managing the landscape: inducing cooperation for the production of ecosystem services. *Ecological Economics* 64, 333–343.
Hagedorn, K. (2008) Particular requirements for institutional analysis in nature-related sectors. *European Review of Agricultural Economics* 35 (3), 357–384.
Hayami, Y. and Ruttan, V.W. (1985) *Agricultural Development: An International Perspective*. The John Hopkins University Press, London.
IFAD (2006) *Community-based Natural Resource Management*. International Fund for Agriculture Development (IFAD), Rome.
Kamar, M.J. (1998) Soil conservation implementation approaches in Kenya. *Advances in Geoecology* 31, 1057–1064.
Kassam, A.H., Friedrich, T., Shaxson, F. and Pretty, J. (2009) The spread of Conservation Agriculture: justification, sustainability and uptake. *International Journal of Agricultural Sustainability* 7(4), 1–29.
Kassam, A.H., Basch, G., Friedrich, T., Shaxson, F., Goddard, T., Amado, T., *et al.* (2013) Sustainable soil management is more than what and how crops are grown. In: Lal, R. and Stewart, R.A (eds) *Principles of Soil Management in Agro-ecosystems*. Advances in Soil Science. CRC Press., Boca Raton, Florida, USA.
Kassam, A.H., Friedrich, T., Shaxson, F., Bartz, H., Mello, I., Kienzle, J. and Pretty, J. (2014) The spread of Conservation Agriculture: policy and institutional support for adoption and uptake. *Field Science Research Reports* 7(4), 1–29.
Lee, D.R. and Barrett, C.B. (eds) (2001) *Tradeoffs or Synergies? Agricultural Intensification, Economic Development and the Environment in Developing Countries*. CAB International, Wallingford, UK.
Lestrelin, G., Nanthavong, K., Jobard, E., Keophoxay, A., Lienhard, P., Khambanseuang, C. and Castella, J.C. (2012) To till or not to till? The diffusion of conservation agriculture in Xieng Khouang province, Lao PDR opportunities and constraints. *Outlook on Agriculture* 41(1), 41–49.
Makini, F.W., Kamau, G.M., Makelo, M.N., Adekunle, W., Mburathi, G.K., Misiko, M., *et al.* (2013) *Operational Field Guide for Developing and Managing Local Agricultural Innovation Platforms*. Kenya Agriculture Research Institute (KARI) and Australian Centre for International Agricultural Research (ACIAR), Nairobi.
Markelova, H. and Mwangi, E. (2010) Collective action for smallholder market access: evidence and implications for Africa. *Review of Policy Research* 27(5), 621–640.
McCarthy, N., Dutilly-Diane, C. and Drabo, B. (2004) Cooperation, collective action and natural resources management in Burkina Faso. *Agricultural Systems* 82, 233–255.
Merrey, D. and Gebreselassie, T. (2011) *Promoting Improved Rainwater and Land Management in the Blue Nile (Abay) Basin of Ethiopia*. NBDC Technical Report No. 1. International Livestock Research Institute (ILRI), Nairobi.
Misiko, M. (2014) *Assessing and Strengthening Capacity for Agricultural Innovation Platforms*. Technical Report. A training workshop held at Hotel Des Mille Collines (Kempinski), Kigali, Rwanda September 15–19, 2014. International Maize and Wheat Improvement Centre (CIMMYT), Addis Ababa.
Misiko, M., Mundy, P. and Ericksen, P. (2013) *Innovation Platforms to Support Natural Resource Management*. Innovation Platforms Practice Brief 11. International Livestock Research Institute (ILRI), Nairobi.

Msuya, D.G. (2013) Farming systems and crop-livestock land use consensus. Tanzanian perspectives. *Open Journal of Ecology* 3, 473–481.

Muchena, F.N., Ondurua, D.D., Gachini, G.N. and de Jager, A. (2005) Turning the tides of soil degradation in Africa: capturing the reality and exploring opportunities. *Land Use Policy* 22, 23–31.

Nederlof, E.S. and Pyburn, R. (2012) *One Finger Cannot Lift a Rock. Facilitating Innovation Platforms to Trigger Institutional Change in West Africa*. Royal Tropical Institute, Amsterdam, The Netherlands.

Nyagumbo, I. (1999) Conservation tillage for sustainable crop production systems: experiences from on-station and on-farm research in Zimbabwe (1988–1997). In: Kaumbutho, P.G. and Simalenga, T.E. (eds) *Conservation Tillage with Animal Traction*. Aminal Traction Network for Eastern and Southern Africa (ATNESA), Harare, Zimbabwe, pp. 107–114.

Ostrom, E. (1990) *Governing the Commons: The Evolution of Institutions for Collective Action*. Cambridge University Press, Cambridge, UK.

Oweis, T. and Hachum, A. (2009) Supplemental irrigation for improved rainfed agriculture in WANA region. In: Wani, S.P., Rockström, J. and Oweis, T. (eds) *Rainfed Agriculture: Unlocking the Potential.* CAB International, Wallingford, UK, pp. 182–196.

Pagiola, S. (1998) *Economic Analysis of Incentives for Soil Conservation*. World Bank, Washington, DC.

Pathak, P., Sahrawat, K.L., Wani, S.P., Sachan, R.C. and Sudi, R. (2009) Opportunities for water harvesting and supplemental irrigation for improving rainfed agriculture in semiarid areas. In: Wani, S.P., Rockström, J. and Oweis, T. (eds) *Rainfed Agriculture: Unlocking the Potential.* CAB International, Wallingford, UK, pp. 197–221.

Pritchett, L. and Woolcock, M. (2003) Solutions when the solution is the problem: arraying the disarray in development. *World Development* 32(2), 191–212.

Rao, C.A.R., Kareemulla, K., Nagasree, K., Venkateswarlu, B. and Kumar, S. (2010) Estimation of economic returns to soil and water conservation research – an ex ante analysis. *Agricultural Economics Research Review* 23(1), 41–46.

Renting, H., Rossing, W.A.H., Groot, J.C.J., Van der Ploeg, J.D., Laurent, C., Perraud, D., *et al.* (2009) Exploring multifunctional agriculture. A review of conceptual approaches and prospects for an integrative transitional framework. *Journal of Environmental Management* 90, S112–S123.

Ribot, J.C. (2002) *Democratic Decentralization of Natural Resources: Institutionalizing Popular Participation*. World Resources Institute, Washington, DC.

Rockström, J., Barron, J. and Fox, P. (2002.) Rainwater management for increased productivity among small-holder farmers in drought prone environments. *Physics and Chemistry of the Earth* 27(11–22), 949–959.

Rockström, J., Karlberg, L., Wani, S.P., Barron, J., Hatibu, N., Oweis, T., *et al.* (2010) Managing water in rainfed agriculture: the need for a paradigm shift. *Agricultural Water Management* 97(4), 543–550.

Rosenzweig, M.R. and Binswanger, H.P. (1993) Wealth, weather risk and the composition and profitability of agricultural investments. *Economic Journal* 103(416), 56–78.

Shiferaw, B. (2006) Poverty and natural resource management in the semi-arid tropics: revisiting challenges and conceptual issues. *SAT eJournal* 2(1), 1–21.

Vogel, H. (1994.) *Conservation Tillage in Zimbabwe: Evaluation of Several Techniques for the Development of Sustainable Crop Production Systems in Smallholder Farming*. African Studies Series, Vol. A11. University of Berne, Switzerland, 150 pp.

Woldeamlak, B. (2007) Soil and water conservation intervention with conventional technologies in northwestern highlands of Ethiopia: acceptance and adoption by farmers. *Land Use Policy* 24(2), 404–416.

World Bank (2008) *Development and Climate Change: A Strategic Framework for the World Bank Group*. Report to the Development Committee, Washington, DC.

World Bank (2012) *Agricultural Innovation Systems: An Investment Sourcebook*. World Bank, Washington, DC.

14 Summing Up

Amir H. Kassam,[1]* Saidi Mkomwa[2] and Theodor Friedrich[3]

[1]University of Reading, UK; [2]African Conservation Tillage Network, Nairobi, Kenya; [3]Food and Agriculture Organization of the United Nations, Havana, Cuba

Reducing soil disturbance by tillage in agricultural land began in the Great Plains in the USA in the 1930s in response to the devastation caused by prolonged drought and tillage. This period became known as the 'Dust Bowl'. Several practices, including stubble mulching, were developed to reduce or eliminate tillage and retain plant residue on the soil surface to alleviate wind and water erosion. This collection of practices led to what became known as conservation tillage, culminating in no-till systems that avoid any soil disturbance by no-till seeding, and maintaining a mulch cover of organic matter on the soil surface. The book *Ploughman's Folly* by Edward Faulkner (1943), an extension agronomist in Ohio, USA, was an important milestone in the development of conservation agriculture practices. Faulkner questioned the wisdom of inversion ploughing and explained the destructive nature of soil tillage. Research in the UK, USA and elsewhere during the late 1940s and 1950s made no-till farming possible. The practice began to be adopted by farmers in the USA in the 1960s and in Brazil, Argentina and Paraguay in the 1970s and in Zimbabwe in the 1980s, with research on no-till farming being published in Zimbabwe and Nigeria in the 1970s and 1980s (Kassam *et al.*, 2014a).

The modern concept of no-till farming – now generally known as Conservation Agriculture (CA) – involves the simultaneous application of three interlinked principles based on locally formulated practices: (i) permanently minimizing or avoiding mechanical soil disturbance (no-till seeding); (ii) maintaining a continuous soil mulch cover with organic matter (crop residues, stubbles and cover crops including legumes); and (iii) growing diverse plant species in the cropping systems that can include annual and perennial crops, trees, shrubs and pastures in associations, sequences and rotations. CA principles can be applied to all land-based rainfed and irrigated

*amirkassam786@googlemail.com

farming systems (annual, perennial, plantation, horticulture, mixed systems including with root and tuber crops and groundnuts, agroforestry, crop–livestock, rice cropping systems).

The three CA practices, in conjunction with good crop, nutrient, pest (weeds, insects and pathogens), water, farm power and energy management, provide an ecological foundation to cropping systems upon which to build sustainable production intensification and land management strategies at the farm and landscape levels. Global literature on CA performance repeatedly shows that such strategies are capable of mobilizing greater crop and land potentials of agricultural production systems while simultaneously facilitating the flow of ecosystem services such as the availability of clean water, carbon sequestration, control of runoff and soil erosion, pollination services and water, nutrient and carbon cycling. Further, it is increasingly becoming clear that CA systems offer greater factor and overall efficiencies and have greater resilience to biotic and abiotic stresses compared with conventional tillage systems. In fact, CA systems are also considered to be climate-smart from an adaptation as well as mitigation viewpoint.

Given these productivity, economic, environmental and social advantages and benefits, the uptake and spread of CA annual cropping systems globally has been increasing at the rate of some 10 million hectares per annum from 2008/09 to 2013. In 2013, CA was spread across 157 million hectares of annual cropland worldwide, and since 2008/09 it had increased by some 47%, from 106 million hectares. In sub-Saharan Africa, the spread of CA increased by some 750,000 ha over the same period, from some 475,000 ha in 2008/09 to 1.2 million hectares in 2013, corresponding to an increase of more than 150%. This figure is considered to be conservative, as it was not possible to update all the area under CA in 2013. We consider that the current area of CA in Africa is around 1.5 million hectares or more.

That most of the 750,000 ha increase in CA area has been with smallholder farmers suggests that hundreds of thousands of farmers, possibly more than a million, are reaping the productivity, economic and environmental benefits of CA. The existence of CA in Africa has been reported from many countries, including South Africa, Zimbabwe, Zambia, Namibia, Malawi, Madagascar, Mozambique, Lesotho, Swaziland, Botswana, Uganda, Tanzania, Kenya, DR Congo, Rwanda, Burundi, Sudan, Ethiopia, Senegal, Burkina Faso, Ghana, Cameroon, Niger, Algeria, Morocco and Tunisia. Much of this recent adoption of CA has occurred as a result of more policy and extension attention and development resources, from donors, governments, non-governmental organizations (NGOs) and national and international research and development organizations, being directed towards the promotion of CA through participatory dissemination and upscaling approaches. This is well reflected across most of the chapters of this book, highlighting area-specific research and scaling initiatives in all the regions of Africa.

As the book indicates, agriculture across Africa is changing for the better, the change being increasingly driven by Africa's own home-grown strategic thinking and action at all levels of decision making and practice about sustainable production intensification, from political to public, private and civil sectors to farming communities in all the regions and in different development circumstances. The 1st Africa Congress on Conservation Agriculture generated the material for this book and presented convincing evidence and examples from across the continent to show that CA is capable of reversing land degradation and building the productivity potential of the agricultural resource base of smallholders for actual production intensification. The Congress also pointed towards the kind of technical, knowledge, communication and service sector support, and therefore capacity development, needed to promote the spread of CA in the farming systems of the different regions in Africa. The Lusaka Declaration (http://act-africa.org/search.php) makes it clear that African research and development stakeholders consider CA also to be climate-smart at the farm and landscape level, as shown by many smallholder farmers, offering an opportunity to build climate-smart and regenerative food and agriculture systems at sub-national and national levels that were simply not possible in the past with conventional tillage agriculture. The key difference is that, in CA systems, degradative components of conventional tilled systems are removed and replaced with core practices that provide a biologically active ecological underpinning to the total production system, including all the components (crop, nutrient, water, pest, farm power and energy management) of productive agriculture, upon which to build an efficient and resilient farming system capable of delivering optimal productivity and ecosystem services.

Africa is already looking to the future with greater confidence in establishing its own effective national and regional strategies for sustainable production intensification and sustainable land management based on CA to address the challenges of food security, climate change, agricultural development and economic growth. Despite the continuing negative pronouncements from some academics from the north about the place for CA in African agriculture development, current scientific and empirical evidence presented in this book and in other books and scientific literature suggests that CA systems offer the best option we have today for the sustainable production of field crops, including annuals and perennials, and of crop–livestock systems in rainfed and irrigated agriculture.

While it is critically important to establish good CA research programmes across Africa, it is equally important to get on with the business of transforming African agriculture through effective formulation and dissemination of CA practices, farmer and stakeholder capacity development, formation of CA farmer groups, networks, cooperatives and associations, promoting the expansion of the private sector service providers (including equipment suppliers and manufacturers) and integrating CA into education and training programmes. At the sector level, it is essential to get all the appropriate components of the national food and agriculture systems to support their respective development sustainably. However, it must not be forgotten that the efficiency and resilience of every local and national food and agriculture system depends on the extent to which production systems or the farming systems are ecologically underpinned for sustainability, including their resilience to biotic and abiotic stresses and shocks. Chapters 1 and 2 make it clear that, conceptually and at the level of farmer practice, CA systems offer the necessary ecological underpinning. In so doing, CA systems not only have the potential to be productive, efficient and resilient but also permit the flow of ecosystem services that are often debilitated or become dysfunctional as a result of continuous degradation of land and the environment that accompanies conventional tillage agriculture.

The increasing recognition being given to CA as being climate-smart is not surprising, as discussed in Chapter 2, nor are the ideas and evidence discussed in Chapters 3 and 4 that CA systems do offer an opportunity to manage agricultural production at a much greater degree of efficiency or higher levels of factor productivity, prompting some to point to CA systems as being able to produce 'more with less' or 'more from the same', and thus empower smallholders to become more profitable and sustainable. Indeed, as improved soil health conditions become established and soil functions related to water, nutrient and carbon cycling become rehabilitated in CA systems, a whole range of productivity, environmental, financial and socio-economic benefits begin to accrue to the producers and their resource base as well as to rural communities and potentially to societies at large.

Chapters 4, 5, 6 and 7 clearly show that benefits from CA can be harnessed in different farming systems in different regions, from southern Africa to North Africa, and, as would be expected, adoption of CA is beset with different mixes of constraints and farmer perception. However, the case of farmers in northern Mozambique (Chapter 6) reveals that even in isolated semi-arid areas without the use of any purchased inputs, production intensification and improvement in productivity and resilience are possible with the locally adapted CA approach. Chapter 6 confirms what many development experts know from experience to be true of traditional cropping systems when these are based on CA practices and include cover crops as illustrated in Chapter 8, or operate on the push–pull technology as illustrated in Chapter 9, or rely on a crop association involving a tree legume as illustrated in Chapter 10. Chapters 8, 9 and 10 show that, while CA in general is considered to be climate-smart and more resilient, additional resilience can be built into CA farming systems by integrating biological processes that can enhance nutrient supply to the crop and provide natural protection against a specific insect pest (maize stem borer) and a specific pernicious hemi-parasitic weed (striga). Similarly, Chapter 11 shows how CA-compliant mechanization not only can raise system productivity but also adds resilience to cropping systems by enabling timely operations to be carried out on larger areas. This is a highlight of the possible opportunities for making smallholder farming attractive to the youth as 'a business that thrives'.

The material contained in this book does not cover in any detail the climate-change mitigation potential of CA. In general, in well established CA systems, benefits include up to 70% decrease in fuel energy or manual labour, up to 50% less mineral fertilizer requirement, 20% or more reduction in pesticide and herbicide use (including the use of no pesticide and herbicide where they are not available or in organic CA systems), some 30–50% less water requirement, and reduced capital outlay on farm mechanization.

All these benefits lower the carbon footprint of agriculture. In addition, it has been shown widely that CA implemented in a proper way sequesters carbon and lowers emissions of greenhouse gases of CO_2, N_2O and CH_4.

What CA can do for small or large producers in semi-arid or humid environments and for agricultural growth and food security in Africa is to usher in the much needed fresh thinking and action about sustainable production intensification, about effective farmer empowerment, about the development of a profitable private service sector, and about relevant education and research support for agriculture development. Indeed, what CA represents is the paradigm change that Africa and the rest of the world need to support to address the major development challenges, including the lack of strategic thinking that the conventional tillage agriculture paradigm has created. CA is an agroecological (farm level) and agroecosystem (landscape level) response to much that is now considered to be unfit for purpose in agriculture, in Africa and in the rest of the world. Changes in agricultural practices that are now called for must be based on the agroecological and biological understanding of sustainable soil management, including the management of coevolved soil and rhizospheric biodiversity, and food webs. The need to understand their productivity, resilience and ecosystem relationships with crops and animals has become more important than ever before. Chapters 12 and 13 illustrate this need in terms of differentiated policy and knowledge support that will be required over the coming years and decades to assist the upscaling and mainstreaming of CA in Africa.

As the book makes clear, CA is most definitely not a quick-fix solution to agricultural development, nor does it offer a silver bullet solution to overcome all constraints, and it is certainly not a panacea solution to all agricultural development problems. However, CA does offer an alternative approach to underpin crop production systems (including with trees and livestock) ecologically so that they are sustainable and resource enhancing and conserving, offering on-farm productivity benefits and landscape-level ecosystem services. As with any farming and land management system, CA has its constraints that must be overcome for large-scale adoption and uptake. However, as indicated earlier, CA is spreading in Africa at a surprisingly faster rate than what is being broadcast by CA antagonists and sceptics, and some academic researchers have even set themselves up as gatekeepers and courtiers. For sure, the establishment of CA practices on small or large farms can be difficult in initial years in some semi-arid areas and on heavy clay soils, or compacted and poorly drained land. Control of insect pests, diseases and weeds can be a concern in some instances and pesticides and herbicides may be required, at least in the initial years. Leaving crop residues and stubbles on fields as mulch would reduce an important source of animal fodder, at least initially, in some areas where livestock play an important role in farm economies. In some instances, such reduction can be temporary because soil health and productivity improvement leads to increased biomass production and additional fodder availability. In other cases, community-based solutions are possible in terms of residue and grazing management to optimize the changing situation. There can be other location-specific socio-economic issues that must be addressed, such as perceived risk of loss of productivity in initial years or possible displacement of paid farm labour. Lack of appropriate equipment and machinery for seeding and fertilizer placement can be problematic.

However, adoption and spread of CA in Africa so far offers lessons that show that the above-mentioned challenges are being overcome by farmers through locally formulated solutions involving a range of public and private stakeholders working together with farmers and their communities along different adoption pathways and transformation. In the coming decades, every effort by all concerned must be made in Africa to support this remarkable ongoing transformation of tillage agriculture to CA in all regions. There are several immediate ways to support and accelerate the widespread upscaling of CA in Africa, as has been described by Kassam *et al.* (2014b), as follows:

- In all new agricultural development projects, include CA as the basis for sustainable production intensification and engage all relevant stakeholders to ensure success.
- Promote and enable CA farmers to work in groups such as farmer field schools, farmer networks and no-till associations to capture economies of scale, reduce failure risks and promote farmer-led innovation of CA systems and practices.
- Revise universities' agriculture curricula to include teaching the next generation of farmers and agriculture development practitioners about CA as an alternative and sustainable way of farming.
- Fund more innovative practical CA research to tackle soil, agronomic and livestock husbandry challenges through universities and research centres.
- Advocate for initial government support in terms of subsidies to make appropriate farm equipment and machinery more readily accessible and to reduce any risks of possible production losses during the initial years of transformation to CA.
- Encourage governments to update their agricultural policies and bring institutional reforms that support the upscaling of CA.
- Develop large-scale programmes that would offer payments to CA farmers working together for their improved stewardship of agricultural lands as observed in terms of improved soil health, ecosystem functions and productivity, and for harnessing societal-level ecosystem services such as carbon sequestration, watershed services for increasing the quantity and quality of water resources, control of soil degradation and erosion and reduction in flood risks and enhancing pollination services.
- Use the fast-growing information communication technologies in Africa, which offer a unique opportunity to convey – in an interactive way – much-needed CA information and knowledge in real time and cost-effectively. Relevant client-demanded content and appropriate platforms need to be developed. These include ePlatforms and eLearning courses.
- At continental level, support the implementation of the CA and climate-change agenda declarations, such as the Malabo Declaration on CAADP

and commitment to accelerate Agricultural Growth and Transformation for Shared Prosperity and Improved Livelihoods (encompassing Vision 25×25). Stronger linkage and involvement are needed amongst regional players: the African Union (AU), NEPAD, CAADP, Regional Economic Communities, national governments, NGOs and the private sector.

Fuller and widespread advantages of benefits offered by CA systems can be taken by African farmers and their rural communities if all stakeholders become involved in facilitating the transformation process, as has been happening in a number of countries in Africa, such as Zambia, Zimbabwe, Malawi and South Africa. However, as the book chapters show, a more structured response to the opportunities presented by CA calls for a realignment of public and private sector agricultural institutions, including research, extension and education, as well as agricultural development policies to enable CA to be the preferred agricultural paradigm choice for building and sustaining productive, efficient and resilient food and agriculture systems at national and regional levels. This is the spirit of the Lusaka Declaration, an outcome of the 1st Africa Congress on Conservation Agriculture held in Lusaka in March 2014, organized by ACT Network. The Congress brought together hundreds of CA stakeholders, including CA farmers from all regions in Africa, to share CA experiences and knowledge and decide on how Africa can take full advantage of the opportunity presented by CA for building resilient farming systems in a changing climate.

The Congress and this book are reliable indicators that Africa is deciding and taking strategic action based on its own understanding of what should be the preferred agriculture paradigm for Africa's future. ACT as a pan-African institution has established an effective continental network of CA stakeholders linked to CA stakeholders internationally to facilitate the much needed transformation of Africa's tillage-based food and agriculture systems to CA-based. This book of national and regional analysis of uptake of CA by smallholders and the required policy and knowledge support for its spread shows that there is now continent-wide support for CA from the key stakeholders in the public, private and civil sectors. The analysis also shows that the principles of CA and their locally formulated adapted practices, along with other production system inputs and practices, appear to offer entirely appropriate and best solutions to smallholder farmers in the African tropics and subtropics, including in winter-rainfall Mediterranean environments, for achieving sustainable production intensification. This is fully backed up by the spread of CA systems globally, of which half the CA area is in the developing countries, offering excellent opportunities for south–south collaboration for the development of CA in Africa. This is not surprising, especially when most of the discussions about the need for sustainable production intensification, climate-smart agriculture, sustainable land management, regenerative agriculture and climate change adaptability and mitigation come to the conclusion that CA systems represent the best practical examples of how to address such needs in the real world in a changing climate.

References

Faulkner, E. (1943) *Ploughmans's Folly*. Univervisty of Oklahoma Press, Norman, Oklahoma, USA.

Kassam, A.H., Derpsch, R. and Friedrich, T. (2014a) Global achievements in soil and water conservation: the case of conservation agriculture. *International Journal for Soil and Water Conservation* 2(1), 5–13, doi: 10.1016/S2095-6339(15)30009-5.

Kassam, A.H., Friedrich, T. and Jat, R.A. (2014b) Summing up. In: Jat, R.A., Sahrawat, K.L. and Kassam, A.H. (eds) *Conservation Agriculture: Global Prospects and Challenges*. CAB International, Wallingford, UK, pp. 375–379.

Index

Page numbers in **bold** refer to illustrations and tables.

acceptance 2, 3
access 71, 200, 206, 228, **230**, 232, **234**, 236, 257
achievements 132–135
action plans 95
actions 256–257, 261
activities 122–123, 223, 225–226, 257, 258–260
actors 102–103, 104, 108–112, **110**, **111**, 122, **220**, 251, 253, 256–257, 258–259
see also non-governmental organization; stakeholders
adaptability 96, 157–158, 196
adaptation 11, 13–16, 25, 26–27, 96, 168
Adapting Conservation Agriculture for Rapid Adoption by Smallholder Farmers in North Africa (CANA) project 131
adoption
benefits 53
constraints 68, **70**, 100, 135, 137, 168
effects 270
experience role 122
factors affecting 68, **69**, 75–96, **92**, 266
farmer types 65, 257–258
impediments 31
increase 107, 220–223
levels 2, **54–56**
obstacles 206
patterns 221
percentages 53, 67
rate **225**
see also uptake
advice 65
affiliation 251
affordability 66, 135, 136, 137, 180, 198–200
Africa Climate-Smart Agriculture Alliance 8
African Conservation Tillage Network (ACT) 5, 53, 180
African CSA Alliance 179
African Drylands Week, 2nd 179
African Heads of State Summit, Malabo 179
African Union (AU) 8, 179
Aga Khan Foundation (AKF) 107, 112–113, 118, 121, 123
Aga Khan Foundation Coastal Rural Support Programme (AKF-CRSP) 107
AGRA-ACT-KARLO **54**
agrarian production systems 216–218, **217**
agribusiness support **230**, **234**
agricultural innovation platforms (AIPs) 247, 250–251, 252, 253–254, **257**, 258, 259–260, **259**
Agricultural Innovation Systems (AIS) 102, 250
Agricultural Knowledge and Information System (AKIS) 102
Agricultural Research Centre for International Development (CIRAD) 5
Agricultural Research Council (ARC) 224, 226, 227
Agricultural Research Trust (ART) 36
Agricultural Technology Clearing Committee (ATCC) 89
Agriculture Climate Change Programme 8
Agro-ecology Based Aggradation Conservation Agriculture (ABACO) 53, 66

Agro-Pastoral Farming System 45
agrochemicals 25
agroecology 2, 10, 20, 46, 215
agroecosystems, managing approach 14, 214
Agroforestry Food Security Programme (2007) Malawi 173–174
agroforestry systems 5, 51, 79, 172–173, 174, 177
  see also trees
agronomic (phenotypic) level 13
aims 228
Alberta Province 25
Algeria 132
Alliance for a Green Revolution in Africa (AGRA) 5, 47, 56
alliances 253–254
  see also partnerships
altitudes 148
animal production
  sustainable 153
  see also livestock
animal traction 53, **54**, **56**, 57, **58**, **59**, 188, **189**, 190, 193
animal/crop conflict (ACC) **70**
animals **58**, **59**, 180, 188, **189**, 190, 193
approaches
  alternative 248, 269
  community-based 103, 113, 269–270
  evolution 101–102
  holistic 153, 247
  innovations 9, 895
  methodological 105–108
  multi-scalar 250
  multi-sectoral 250–251
  participatory 4, 89, 95, 100, 137, 250, 257
  segmentation 228–229
  smallholders system 227–231
  tillage–seed–fertilizer–pesticide–credit 2
  typology approach 228
  see also innovation systems
Arab Authority for Agricultural Investment and Development (AAAID) 130, 132
area, (ha) **52**
areas, Africa CA practised 3–4, **4**, 266
assets 186, **231**, **235**
Associação Meio Ambiente (AMA) 114
Association pour la Promotion d'une Agriculture Durable APAD (Tunisia) 5
associations 115–116, 171
associations (intercropping) 78, **172**
attitudes, changing 94–96
AU-NEPAD 7, 8
Australian Centre for International Agricultural Research (ACIAR) 131
awareness 43, 70, **220**, **225**, 237

back-filling 88
banana–coffee-based farming systems 46
barley 51, 65, 128, 129, 130, 217
basin system 36, 82–85, 86, 88, 93
beliefs and attitudes (BA) **70**
benchmark farmers 114–115
benefit–cost ratio (BCR) 63, 64, **64**, 68, 156
benefits
  additional 146, 155
  community levels 255–256
  crop-related 210, 248, 257
  environmental 3
  extending 246–261
  financial 3
  intercropping 81
  linking across scales 255
  long-term 19
  member 253
  monetary 225
  net **63**, 141
  perceived 117, 119
  push–pull technology 159, 160–162
  range 268
  shorter-term 19, 255
  social 248
  see also ecosystem; incomes; sustainability; timeliness; weeds, reduction; yields
Bergville smallholder farming systems **230–231**, 231, **233–235**, 237–241, **240**
best practices 114
Biggs and Matsaert framework 108
biodiversity conservation 3
biomass 51, 57, 93, **107**, 144, 148, 168, 170, 175
  see also crop residues
Bonn Challenge 179
Brachiaria cv. *Mulato* 158
Building Resilience to Climate Change (BRCC) 91
bulk packages 66
Bungoma smallholders 256
Bungoma South Farming Innovation (BUSOFIP) 252, 253, 256, 257, **257**, 261
Burundi **52**
business person 209

C/N ratio 144
CA inputs (CAI) **70**
CA for NRM (CA4NRM) smallholder framework components 258–259
CA Western Cape 223
CA-FIP 224, 231
Cabo Delgado, Mozambique 100–123
canavalia, *Canavalia ensiformis* 49
canola, *Brassica* spp. 49

capabilities 228
capacity 21–22, 23, 43, 68, 70, 224
capital 37, 186, 199
carbon 25, 50, 173, 176, 196, 211
see also soil organic carbon
carbon dioxide (CO_2) 177
case studies 33, 43, 53–64, 100–123, 214, 253–254
cattle 148
see also animals; livestock
cereal-based production systems 46, 128, 129, 132, 135, 152
see also barley; maize; wheat
cereal-livestock productivity 156
CFU/COMESA Conservation Agriculture Regional Project (CARP) 90
challenges 2, 4, 92–94, 135–136, 219, 227–228
characteristics, conservation agriculture farmers **52**
characterizing 16, 24
chemicals 25, 155
chisel-type furrow openers 190, 191–192, 210, 218
choice of crop enterprise (CCE) **70**
choosing, right system 148
climate
Mediterranean, characteristics 127
parameters 10–11
pattern 16
plant growth determinant 42–43
resilience 64
resources 16
smartness 3, 6, 10, 20, 24, 151–163, 167, 169, 175–177, 268
sub-humid, characteristics 106
variability 15, 16
defined 13
see also drought
climate change
adaptability 11, 196
adaptation strategy 168
adoption constraint 70, **70**
challenges 3
compounding effects 152–153
impacts 44, **220**
mitigation 25, 268–269
stresses and shocks 194
yield fluctuations cushioning **64**
Climate Change Programme 179
climate-smart agriculture (CSA) 3, 6, 20, 151–163, 167, 169
Climate-smart technology 10, 159–160, 175–177, 268
collaboration 96, 178
collectives 246, 252, 258
commercial systems 46, **52**, 65, 185–187, 215, 220–222, 223–224, 236–241
common goal 68, **70**
Common Market for Eastern and Southern Africa (COMESA) 5, 90, 177–178
communication 103
see also promotion; training
community-based approaches 103, 113, 269–270
community-based extension (CBE) initiatives 103
see also farmer field schools
community-based organization (CBO) 253, 254, 256–257, 261
companion plants 154, 155, 156, 158, 159, **159**, 160
comparison
CA versus conventional ridge tillage (CRT) 79–89
NT versus CT 32, 130, 133, 134
competition 170
competitiveness, farming practices **220**
compost 107, 108
Comprehensive Africa Agriculture Development Programme (CAADP) 7–8, 43, 179
concepts 15, 47, 93, 102, 167, 261
conferences 5, 43, 140, 210, 223, 224, 267, 271
conflict 13, **70**
conservation
defined 105
evergreen 167
explained 1–2
measures 79
Conservation Agriculture for Natural Resources Management (CA4NRM) 258–259
Conservation Agriculture for North Africa (CANA) 131, 132, 136
Conservation Agriculture Regional Project (CARP) 90
conservation agriculture technologies (CATs) 226
Conservation Agriculture with trees (CAWT) 149, 167
Conservation Farming Unit (CFU) Zambia 5, 86, 90, 170–171
conservation tillage (CT) 132–133
constraints
adoption 68, **70**, 100, 135, 137, 168
biotic 154
innovation development 102
major, smallholder 222
mechanization 198, 211
rural development 216, 250
sustainable agriculture 216
technology uptake **54–56**
constructivism 215

constructs, social 214
Consultative Group to Assist the Poor (CGAP) 228–229
contacts 120
contention 100
  see also conflict
contracts 66
  see also services
control plots, monitored 237
conventional agriculture (CV) 107, **134**
conventional ridge tillage (CRT) 76, 79–89, **91**
conventional systems 11, 42, **85**, **86**, 218, 247–248
conventional tillage (CT) 11, 13, 15, 107, 129, **135**, **220**
Cooperative for Assistance and Relief Everywhere (CARE) 5
cooperatives 66, 115
coping 15, 16
cost–benefit analysis 59, 63–64
costs
  CA versus CRT 81–82, **83–84**
  inputs 59, 78, 136
  labour 67, 85–88, **87**, **88**, 94
  machinery 36, 131, 136
  management **55**
  pressure, conventional tillage **220**
  reduction 177–178, 225
  saving 248
countries 3–5, **52**, 127, **135**
cover crops 16, 49, **54–56**, 68, 142–149, **189**, 195, 210, 221
cover use 247
cowpea, *Vigna unguiculata* 49, 216–217
credit **188**
crimper, crop 189
crisis 2, 131
crop residues
  benefits 78
  burning 42, 93
  cereal-straw residue 136
  cover 192
  implements **54–56**
  management 210, 222
  mulch component **107**, 168
  retention 77, 78, 128, 170, 221
  spreaders 188–189, **188**
  use 65, 133
Crop–Livestock Conservation Agriculture (CLCA) 132
crop–livestock integration 128, 132, 137
crop–livestock production systems 128
cropping ratios 197–198
cropping systems
  alley 145
  diversification 78, 195
  diversity 105
  dryland 215
  improved 197–198
  level 13
  maize yields **90**
  mixed 108, 154
  monocropping 16, 128
  multiple 160
  non-diversified 128
  practices, conventional 218
  sequences (relay cropping) 78
  use 258
  Western Cape 221
  see also implements; intercropping; mechanization; rotation systems
crops 12, 51, **135**, 156, 216–217, **231**, 236, 247
  see also barley; maize; sorghum; wheat
Crotalaria spp. 49
cultivars, specially bred input-responsive 14
Cultivating Good Water programme partnership 24
cultivation methods, South Africa 217
cultivators 218
cultivators, collaboration 256
cultural contexts **134**, **135**

data 58–59, 104, 105, 224, 226
  see also surveys
dealership systems 201
decision makers 70
decision making 239
degradation 15, 22, 42, 43, 44, 45, 196, 247, 248, 254
demand, creation 208
demonstration 95, 261
Department of Agricultural Research Services (DARS) 89
Department of Agricultural Research and Technical Services (DARTS) Malawi 172
Department of Agriculture, Forestry and Fisheries (DAFF) 226
Department for International Development (DFID), UK 5
Department of Rural Development and Land Reform (DRDLR) 226
design, critical part 208
Desmodium spp. 49, 155, 156, 158–159
development 95, 101–102, 128, 163
dialogue 261
diammonium phosphate (DAP) fertilizer 59, **60**
dibble stick **105**, **106**, **107**, 190
diet 151
diffusion 96
direct-seeding 129, 130, 131, 132, 135, 136

disasters 2
disc harrows 218
disc openers 129, 190
disc planters 222
disc-type furrow openers 192
discovery learning 232
discs 132, 190–191, **191**, 210, 218
diseases 130, 136
disaggregation of farmers by wealth 124
dissemination 102–103, 163
distribution **52**, 201
diversification 1, 16, 17, **18–19**, 77, 78, 170, 195, 259, 261
diversity 105, 222, **225**, **235**, 236
Dolichos lablab, lablab beans 49, 144
donkeys **54–56**
donors 89
double-disc openers 190
double-storey production 170
draught power 180
    see also animal traction
drills
    affordability 135, 137
    direct-seeding 129, 130, 131, 132, 135, 136
    double-disc multi-purpose 132
    low-cost 131, 136
    prototype 129, 130
    requirements 129–130
    unavailability 137
drivers 4, 36, 43, 53, 66, 93, 94
drought 44, 70, 142, 148, 158, **159**, 176
dry season problems 147
drylands 159, 168, 248
Dust Bowl 265

East Africa (EA) 41–71
East African Community (EAC) 5
Eastern Free State (EFS) South Africa 224
ecologically-based agricultural paradigm 26
economic analysis 31–32, 60, **60**
economics 131, 174–175, 251
economy, cash 159
ecoservice systems **220**
ecosystem 3, 14, 17, **18–19**, 19, 20, 158, 219, **220**
education 267, 269, 271
    see also learning
educators
    peer 163
        see also referents, social
efficiency 30–33
elevations 148, 149
employment 152
empowering 41
empowerment 30, 41–71, 250–251
Empresa de Pesquisa Agropecuária e Extenção Rural de Santa Catarina (EPAGRI) programme 140
enclosure **188**
endophytes, symbiotic 12
energy requirement, mechanized systems 193
enhancement 21–22, 259
enterprises, disconnected **256**
entry points 6, 68, 123, 134, 185–187, 248
environment
    conditions 215–216
    degradation 43, 44
    impact of tillage-based agriculture 2
    marginal 197–198
    performance **231**, **235**
    risk-prone 215–216
    services 246
    sustainability 151–163
equipment
    availability 223
    mechanized 195–196
    small 247
    special 190
    specifications 209–210
    subsidy system 208
    unaffordable 198–200
    unavailability 200–202
    unsuitability 202–203
    see also implements; machinery
erosion 22–23, 35, 50, 196, 210–211, 218, **219**, 246, 248
    see also degradation
European Union (EU) 5, 47, 53, **54**, **55**, **56**, 66
evapotranspiration rates 215
EverGreen Agriculture (EVA) 167–180
EverGreen Conservation Agriculture (ECA) 167–180, **168**, **171**
experiences 129–132, 225
experiments 53–60, 226
expertise 2
export earnings 152
extension 65, 93, 95, 123, 171, 204
    see also farmer field schools

facilitators, village-level 122
Faidherbia albida 50, 79, 167–180, **172**
failures 102
fallows 65, **133**, 137, 173
family 46, 76
Farm Mechanization and Conservation Agriculture for Sustainable Intensification (FACASI) 47, **56**
farm-led experimentation (on-farm trial) process 237
farm-scale shorter-term rewards 258–259

farmer field schools (FFS) 4, 53, 66, 103, 112, 113, 115, 118, 122, 231, 253
Farmer Innovation Programme (FIP) 224
Farmer-centred Innovation Systems 214
farmer-managed natural regeneration (FMNR) 179
farmers
  benchmark 114–115
  conventional 95, 107
  days 223
  experienced 114–115
  groups 163
  key innovators 232
  labour estimates **87**
  organizations 124
  perceptions 116–124
  practising CA 45–46, 51–53, 90, **91**
  types **52**, 66, **217**
farmers' perception 100
farmers' practice (FP) 53, 57, 62, 63
farming systems, major 45–46
farming systems research (FSR) model 101, 102
Faulkner, Edward 265
fears, potential 123
feed 128
females 76, **162**
fertilizer-tree technology (FTT) 179
fertilizers
  biological, production 177
  cost-effective use 78
  crop response 23
  equipment **190**
  inorganic, promotion 53
  shrubs species 169
  subsidies 174, 177–178
  treatments 130
  tree species 149, 169, 172
  unaffordability 171
  use 34, 35
field crops 216–217
  see also barley; cowpea, *Vigna unguiculata*; grain; groundnut yields; lucerne, *Medicago sativa*; maize; oats; rye; sorghum; sunflower, *Helianthus annuus* L.; wheat
field days 95
Field Masters Ltd 66
finance 3, 136, 206
First African Congress on Conservation Agriculture (IACCA) 43, 267, 271
flexibility 88–92, 95
fodder 155, 158, 159, 161, 169, 170
food
  crop systems 175
  insecurity 2, 44, 152
  legumes 128
  market 42
  production 41, 216
  security 42, 43, 94, 158, 160, 215–216, 226, **231**, **235**
Food and Agriculture Organization of the United Nations (FAO) 5, 167–168
forage 49, 136, 156
forecasting 13
foreign aid dependency 75
forest management 246
  see also agroforestry systems; trees
four-wheel tractor (4WT) 191
frameworks 3, 108, 226, 256, 258–260, **259**
French Agency for Development (AFD) 132
French Agricultural Research Centre for International Development (CIRAD) 131–132
French cooperation agencies 131–132
fuel 35–36, 175
funding 5, 89
furrow openers 129, 189, 190–191

G8 66
gender 161, **162**, 205, **230**, **233**
  see also women
genetics 13, 16–17
German Development Cooperation (GIZ) 5, 47
germination, suicidal 155
germplasm, new 258
Gliricidia spp. 79, 173, 174
global environment for CA 3
Global Environment Fund (GEF) 131
glyphosate 50, 79
good agricultural practices 11, 113–114
good agronomic practices 53
good practices **77**, 90
governance structures 250
government, devolved 250
Government of Malawi (GoM) 75
grain 60–62, 133, 153, 173, 216–217
Grain SA 223–224, 231
Grain SA CA-FIP study sites 226
grazing 130, 215, 221
green manure cover crops (GMCC) 35, 139–140, 142–149
green manure cover crops (GMCC) species
  see mucana, *Mucana* spp.; ratooned pigeon peas, *Cajanus cajan*
Green Moroccan Strategy (*Plan Maroc Vert* (PMV)) 131
Green Revolution 15, 25, 26, 31
greenhouse gas (GHG) emissions 25, 196
gross domestic product (GDP) 152
gross margins 81–82, **83–84**
ground cover 81

groundnut yields 76, **81**, 216–217
groups
  activities 122–123
  delivering agricultural extension 163
  innovation platforms 115–116, 223
  mechanization assets 186
  organized 250
  ownership 186
  problem solving 227
  successful 223
  whole-village mobilization role 173
  working together 226–227
growing season 128, 152–153
growing seasons days (GSDs) 44
guidelines development 96
Guiera senegalensis shrub 175

hand hoe **54–56**, 247
harvest operation 189
harvest techniques 188
Heads of State and Government Orientation Committee (HSGOC) 8
health 160
hectares, CA practice 3–4, 26, 139, 140
Heifer International (HI) 163
Heifer Project International (HPI) 163
herbicide effects (HE) **70**
herbicides
  cost 136
  effects **70**
  efficiency 66
  implements **54–56**
  improper use 136
  knowledge 254
  labour saving 36, 193
  optional spraying 66
  pre-emergence 130
  pre-sowing 135
  recommended option 49–50
  systemic 79
  use 78, 248, 258
Hi-Bred Seeds Corporation 178
high external input systems (HEI) 217–218
Higher School of Agriculture of Kef (ESAK) 132
Highland Farming Systems 46
hiring 66
HIV-AIDS pandemic 222
hoes 129, 190
holistic approach 153, 247
homogenous 228
horsepower (HP) 129, 200, 201
households 68, **69**, 76, 105
hunger 152
hydroelectricity 23

image, farming **220**
impacts 2, 44, **220**, 224–225
implementation 75, 123, 163, 203, 231–232, 250, 251
implementers 227
implements **54–56**, 180, 185, 210, 247
  see also equipment
importer 42
imports 198, 206–207, 216, 223
incentives 68, 251–252
incomes 31, 42, 94, 161, **231**, **235**
indicators 223, 224, **225**, 239, 241, **241**, 271
indigenous CA systems 46
infiltration 21, 22, 23, **85**, **86**, 134
influence, large-scale farmers on smallholder farmers 64–66
information 92–93, 104, 115, 223, 229, 252
information and communication technologies (ICT) **110**, 114, 229
initiatives 5–6, 51, 177–178, 222, 223–224, 225–227, 246
innovation platforms (IPs) 103–104, **220**, 223, 224, 226, 237, 246–261
innovation systems (IS) 24, 27, 100–123, 214–215, 224, 227, 231, 232, 236, 237
innovations
  approaches 95
  brokers 104
  climate-smart technological innovation 157–163
  development constraints 102
  farmer perception 122
  farmer-led participatory approach 89
  institutional 248, 259–260
  knowledge intensive 123
  matrix 108–112, **110**
  space 123
  technical 247
input–output system concept 15
inputs
  affordability 66
  challenge 66
  external **135**, 218
  levels 33, 195, 218, **225**
  loan 66
  mechanization 206–210
  misconceptions 93–94
  production 14–15
  (seed) supply companies 114
  subsidized 76, 103
  types 20, 37, 59, 136
  unaffordable 157
  unaffordable/unavailable 136
  see also fertilizers; fuel; herbicides; labour; nutrients; pesticides; seeds; water

INRA-Settat 130
insect pest–predator protection dynamics 19
insect pests 152, 154, 156
Institut National de la Recherche Agronomique INRA (National Agronomic Research Institute) 129, 130, 131
Institut Technique des Grandes Cultures (ITGC) 132
institutions 31, 33, 75, 122, 248, 249–251, 254, 261
insurance 253
Integrated Agricultural and Rural Development Project (*Projet de Développement Agricole et Rural Intégré*, PDARI) 132
Integrated Crop–Livestock Conservation Agriculture (CLCA) for Sustainable Intensification of Cereal-based Systems in North Africa and Central Asia project 132
integrated crop–livestock system (IC–LS) 50–51
integrated pest management (IPM) 153, 154
integrated soil fertilizer management (ISFM) 153, 154, 174
integration 17, 27, 50–51, **56**, 153, 248–249
intensification 16, 25, 31, 42, 50, 153, 247, 249
interactions **113**, 231
intercropping 35, 78, 148, 160, 169, 170, **172**, 173
intercrops 49, 77, 154, 155, 158, **159**, 160
  see also pigeon peas
intermediaries 186
International Centre for Agricultural Research in the Dry Areas (ICARDA) 131, 161
International Centre of Insect Physiology and Ecology (*icipe*) 151, 153–154, 158, 163
International Convention on Biodiversity 3
International Crops Research Institute for the Semi-Arid Tropics (ICRISAT) 35
international food security assessment report (2013–2023) 44
International Fund for Agricultural Development (IFAD) 5
International Maize and Wheat Improvement Centre (CIMMYT) 5, 35, 76, 79, **80**, **81**, **82**, 89, 96, 178
interplanting 79
interventions 5, 46, 47, 217, 222, **233–235**, 236
invasive alien organisms **225**
investments 31, 175, 258–259, 261
  see also returns
irrigation furrows 189
issues 236
  see also fertilizers; herbicides; inputs; pesticides; soil, disturbance; soil, quality; water use efficiency
Itaipu 24

jab planter **54–56**, **59**, **106**, **107**, 190
jack beans, *Canavaliaensiformis* 148–149

Kenya **52**
Kenya Network for Dissemination of Animal Traction Technologies (KENDAT) **54**, **56**
Kenyan Agricultural Research Institute (KARI) 47, 49, **54**, **55**, **56**
Kieni Innovation Platform 253
knife roller **189**, 210
knowledge
  adoption constraint 70, **70**
  base 96
  development 24, 30
  dissemination 71, 96, 270
  flows 122
  intensive 37, 68
  land use 57
  limited 149, 203–204
  new scientific 26
  sharing 163
  smallholder 261
  systems 20
Kuhn NT seed drill 131
Kulima 114
KwaZulu-Natal KZN 224
Kyoto Protocol 3

lablab beans, *Dolichos lablab* 143, 144, 148
labour
  access **230**, **233**
  costs 67, 81–82, 85–88, **87**, **88**, 194
  demands 141, 145
  family 76
  farmers estimates **87**
  force, crop cultivation percentages 44
  inputs **82**
  management **55**
  market 42
  potential fears 123
  productivity 197
  reduction 122, 143
  returns 157
  saving 36, 193, 248
  use 35–36
  values 59
Laikipia County, Kenya 53–64

land
  degradation 22–23, 45, 128, 152, 153, 218–219
  enclosure **188**
  holdings 75, 152, 205
  management 34
  preparation 48, 79, 134, 175, 185, **185**, 187–189, 210
  restoration 179
  size **230**, **233**
  tenure system **188**
  use 11–12, 57, 246
Land Management and Environment Programme (LAMP) 51
land-based production systems 17–18
land-to-labour ratio 65
Landcare 178, 226
landless 216
landscapes 179, 215, 246, 258, 260
Lead Farmer Networks 4
leadership 96, 259
learning 65–66, 105, 109, 121, 122, 124, 143, 231, 232, 239, 261
least significant difference (LSD) test 59, **90**
Legume Research Network Project (LRNP) 47
legumes **18–19**, 78, 79, 144, 145, 148, 155, 160, 173, 195
  see also Desmodium spp.; *Gliricidia* spp.; *Sesbania* spp.; tephrosia, *Tephrosia* spp.
Lengetia Farm 65, 66
lessons 214, 236–241
ley cropping 222
Likert-type scale 117
linkages, weak between research, extension, training, manufacturing and farmers 204
litter layer 143
  see also mulch
livelihoods **231**, 232, **235**, 248, **256**, 259
livestock 50–51, **56**, **171**, **188**, 236
  see also crop–livestock integration
local-level
  equipment manufacture 207–208, 211
  government organizations 114
  organization **230**, **234**, 250
  solutions 123
low external input sustainable agriculture (LEISA) 218
low-input management practices 33
lowland areas 149
lucerne, *Medicago sativa* 49, 216–217
Lusaka Declaration 267, 271

machinery 25, 36, 128, 129, 132, 206–207
  see also mechanization
machines 180
Machinga Agricultural Development Division (ADD) **80**, **81**
macro-economic objectives 216
Magoye ripper-planter 191
Mainstreaming Framework 226
maize
  data analysis 59
  harvest decrease 216
  hybrids 178
    see also genetics
  intercropping 148
  on-farm trials 79, **80**
  production, Zambia 170
  production focus 46
  production viability 225
  yields **61**, 62, **80**, **90**, **91**, **159**, **172**
Maize Innovation Platform MIP (of Nyagatare, Rwanda) 252, 253, 255, **257**, 258
Maize Mixed Farming System 45
Maize Trust of South Africa 224
Malabo Declaration 8, 179
Malawi 75–96, 171–174
Malawi Agroforestry Food Security Programme 177
Malawi hoe 88
Malawian Smallholders Farmers' Association (NASFAM) 173, 178
malnutrition 41
management, strategies, improved 154
manual forms of CA **106**, 123
manufacturing 201, 207–208, 211, 223
marginal rate of return (MRR) 159
market-led CA 71
marketing difficulties 202
markets 66, 102–103, 139, 185–187, 228, **230**, **235**
materials 140, 173, 207, 208
matrix 104, 108, **110**, **111**
measuring **225**
mechanization
  adoption constraints 198, 211
  contributing resilience 183–211
  evergreen agriculture accommodation 171
  inputs
  intensification drive 16
  level **230**
  needs 65, 187–194
  options 70
  profitable 4
  promotion 184
  requirements 210
  service providers 211
  sustainable 185, 211
  see also machinery

media 114, 122
Memorandum of Understanding (MOU) **109**, 112, 121
mesofauna 195
messages 95, 137, 204
methodological approach 105–108
methods 252–253
Metuge District, innovation system 108–116
micro-dosing 35
micro-pits **106**, **107**
microorganisms 12, 17, 194, 195
 see also mesofauna
migration 184
Millennium Development Goals (MDG) 152
Ministry of Agriculture 173
Ministry of Agriculture, Irrigation and Water Development (MoAIWD) 93, 95
misconceptions 93–94
mixed messages on the roles and effects of herbicides 204
mixed-crop livestock systems 152
mobilization, community 178
modelling 215, 224–225, 236–241
models 95, 178, 224, 227–231
moisture stress 215
monitoring 232, 237, 239, **241**
monoculture 222
Morocco 129, 131, 135
motivation 105, 117, 118, **118**, 120, **120**, 122, **220**, 222
mouldboard plough 218, 247
mucana, *Mucuna* spp. 49, 143
mulch 1, **18–19**, 46, 107, **107**, 140, 142, 143, 144, 196
 see also cover crops; green manure cover crops; crop residues
mulching 142
Mycorrhiza 12

Napier grass 155, 156
National Adaptation Programmes of Action (NAPA) 8
National Agricultural Research System (NARS) 103, 104
National CA Task Force (NCATF) 96, 226
National Faidherbia Programme 175
National Field Crops Institute (INGC) 132
National LandCare Programme (NLP) 226
national security concern 225–227
natural resource base 11, 13, 247
 see also landscapes; soil
natural resources 50, 128, 196, 231
Natural Resources Management (NRM) 50, 246–261, **259**

net mitigation outcome 25
net present value (NPV) 157, 225
networking 124, **220**
networks 104, 108, **111**, 112, 121, 220–223
New Partnership for Africa's Development (NEPAD) 6–7
New York Summit on Climate Change 179
nightfall, first hour (scotophase) 155
nitrogen
 fertilizer 35, 169–170, 177
 fixing **18–19**, **55**, 146, 149, 156, 169, 173, 179, 195
  see also Gliricidia spp.; pigeon peas; *Sesbania* spp.; tephrosia, *Tephrosia* spp.
 mulch lack 144
nitrous oxide (N_2O) 50
no tillage, or no-till (NT) **18–19**
no-till (NT)
 adoption 135–136
 after 2 years **135**
 area 131, 132
 clubs 223, 224
 direct seeding equipment 190
 dissemination and adoption 131
 equipment 68, 131, 196
 implements 190–191, 222, **225**
 principle 93, 168, 170
 production 128, 129
 programme 140
 projects 132
 promoted 132
 seeding 17
No-Till CA Associations 4
No-Till Club, KwaZulu-Natal 223
no-till with dolichos lablab (NDL) 53, 57, 60, 63
no-till with dolichos and pigeon peas (NDP) 53, 57, 62, 63, 64
no-till with no cover crops (NCC) 57
no-till with no cover crops (NNC) treatment **57**, 62, 63
no-till with pigeon peas (NPP) 53, 57, 60, 63
non-adoption, reasons 116
non-affordability 180
non-commercial **233–235**
non-governmental organization (NGO) 113–114, 121, 163, 180
non-users 116–118, **116**
North Africa 127–137
Norwegian Agency for Development Cooperation (NORAD) 5
NT *see* no-till
nurturing environment 207
nutrients
 availability 134
 deficiency 34–35

levels 33
lost 143
poor response 22
preservation 248
productivity, improved 197
recycling 50
sources **18–19**, 144, 146, 195
uptake efficiency 23
use efficiency 197
Nyagatare Maize Investment Group Ltd (NYAMIG) (of MIP, Rwanda) 253, **257**

oats 216–217, 221
oils 49
oilseeds 128
Oldrieve, Brian 82
openers 129
Orange Free State Cooperative (OVK) 224
organic matter 1, 17, **107**, 152, 236
organizations 113, 121, 124, 163, 178, 250, 261
see also community-based organization
Ottosdal No-Till Club, North West Province 223, 224
Oued Zem platform 131
out-degree centrality 112
outcomes 19, 236–237, 258–259
output 14–15
outreach services 65, 66
oxen **54–56**

Parana Basin III in Brazil 24
parasites 143
participatory approach 4, 95, 100, 131, 137, **240**, 250, 257
participatory innovation development (PID) 231
participatory monitoring and evaluation (PM&E) 232, 239
partners 5, 75, 172–173
Partnership to Create an EverGreen Agriculture 180
partnerships 6–7, 24, 103–104, 105, 112, 114, 162–163, 180
Pastoral Farming System 46
pasture-based CA systems 26
Payment for Ecosystems Services (PES) 239
perceptions 100–123, **117**, **119**
performance 20, 32, 47, 95–96
pesticides 107–108, 193
pests 117, 152, 153, 154, **225**
phosphate 205
phosphorus 144
pigeon peas 49, 53, 57, 60, 63, 139, 144, 173, 174
pioneers 46–47
plantation crops 46
planters 190, 191, **192**, 222, 223, **225**
planting
basin system 93
basins 82–85, 93
implements 76, 210
methods 89
pits 175
rains 88
relay systems 49, 173
seeding technology 48
spacing 79
technical features 189–192
techniques 139
technology **54–56**
zero tillage 66
plants 158–159
diversity 236
emergence 129
growth constraints 127
intercrops 158, **159**
nutrients supplying **18–19**
scores **240**, **241**
spacing 89
platforms 6, 131, 159, 237, 251, 252, 253, **257**, 258
see also agricultural innovation platforms; innovation platforms
plough pan 23
ploughing, conventional **85**, **86**
Ploughman's Folly 265
policies
aspects 47
effect 32
implications 241
instruments 251
makers 70
promote wheat production 128
support 24–25, 43, 121, 122, 247, 259, 269
poor 104, 108, **118**, **119**, **120**, 121, 122, 173–174, 200, 228
population 75, 184
post Millennium Development Goals (MDG) targets 152
poverty 2, 42, 43, 75, 151–152, 222, **231**, **235**, 248, 250
Land and Agrarian Studies (PLAAS) 228
power 23, **56**, 184–185, 193–194
power take off (PTO) 129
practices
acceptable variations 258–259
adaptable, developing 163
adaptive 163
complimentary 78–79
conventional practices 218
core 17

practices (*continued*)
endorsed 92
farming, adaptation 11, 13, 16, 26–27
integrate 20
locally formulated 2
ridging 75
see also principles
precarious systems 183
precipitation 65, 127
see also rainfall
predicting weather and extreme climatic events 11
prices 59, 199, 216
principal component analysis (PCA) 105
principles
based on locally formulated practices 265–266
basic 77–79, 95
first 17, 105
implementing 75–97, **77**
market development 186
no-till conservation 170
practices integration 19–20
provision 155–156
second principle 17
smallholder framework component 258–259
third principle 17
see also species, diversification
three interlinked 27, 43, 48, 53, 105, 155–156, 214
see also diversity; no-till; soil cover
translation into practices 2
see also practices
priorities 31, **230**, **233**
private sector 6, 114
problems
key 248
see also degradation; erosion
production systems 11–12, **18–19**, 21, 33, 51, 128, 136, 199
productivity 43, 68, 75, 140–141, 151–152, 219, **230**, **233**, 255
see also labour; yields
profit 30, 32, 41–71, 142, 145
Programme of Technology Development for Arid Regions 130
programmes 130, 149
Project of Integration of Climatic Changes into the Green Moroccan Strategy (PICC-PMV) 132, 135
projects 47, 51, 53, 132, 178, 222, 227, 253, 255, 256, 258, 261
Projet de Développement Agricole et Rural Intégré (PDARI) project 132
Projet d'Integration des Changements Climatiques dans le Plan Maroc Vert (PICC-PMV) project 132, 135
promoters 113
promotion 5, 46–47
provisioning ecosystems services 14
purchasing power, low 198
push–pull technology 151–163, **154**, **157**

quality 66
questionnaire 66–67

radishes 221
rainfall
amounts 62, 158
Cabo Delgado 106
characteristics 12
cumulative rainfall **61**
decrease effect 63, 70, 80, 94
event 134
increases 42–43
inefficient use 34
line 215
Morocco 133–134
North Africa 127, 128
South Africa 215–216
unpredictable 153
use 34
variability 216
see also precipitation
rainfed regions
Africa 42, 167, 267
agroclimatic characteristics 24
CA principles application 265
Cabo Delgado 105
dependence 43
East Africa 44
growing period 198
North Africa 127, 128, 129
production systems 15, 17
uncertainties 158
rains 134–135
rainwater, capture and storage 176
randomized complete block design (RCBD) 53
ratooned pigeon peas, *Cajanus cajan* 144
recommendation 86, 96
referents, social 116–124, **118**, 120, **120**, 122, 123
reflexivity 163
regeneration, natural 248
Regional Economic Communities (RECs) 6
regional environment for CA 3
Regional Land Management Unit (RELMA) 5, 47, 51

rentals 66
repair services 198–200
repellents 155, 158
repetition 221
requirement, key 24
research 24, 222, 225, 226, 227, 237, 267, 269, 270
research and development (R&D) 130, 155
resilience
adding 194
agroecosystems 162
building 11, 12, 16, 24–25
contributing 183
defined 44, 194
developing 151
drought 70
ecological foundation 194
enhancement 8, 259
environmental 51, 137
increase 175
main pillars 137
mechanization conferred 210
potential 157–158
precarious systems 183–211
stresses 12
weatherproofing objectives 13, 16–17, 20, 21, 25
resistance to change 94
resource saving, integrated 248–249
resources
appropriate interventions category **233**
availability 228
biological 175
conserving, multifunctional 154
key 247
local 141, 142, 145
rehabilitating and conserving natural resources 194–196
scarcity 248
segmentation category **230**
use efficiency **220**
see also compost; crop residues; green manure cover crops; labour; natural resources; rotation systems
respondents, profiles 67
results, slow 147
returns 23, 59, 157, 159, **225**, 252
revenues **83–84**, 157
reward systems 176–177
ridging 75, 86, 87, 93, 95
rip-on-row 221
rippers **58**, 188, 191, 210, 218
ripping **55**, 221–222
risk 34–35, 44, 195, 215–216, **225**
role models 223
see also referents, social
rollers **189**, 210
rotation systems
annual 221
associations **77**
cropping diversification 78
effect on yield **133**
fallows incorporating nitrogen-fixing trees 173
inputs unaffordable/unavailable 136
maize–soybean–wheat 221
percentage usage 222
planting systems 49
practice constraints 128
sequences diversifying 156
technology **54–56**, 65
wheat and wheat–fallow 133
wheat–barley–fallow or wheat–chickpea 130
row spacing 81
Royal Norwegian Government-Initiated CA support programme in Zambia 5
Royal Tropical Institute RTI
rules 249, 259
runner beans, *Phaseolus coccineus* 144, 148
rural development 246, 250
Rural Institutions 237
rural people 216
rural shops 114
rural youth 43
Rwanda **52**, 258
rweya–kibanja system 46
rye 216–217, 221

Sahel, evergreen agriculture 174–175
sales commissions 115
sandy soils 221–222
Sasakawa Global (2000) 47
Save and Grow guidelines 186
savings 135
scalability 140–142, 187
scaling 2, 75, 89, 94–96, 169, 179, 219, 232, 253, 257–258
scores 239, **240**, **241**
scoring techniques 239
scrapers 210
seasonality 34, 128, 203, 215, 216
sector 12
sector timelines 104
seeders 57, **60**, 191–192, 196
seeding
direct seeding 77, 82–85, 210
drills 129, 130, 131, 132, 135, 136
mechanical 188
technical features 189–192
technology choice 48
seeds 49, 78, 103, 104, 114, 136

segmentation **230–231**
self-initiation 120
self-initiative 122
self-observation 120, 122
self-organization **220**
self-perpetuating learning systems 215, 231
Selian Agricultural Research Institute (SARI) 47
semi-commercial **233–235**
services
animal traction planters 193
commercialization 185–187
delivery 103
entrepreneurial service provision 71
environmental service 246
extension 204
hiring 186
providers 186, 208–209, 210
provision 186–187, 209
societal 3
specialist, provision 208–210
support 66
Serviços Distritais das Actividades Económicas SDAE (District Services of Economic Affairs) 108, **109**, **110–111**, 112, **113**, 114, 121–122
Sesbania spp. 79, 174
shallow weeders **58**, 66
shocks 195
short-term payoffs 248
shrubs 173
see also *Sesbania* spp.; tephrosia, *Tephrosia* spp.
simulation packages 224
sites 79
skills 203–204, 250, 261
slashers 210
Smallholder Farmer Innovation Programme (SFIP) 229
smallholder systems
agriculture, functions 246
conservation agriculture 183–211
farmers 30, 41–71
farms, size 128
framework **259**
initiatives 225–231
organization 261
production sector 222–223
size 51, **52**, 65
subsistence farming 216
types 222
social issues/incentives to unlock CA adoption 68
social network analysis (SNA) 104, 105, 108–112, 121
socio-economic potentials 10
soil
biological processes 19, 21
compaction 23, 35, 195, 196, 210–211, 222
condition 194
conservation technologies 65
de-compaction 248
development, water functions 22–23
disturbance 1, 17, 22, 76, 77–78, **77**, 87–88, 129, 134, 236, 247
drainage 21
fertility 42, **55**, 107, 155, 169, 174, 258
formation 11, 22
health 21, 222, **225**
loss 219, **225**, 248
management 11, 34, 247
mining 128
moisture-holding capacity 21–22, 23
nutrients 146
openers 196
properties, improve 78
protection 160–161
quality 134, 136, **220**, 236
rehabilitation 196
restoration 237
scores **240**
water 12, 21, 128, 215
see also degradation; erosion
Soil Conservation and Agroforestry Programme for Arusha (SCAPA) 51
soil cover 49, 57, **70**, 78, 170, **188**, 192
Soil Fertility Initiative 5, 47
Soil Health Tool **225**
soil organic carbon (SOC) 134, 136, 218, 224, **225**
soil organic matter (SOM) 22, 35, 128, 146, 224, **225**
soils 130, 134, 144, 152, 170, **225**
solutions 139, 153–157, 219, 254–255, 269–270
sorghum 216–217
South Africa 214–242
South African Farm Implement Manufacturers (SAFIM) 223
Southern African Development Community (SADC) 5
sowing
direct 132
see also seeding
soybean 216–217
spacing 79, 81, 86, 88, 89
spares 198–202
spear grass, *Imperata cylindrica* 143
species, diversification 1, 16, 17, **18–19**, 170, 259, 261
specificity, local 227
sprayers 193, 210
spreaders, crop residues 188–189
spreading, CA 26, 266, 269, 270
stability 162, 194

stakeholders 4–6, 27, 46, 68, 104, 206, **206**, 226
stakes (cuttings) 149
standard error of the difference (SED) 80, 81, 91
standardization 16
state role 102–103
Statistical Package for Social Sciences (SPSS) 67
stem borer 154, 155, 158, 159, 160
stewardship 250
stover 59, 63
straw chopper 192
strawspreaders 210
stresses 12, 23, 194, 195
Striga spp., witchweed 49, 143, 152, 154, 155, 156, 158, 160
structures, new CA 223
studies 33, 43, 53–64, 100–123, 214, 228, 253–254
Stulwane learning group **240**
Sub-Saharan Africa (SSA) 100, 152
subsidy 103, 177
subsistence agriculture dominance 199–200
subsistence farming 44
subsoilers 210
success 177, 219, 224, **225**
sunflower, *Helianthus annuus* L. 49, 216–217
support 5, 24–25, 43, 136, 206–207, **230**, 232–236, **234**, 247
surveys 66–68, 104, 105
sustainability
  adaptable practices, developing 163
  agricultural systems, mainstreaming 214
  developing adaptable practices 163
  development 25, 51, 247
  dryland farming systems 217
  ecological foundation 20
  farming 41–71
  intensification 30–37, 43, 51, 153, 248, 258
  intensive 30, 34–37, 43, 51, 153, 248, 258
  issue 122
  land management practices 175
  mechanized 185, 211
  NT development 13
  production **18–19**, 20, 153
  subsidy pathway 177–178
Sustainable Agricultural Production Program **91**
Sustainable Agricultural and Rural Development (SARD) 47, **54**, **55**, **56**
sustainable farming 41
Sustainable Intensification of Maize-Legume Systems for Food Security in Eastern and Southern Africa (SIMLESA) 89, 253, 255, 256, 258, 261
Swedish International Development Agency (SIDA) 47, 51
synchronization 147–148
synergies 17–18
System of Rice Intensification (SRI) 17
systems, adaptive 227

Tanzania **52**
targeting 178, 256–257
Technical Cooperation Projects (TCPs) 5, 47
Technical Institute for Field Crops (ITGC) 132
techniques 139
technologies
  adoption 120
  beneficiaries 100, 122, 136, 173
  climate-smart push–pull 151
  colonial period 137
  Green Revolution 31
  main 65
  new 128
  promotion 132
  push–pull 47, 151–163, **154**, **157**
  situation analysis **54–56**
  transfer, import-based 198
  trees 149
temperature 127, 152–153, 158, 176, 215
temporary measures 23
tephrosia, *Tephrosia* spp. 79, 148, 169, 174
testing 236
Third World Congress on Conservation Agriculture 5
three crop sequence 108
ties 108
tillage 22, 23, 26, 34, 130, 133, **133**, 136, 137
tillage-based farming 194
tillage-based mechanical management 21
tillers 218
timeline, CA innovation system set-up **109**
timeliness 13, 184, 197, 219
timing 13
tine furrow openers 190, 191, 218
tools 93–94, 180
topsoil 22, 219
Total LandCare (TLC) 76–79, **77**, 90, **91**, 95, 96, 97, 172
traction **233**
tractor-drawn implements 218, 222
tractors **54–56**, 129, 180, 185–186, 190, 191, 193–194, 200, 201, 250
trade-offs 50
training 70, 93, 95, 96, 112, 114, 163, 173, 204, 208, 209–210, 267
transfer of technology (TOT) model 101
transformation 6, 19, 26, 267
transition, conventional to CA system model 224
trap crops 154, 155, 158
trap plants 155, 156, 158, **159**, 160

treatments 53, **60**, 61–62, **61**, **63**, **64**, 68, 79
    see also farmers' practice; fertilizers; no-till
tree-based perennial CA 26
tree-crop association 170
tree-crop diversification 170
trees 50–51, 143–144, 149, 167–180, **168**, **171**
    see also agroforestry systems; *Faidherbia albida*; fertilizer-tree technology; intercropping
trials 79, 133, 134, 224, 226, 237, 239, **240**
Tunisia 129, 131–132
two-wheel tractors (2WTs) 186, 191, 196
typology approach 48, 228

Uganda **52**
Umokazi 113–114
UN Convention on Combating Desertification 179
unaffordability 137, 198–200
unavailability 137, 200–202
underpinning, ecological 12
understanding 92–93, 237
    see also knowledge; learning
unemployed 216
United Nations Convention to Combat Desertification 3
United Nations Framework Convention on Climate Change 3
United States Agency for International Development (USAID) 31, 90
unskilled extension workers 204
upscaling 43, 68, 71, 95–96, 177, 178, 179, 180, 270–271
uptake 94, 156, 160, 167, 174, 266
    see also adoption
users 119–121

validation plot 53
value chain 46, 238–239
vegetation crusher 189
verification, in-field 236
vetch 221
vetiver grass **77**, 79
views, gathering 104
Village Development Organization (VDO) 113
village-level facilitators 122
vision (25×25) 179
visual soil assessment (VSA) **240**, **241**
vulnerability 43, 44

water
    climate-smartness 20, 24
    conservation 20, 65, 136
    cycling 21–22
    efficient 25
    filtration 21–22, 173
    harvesting 129
    holding capacity 224
    improved water use efficiency and productivity 197
    infiltration 22, 23, 34–35, **85**, **86**, 134
    management 246, 247
    quality decline 216
    regimes **220**
    related aspects 11
    reserves 23
    retention 248
    services 11, 21–22, 23
    storage 34–35, **133**
    stress 43
    use 136, 197, **225**
    see also weatherproofing
water use efficiency (WUE) 136, **225**, 236
watershed level 24
way forward 205, 260–261
weak artisan, blacksmith and craft workshops 201–202
wealth 105, 108, 118, **118**, 119, **119**, **120**, 122
weather 10–11, 16, 42–43, 194, 215, 216, **220**
weatherproofing 10–27
weed management (WM) **70**
weeding 79, 168, 210
weeds
    burned 93
    control 35, 79, 135, 146
    and cover crop maintenance, technical features 192–193
    local solutions 123
    management 21, 36, 38, 49–50, **55**, **56**, **70**, 190–191, 192–193, 210
    pressures 152
    reduction 142, 143, 155, 248
    suppression 122, 173
    wipers 193
    witchweed 49, 143, 152, 154, 155, 156, 158, 160
weeping lovegrass, *Eragrostis curvula* 222
wellbeing **231**, **235**
wheat 65, 128, 133, **133**, 216–217
wheeled tractor (WT) 186, 191, 196
win–win outcomes 250
wind speeds 176
windbreak 176
witchweed, *Striga* spp. 49, 143, 152, 154, 155, 156, 158, 160
women 41, 53, 76
World Agroforestry Centre (ICRAF) 5, 172–173, 178
World Wildlife Fund (WWF) **110–111**, **113**, 114

yields
  assessment 79–81, **80**, **83–84**
  attainable 32, 33, 38
  CA effects 37–38, 142, 219, 224, **240**, **241**, 248
  data 59
  desmodium effect 155
  economically attainable 32, 38
  farmer perception 122
  GMMC effect 143–145
  improvement 136
  increase 61, 68, 173, 174–175
  increases 80, 172
  potential, realistic measure 33
  reduction factors 32
  results 168
  treatments effect 63
  trends **220**
  variation **62**
youths, rural 184

zaï technique of cultivation 175
Zambia, Conservation Farming 36
Zambia Champions *Faidherbia* 170–171
Zambian *chaka* hoe 88
zero ripping, implements **54–56**
zero-tillage **54–56**, 66, 140, 167, 177